TRANSGENIC AND KNOCKOUT MODELS OF NEUROPSYCHIATRIC DISORDERS

Series Editors:
Ralph Lydic and
Helen A. Baghdoyan

Transgenic and Knockout Models of Neuropsychiatric Disorders, edited by *Gene S. Fisch and Jonathan Flint*, 2006

The Orexin/Hypocretin System: *Physiology and Pathophysiology*, edited by *Seiji Nishino and Takeshi Sakurai*, 2006

The G Protein-Coupled Receptors Handbook, edited by *Lakshmi A. Devi*, 2005

Attention Deficit Hyperactivity Disorder: *From Genes to Patients*, edited by *David Gozal and Dennis L. Molfese*, 2005

Genetics and Genomics of Neurobehavioral Disorders, edited by *Gene S. Fisch*, 2003

Sedation and Analgesia for Diagnostic and Therapeutic Procedures, edited by *Shobha Malviya, Norah N. Naughton, and Kevin K. Tremper*, 2003

Neural Mechanisms of Anesthesia, edited by *Joseph F. Antognini, Earl E. Carstens, and Douglas E. Raines*, 2002

Glutamate and Addiction, edited by *Barbara Herman*, 2002

Molecular Mechanisms of Neurodegenerative Diseases, edited by *Marie-Françoise Chesselet*, 2000

Contemporary Clinical Neuroscience

TRANSGENIC AND KNOCKOUT MODELS OF NEUROPSYCHIATRIC DISORDERS

Edited by

GENE S. FISCH, PhD

*Department of Statistics, Yeshiva University
and Department of Psychology, CUNY/Lehman College
New York, NY*

JONATHAN FLINT, MD, MRCPsych

*Wellcome Trust Centre for Human Genetics
and Department of Psychiatry, University of Oxford
Oxford, UK*

HUMANA PRESS ✻ TOTOWA, NEW JERSEY

© 2006 Humana Press Inc.
999 Riverview Drive, Suite 208
Totowa, New Jersey 07512

humanapress.com

All rights reserved.

No part of this book may be reproduced, stored in a retrieval system, or transmitted in any form or by any means, electronic, mechanical, photocopying, microfilming, recording, or otherwise without written permission from the Publisher.

All authored papers, comments, opinions, conclusions, or recommendations are those of the author(s), and do not necessarily reflect the views of the publisher.

The content and opinions expressed in this book are the sole work of the authors and editors, who have warranted due diligence in the creation and issuance of their work. The publisher, editors, and authors are not responsible for errors or omissions or for any consequences arising from the information or opinions presented in this book and make no warranty, express or implied, with respect to its contents.

Production Editor: Melissa Caravella

Cover design by Patricia F. Cleary

Cover Illustration: From Fig. 2 in Chapter 7, "How Can Studies of Animals Help to Uncover the Roles of Genes Implicated in Human Speech and Language Disorders?" by Simon E. Fisher.

For additional copies, pricing for bulk purchases, and/or information about other Humana titles, contact Humana at the above address or at any of the following numbers: Tel.: 973-256-1699; Fax: 973-256-8341; E-mail: orders@humanapr.com; or visit our Website: www.humanapress.com

The opinions expressed herein are the views of the authors and may not necessarily reflect the official policy of the National Institute on Drug Abuse or any other parts of the US Department of Health and Human Services. The US Government does not endorse or favor any specific commercial product or company. Trade, proprietary, or company names appearing in this publication are used only because they are considered essential in the context of the studies reported herein.

This publication is printed on acid-free paper. ∞
ANSI Z39.48-1984 (American National Standards Institute) Permanence of Paper for Printed Library Materials.

Photocopy Authorization Policy:

Authorization to photocopy items for internal or personal use, or the internal or personal use of specific clients, is granted by Humana Press Inc., provided that the base fee of US $30.00 per copy is paid directly to the Copyright Clearance Center at 222 Rosewood Drive, Danvers, MA 01923. For those organizations that have been granted a photocopy license from the CCC, a separate system of payment has been arranged and is acceptable to Humana Press Inc. The fee code for users of the Transactional Reporting Service is: [1-58829-507-9/06 $30.00].

Printed in the United States of America. 10 9 8 7 6 5 4 3 2 1

e-ISBN 1-59745-058-8

Library of Congress Cataloging-in-Publication Data

Transgenic and knockout models of neuropsychiatric disorders / edited by Gene S. Fisch, Jonathan Flint.
   p. ; cm. -- (Contemporary clinical neuroscience)
 Includes bibliographical references and index.
 ISBN 1-58829-507-9 (alk. paper)
 1. Mental illness--Animal models. 2. Mental illness--Genetic aspects. 3. Transgenic mice--Genetics.
 [DNLM: 1. Mental Disorders--genetics. 2. Mental Disorders--physiopathology. 3. Animal Experimentation. 4. Disease Models, Animal. 5. Mice, Transgenic--genetics. 6. Psychology, Experimental. WM 140 T772 2006] I. Fisch, Gene S. II. Flint, Jonathan. III. Series.
 RC455.4.A54T73 2006
 616.8--dc22
                                      2005020469

PREFACE

Since the composition and structure of DNA was first revealed more than half a century ago, advances in genetics have been nothing short of remarkable. Fifty years later, we have seen an exponential increase in the amount of research generated in genetics at all levels of analysis, including the discovery of genes that produce a broad range of diseases and disorders, as well as the completion of the human genome project. In addition, the genomes of a number of model organisms, including both mouse and rat, have been deciphered. Consequently, development of research in the neurosciences at the behavioral, biological, and physiological levels has accelerated in the last 20 years. Not only are we now in a position to develop meaningful genotype–phenotype relationships for complex neurobehavioral and neuropsychiatric disorders, we may also be at a point where we can create animal models of human psychiatric illness, determine the pathophysiology of these disorders, and perhaps advance genetic and other forms of molecular therapy.

The creation of transgenic and knockout mice and, in the last decade, the development of transgenic and knockout mouse models of genetic disorders associated with neurobehavioral and neuropsychiatric dysfunction, has been extraordinary. In recent years, mouse models have been created for cognitive and learning impairment (e.g., fragile X syndrome and nonsyndromal X-linked mental retardation), for pervasive developmental disorders (Rett syndrome), for psychosis and schizophrenia (del22q11), and for anxiety and depression.

Considering that the use of infrahuman animals in experiments to study the principles of behavior and rules of neuroscience is a relatively recent phenomenon, it is important to determine whether, or to what extent, infrahuman animals model human disabilities. This is particularly the case for researchers who study complex neurobehavioral and neuropsychiatric disorders such as schizophrenia, depression, mental retardation, and autism.

What, then, are the necessary and sufficient criteria to demonstrate that a given infrahuman organism is a proper model for neuropsychiatric dysfunction? Given the current state of transgenic and knockout technology, neuroscience, and means by which to evaluate animal behavior, we are now in a position to examine transgenic and knockout mouse models and the extent to which they mimic human illness. The goal of *Transgenic and Knockout Models of Neuropsychiatric Disorders* is to provide the reader with a clear and comprehensive assessment of how and whether genetic abnormalities produced in transgenic and knockout mouse models manifest neuropsychiatric disorders.

To accomplish this, we have divided the text into three main sections. In Part I, Chapter 1, we introduce and provide an overview of the history and assessment of neuropsychiatric disorders, and the controversial notion of continuity of species of human and infrahuman behavior articulated in Darwin's theory of evolution. In Chapter 2, we examine attempts to utilize infrahuman animals in experiments designed to improve our understanding of human behavior and psychiatric disease. Chapter 3 presents an epistemological argument concerning the problem of language in transgenic and knockout animal models, and whether nonhuman models of complex neuropsychiatric dysfunctions involving language are feasible. Chapter 4 presents the counterargument, noting how operant conditioning procedures can be used with infrahuman animals and transgenic and knockout mice in particular to study complex behavioral disorders.

In Part II, we present current research on transgenic and knockout models leading to the analysis of neurocognitive dysfunction—neurobiological disorders producing learning disabilities and mental retardation. Chapter 5 shows how mouse models can be used to examine the pathogenesis of human neurological diseases, particularly polyglutamate disorders. Chapter 6 provides a comprehensive review of several transgenic and knockout mouse models of mental retardation, noting the similarities and differences of neurocognitive development and function in humans compared to their mouse counterparts. Chapter 7 examines speech and language function, and how one gene associated with speech and language impairment and conserved evolutionarily in infrahuman organisms may be useful in examining human speech disorders using mouse models. Chapter 8 examines mouse models of genetic disorders associated with a high risk of developing autism, and the difficulties of developing mouse analogs of human disorders that involve cognitive, social, and speech and language dysfunction.

Part III focuses on neuropsychiatric dysfunctions: psychosis and schizophrenia, and the mood disorders: anxiety, depression, and bipolar disorder. Chapter 9 discusses the difficulties in developing a mouse model for schizophrenia imposed by the constraints involving diagnosis, brain development, and the polygenic etiology of the disorder. Chapter 10 investigates the use of infrahuman animals to examine the cognitive and negative affect features of psychosis, and how NR1-deficient mice can be used as a model. Chapter 11 reviews anxiety and fear in mutant mice in an ethological setting, and the effects of drugs in modifying their behavior. Chapter 12 examines a variety of mutant mouse models in several animal experimental paradigms to assess anxiety and depression, as well as the effect of strain differences on behavior. Chapter 13 presents an overview of neurochemical, neuroendocrine, and behavioral changes in bipolar disorder, and the available mutant mouse models in which some aspects of the disorder are emulated.

We hope our readers find this comprehensive overview of animal models of neuropsychiatric disorders to be a useful tool in their research and classrooms.

Gene S. Fisch, PhD
Jonathan Flint, MD, MRCPsych

Contents

Preface v
Contributors xi

I Introduction and Overview

1. Transgenic and Knockout Models of Psychiatric Disorders: *Introduction, History, and Assessment* 3
   Gene S. Fisch

2. Transgenic Mouse Models and Human Psychiatric Disease ... 25
   Jonathan Flint

3. Transgenic and Knockout Mouse Models: *The Problem of Language* 45
   Linda J. Hayes and Diana Delgado

4. If Only They Could Talk: *Genetic Mouse Models for Psychiatric Disorders* 69
   Trevor Humby and Lawrence Wilkinson

II Transgenic and Knockout Models of Neurocognitive Dysfunction

5. Spinocerebellar Ataxia Type 1: *Lessons From Modeling a Movement Disease in Mice* 87
   Harry T. Orr

6. Mouse Models of Hereditary Mental Retardation 101
   Hans Welzl, Patrizia D'Adamo, David P. Wolfer, and Hans-Peter Lipp

7. How Can Studies of Animals Help to Uncover the Roles of Genes Implicated in Human Speech and Language Disorders? 127
   Simon E. Fisher

8. Animal Models of Autism: *Proposed Behavioral Paradigms and Biological Studies* 151
   Thomas Bourgeron, Stéphane Jamain, and Sylvie Granon

III Transgenic and Knockout Models of Neuropsychiatric Dysfunction

9   Genetic Mouse Models of Psychiatric Disorders:
    Advantages, Limitations, and Challenges 177
    Joseph A. Gogos and Maria Karayiorgou

10 Animal Models of Psychosis 193
    **Stephen I. Deutsch, Katrice Long, Richard B. Rosse,
    Yousef Tizabi, Ronit Weizman, Judy Eller,
    and John Mastropaolo**

11 Animal Models of Anxiety: *The Ethological Perspective* 221
    **Robert Gerlai, Robert Blanchard,
    and Caroline Blanchard**

12 Modeling Human Anxiety and Depression
    in Mutant Mice 237
    Andrew Holmes and John F. Cryan

13 Mutant Mouse Models of Bipolar Disorder:
    Are There Any? 265
    **Anneloes Dirks, Lucianne Groenink,
    and Berend Olivier**

Index 287

CONTRIBUTORS

CAROLINE BLANCHARD, PhD • *Department of Genetics and Molecular Biology, John A. Burns School of Medicine, Honolulu, HI; Pacific Biomedical Research Center, University of Hawaii at Manoa, Honolulu, HI*
ROBERT BLANCHARD, PhD • *Department of Psychology and Pacific Biomedical Research Center, University of Hawaii at Manoa, Honolulu, HI*
THOMAS BOURGERON, PhD • *Department of Neuroscience, Human Genetics and Cognitive Functions, Institut Pasteur, Paris, France*
JOHN F. CRYAN, PhD • *Department of Pharmacology, School of Pharmacy, University of College Cork, Cork, Ireland*
PATRIZIA D'ADAMO, PhD • *Institute of Anatomy, Division of Neuroanatomy and Behavior, University of Zurich, Zurich, Switzerland; DIBIT, San Raffaele Scientific Institute, Milano, Italy*
DIANA DELGADO, MA • *Department of Psychology, University of Nevada, Reno, NV*
STEPHEN I. DEUTSCH, MD, PhD • *Department of Veterans Affairs Medical Center, Washington, DC; Department of Psychiatry, Georgetown University School of Medicine, Washington, DC*
ANNELOES DIRKS, PhD • *Department of Psychopharmacology, Utrecht Institute of Pharmaceutical Sciences and Rudolf Magnus Institute, Utrecht University, Utrecht, The Netherlands*
JUDY ELLER, BA • *Department of Veterans Affairs Medical Center, Washington, DC*
GENE S. FISCH, PhD • *Department of Statistics, Yeshiva University and Department of Psychology, CUNY/Lehman College, New York, NY*
SIMON E. FISHER, DPhil • *Wellcome Trust Centre for Human Genetics, Nuffield Department of Clinical Medicine, University of Oxford, Oxford, UK*
JONATHAN FLINT, MD, MRCPsych • *Wellcome Trust Centre for Human Genetics and Department of Psychiatry, University of Oxford, Oxford, UK*
ROBERT GERLAI, PhD • *Department of Psychology, University of Toronto at Mississauga, Ontario, Canada*
JOSEPH A. GOGOS, MD, PhD • *Department of Physiology and Cellular Biophysics, Columbia University, College of Physicians and Surgeons, Center for Neurobiology and Behavior, New York, NY*
SYLVIE GRANON, PhD • *Department of Neuroscience, Receptors and Cognition, Institut Pasteur, Paris, France*

LUCIANNE GROENINK, PhD • *Department of Psychopharmacology, Utrecht Institute of Pharmaceutical Sciences and Rudolf Magnus Institute, Utrecht University, Utrecht, The Netherlands*

LINDA J. HAYES, PhD • *Department of Psychology, University of Nevada, Reno, NV*

ANDREW HOLMES, PhD • *Section on Behavioral Science and Genetics, Laboratory for Integrative Neuroscience, National Institute on Alcohol Abuse and Alcoholism, National Institutes of Health, Bethesda, MD*

TREVOR HUMBY, PhD • *Laboratory of Cognitive and Behavioural Neuroscience, The Babraham Institute, Babraham Research Campus, Cambridge, UK*

STÉPHANE JAMAIN, PhD • *Department of Molecular Neurobiology, Max-Planck Institute for Experimental Medicine, Göttingen, Germany*

MARIA KARAYIORGOU, MD • *Laboratory of Human Neurogenetics, The Rockefeller University, New York, NY*

HANS-PETER LIPP, PhD • *Institute of Anatomy, Division of Neuroanatomy and Behavior, University of Zurich, Zurich, Switzerland*

KATRICE LONG, BS • *Department of Veterans Affairs Medical Center, Washington, DC*

JOHN MASTROPAOLO, PhD • *Department of Veterans Affairs Medical Center, Washington, DC*

BEREND OLIVIER, PhD • *Department of Psychopharmacology, Utrecht Institute of Pharmaceutical Sciences and Rudolf Magnus Institute, Utrecht University, Utrecht, The Netherlands; Department of Psychiatry, Yale University School of Medicine, New Haven, CT*

HARRY T. ORR, PhD • *Department of Laboratory Medicine and Pathology and Institute of Human Genetics, University of Minnesota, Minneapolis, MN*

RICHARD B. ROSSE, MD • *Department of Veterans Affairs Medical Center, Washington, DC; Department of Psychiatry, Georgetown University School of Medicine, Washington, DC*

YOUSEF TIZABI, PhD • *Department of Pharmacology, Howard University College of Medicine, Washington, DC*

RONIT WEIZMAN, MD • *Department of Psychiatry, Sackler School of Medicine, Tel Aviv University, Tel Aviv, Israel*

HANS WELZL, PhD • *Institute of Anatomy, Division of Neuroanatomy and Behavior, University of Zurich, Zurich, Switzerland*

LAWRENCE WILKINSON, PhD • *Laboratory of Cognitive and Behavioural Neuroscience, The Babraham Institute, Babraham Research Campus, Cambridge, UK*

DAVID P. WOLFER, MD • *Institute of Anatomy, Division of Neuroanatomy and Behavior, University of Zurich, Zurich, Switzerland*

I Introduction and Overview

1
Transgenic and Knockout Models of Psychiatric Disorders
Introduction, History, and Assessment

Gene S. Fisch

Summary

Humans have long distinguished themselves from infrahuman organisms. Modern notions of humans and infrahuman animals, however, date from the mid-19th century and are essentially derived from Darwin's *The Origin of Species*. Although Darwin differentiated between acquired habits in humans and inherited instincts in animals, his theory of evolution embraced the notion of continuity of species. Late-19th and early-20th century animal psychologists debated this issue; particularly, whether animals had minds, and, if so, whether they were capable of the same thought and emotion evinced in humans. To avoid the problems created by mentalism and consciousness in animal behavior, many leading psychologists of the time adopted the mechanistic assumption, as extant among the British associationists. Others, however, subscribed to the belief that animals were capable of problem solving that went beyond the simple conditioning paradigm incorporating the principle of reinforcement. As a formal approach, operant conditioning was instrumental in providing many important results and was an effective epistemological framework in which to view animal and human behavior. For many psychologists, however, behaviorism seemed limited to what could be inferred from bar pressing and key pecking. As enthusiasm for behaviorism ebbed, cognitive psychology began to assert its influence, restoring the concepts of mentalism to and consciousness in animals, and directing its research efforts to intelligence and problem solving. At approximately the same time, other related areas in science—neurobiology, genetics—converged onto issues related to learning and memory, the result of which was the emergence of cognitive neuroscience. The revolution in genetics brought about by the discovery of the structure of DNA, along with the discovery of genetic abnormalities associated with learning disabilities, accelerated research into genetics factors that produced mental retardation and psychopathology. Cognitive psychology and cognitive neuroscience provided the epistemological justification for using animal models to explore various forms of human cognitive impairment. Studies in neuroscience using recently developed behavioral procedures have identified brain structures associated with certain features of learning and emotion. Successful development of knockout and transgenic technologies, followed by the creation of mouse models of genetic disorders, were used to identify many of the neurobiological and neurophysiological functions associated with neurobehavioral disorders. However, mice are not

From: *Contemporary Clinical Neuroscience: Transgenic and Knockout Models of Neuropsychiatric Disorders*
Edited by: G. S. Fisch and J. Flint © Humana Press Inc., Totowa, NJ

humans, and the question regarding whether studies of animal behavior are relevant to human thought, problem solving, intelligence, and emotions has not been resolved.

Key Words: Animal models; experimental psychology; behaviorism; cognitive psychology; genetics; neuroscience; psychiatric disorders; mouse models.

1. INTRODUCTION AND EARLY HISTORY

What is the nature of man? And how shall we know it?

—Josiah Royce, The Nature of Man *(1)*

It is often remarked that psychology has a long past and a short history, an aphorism suitable to Royce's epigraph. Humans have long distinguished themselves from infrahuman organisms using various rationales. From Aristotle's time until that of Descartes, living creatures were assigned to one of three ordered categories: plants, animals, and humans. Descartes, whose meditations on the nature of existence emerged from the 17th century rationalist tradition, considered infrahuman species, animals, as machines ("automata"), because all of their behavior could be reduced to simple reflex reactions. Although animals may also have sensations and a form of consciousness, he did not consider this equivalent to human thinking, nor did he think it was possible to demonstrate these differences with complete certainty. Humans could also respond reflexively, but ultimately were guided by a "rational soul," the organ for which was located in the brain, that distinguished them from other animals.

Early theological and pre-18th century ideas also differentiated humans from other animals, as expressed by the theory of special creation ("every living creature after its own kind"). Beginning at approximately the mid-18th century, during the period of the Enlightenment, disputes arose among various Protestant factions regarding the nature of human and infrahuman beings. Were humans distinct and superior to other animals, as expressed in theology? Were humans like other animals, like automata, as Descartes had said? Or, were animals neither automata nor reasoning creatures, but something in between?

Although evolutionary notions have been traced back to Thales (640–546 BC), the roots of modern theories probably took hold in the mid- to late-18th century and are found in the writings of Erasmus Darwin (1731–1802) about the transmutation of species. A more fully articulated evolutionary model would be developed later by his grandson, Charles Darwin (1809–1882), with particular reference to habit and instinct in humans and animals. The concepts of habit and instinct had been explored initially almost a century earlier by the late-18th century philosophers of France and Germany, such as La Mettrie and Reimarus, the latter of whom argued that certain features of human behavior—sensation, attention, recognition, and some image associations—could be elements of animal life as well. However, abstract thought and reflection distinguished humans from the other animals. Moreover, skills demonstrated by some animals, e.g., nest building or web spinning, were instinctive, not learned.

Shortly after Erasmus's death, Lamarck proposed that acquired characteristics could be inherited, a notion also developed by several of the 18th century philosophers and which persisted into the 20th century. For example, the psychologist, William McDougall *(2)* published a paper in 1927 in the *British Journal of Psychology* in which he claimed to have bred successive generations of white rats to make a special discrimination.

During the early- to mid-19th century, Alfred Wallace and Charles Darwin developed another, competing, theory. As articulated by Darwin *(3)* in 1859 in *The Origin of Species*, the Darwin-Wallace theory of natural selection was based on three empirical observations:

1. Organisms vary, and these variations are inherited, at least in part, by their offspring.
2. Organisms produce more offspring than could possibly survive.
3. On average, those variants which adapt best to their environments have the better chance of survival and propagation.

Later, in 1872, in *The Expression of the Emotions in Man and Animals*, Darwin *(4)* stated, "not only has the body been inherited by animal ancestors, but there is continuity in respect to mind between animals and humans." Humans and animals have homologous structures, structures derived from a common source. There are, in addition, analogous behaviors in humans and animals. For example, Darwin claimed that the human sneer was a remnant of the animal's preparatory biting response, whereas fist-clenching was a vestigial activity of clawing. Animals and humans were similar in other ways. Both were constituted of innate and acquired behaviors, both could demonstrate imitative behavior, curiosity (exploratory behavior), and, according to Darwin, imagination and reasoning.

In certain other aspects of behavior, however, Darwin did differentiate between humans and infrahuman animals, between inherited traits in animals and learned or acquired features in humans. In 1871, in *The Descent of Man*, Darwin *(5)* distinguished between animal and human societies, including the role played by instincts in animals. In particular, he stated that communication in animals was innate, whereas, in humans, it was learned. Habits, as different from instincts, may be acquired and modified and made more complex; but, unlike the Lamarckian and late-18th century philosophers' theories, Darwin stated that habits could not be inherited from one generation to the next. He wrote that instincts may undergo variation, but unlike habits, may be inherited. Darwin declined to define instinct, stating, "everyone understands what is meant." He did state, however, that instinct impels organisms to perform acts without experience and without knowledge of purpose. Parenthetically, the study of instinctive behaviors would later be embraced by the ethologists of the early mid-20th century.

Darwin's writings addressed the arguments that differentiated humans from other species, and altered and updated the Cartesian point of view. His writings became the impetus for the development of comparative psychology and animal experimental psychology. A year after the *Expression of Emotion in Man and Animals* appeared in print, Douglas Spalding *(6)* published his results from a series of experiments aimed at identifying instincts in birds. In his studies, chicks were either blinded or hooded soon after birth, but were able to flap their wings and fly and to peck at food to eat, from which, Spalding concluded that flying and pecking were instincts, not learned behaviors. In 1884, John Lubbock *(7)* published *Ants, Bees and Wasps*, his experimental studies of individual and social behaviors of insects. Lubbock was the first to introduce the maze to study animal intelligence systematically, anticipating William Small's efforts by more than a decade.

Unlike Spalding and Lubbock, George Romanes *(8)* provided only anecdotal evidence of animal and human behaviors from an anthropomorphic perspective. In his 1882 text, *Animal Intelligence*, he endorsed Darwin's theory of continuity of species between humans and animals. Like Darwin, Romanes assumed continuity manifested itself not only physiologically, but psychologically as well. According to Romanes, animals were not automata, as Descartes had proposed, but had minds similar to humans, because they were capable of making associations and adjustments in learning from their own experiences. The difference between the minds of animals and those of humans lay in the function of the degree to which they were separated on the evolutionary scale. A hierarchy could be established according to the number and extent of the associations animals and humans could make:

1. Simple ideas, such as sensations and perceptions, that are common to humans and animals.
2. Complex ideas, such as compound associations, that affect some animals and humans.
3. Abstract and general thinking, which are unique to humans.

Expressions of affection, sympathy, jealousy, and rage in animals were analogous to those exhibited in humans, and were sufficient to infer mental states in animals corresponding to those observed in humans. Despite the highly suppositional inference, Romanes was convinced that it was valid and that it was the only conclusion that could be drawn. Its validity had special meaning for the evolutionist, in that there must be psychological as well as physiological continuity throughout the animal kingdom.

As chair of Experimental Psychology at Harvard, William James incorporated Darwin's evolutionary theory into the mainstream of psychology. Specifically, he argued that function within species evolved as it adapted successfully to the environment. Consciousness was a function of mind, the organ, which developed and adapted to the complexities of the environment.

The counter-revolution to mentalism and anthropomorphism in comparative psychology arose in the waning years of the 19th century, after C. Lloyd Morgan *(9)* made an appeal to the following rule: "In no case may we interpret an action as the outcome of the exercise of a higher psychical facility, if it can be interpreted as the outcome of the exercise of one which stands lower on the psychological scale" *(9)*.

Morgan derived his rule after constructing three models by which one could compare animals with humans *(9)*. Model 1 asserted that animals and humans were alike in all ways, but animals were without "higher faculties." Model 2 assumed that animals and humans were alike in all ways, but animals were not as highly developed overall. Model 3 suggested that there were many opportunities for error in estimating variations between humans and animals. The first two models were highly anthropomorphic, whereas the third, which Morgan adopted, was closest to his canon.

Interestingly, in 1894, Wilhelm Wundt *(10)* expressed similar thoughts in his *Lectures on Human and Animal Psychology*, despite his being the architect of introspectionism. He noted that comparative psychologists had adopted the point of view of the comparative anatomists and physiologists, i.e., to chart the evolution of mental life in animals and apply that representation directly to humans. Wundt stated that investigators of his time were apt to describe animal behavior as acts of understanding, but that their inferences were more likely to reflect those of the observer than of the animal. He presented an example of the problem of drawing inferences from animal activity *(10)*. Huber, an acute observer of the social behavior of ants, removed an ant from the colony and replaced it some time later, and interpreted the "friendly" responses made by other ants to it on its return as a form of memory. Lubbock, in contrast, performed a similar experiment by removing ant larvae and raising them apart from the colony for several weeks. After their return to the colony, he also noted particularly welcoming responses made by the other ants. Lubbock, however, interpreted these acts as recognition not memory. Wundt recounts the story to invoke Occam's razor: when given the choice of two or more competing theories, choose the one that provides the simplest explanation. Wundt does allow for a form of animal cognition, which he describes as the "assimilative effect of previous impressions upon ones of the same character," and that among certain animals, e.g., cats, dogs, and elephants, their memories are capable of retaining specific items for relatively long periods.

2. ANIMAL MODELS AND THE 20TH CENTURY

Edward L. Thorndike, a student of William James at Harvard University, adopted a position akin to that of Wundt. He observed that anecdotes involving supra-ordinate intelligence in animals were the stories that become widely known. He noted that thousands of abandoned dogs and cats die or were killed in the streets; but should one such animal manage to travel from the Bronx to Brooklyn and happen onto its former master, the story would be printed in all the newspapers. In all such instances, whether observed firsthand or reported by others, Thorndike *(11)* noted that these anecdotes suffer from three important defects:

1. Often it is the single case that is reported.
2. The observation is not repeated, nor are the conditions under which the observation is made well-controlled.
3. The previous history of the animal is not well-known.

Thorndike's remedy for these shortcomings was to perform experiments. He developed what he called the "puzzle box," using it to measure how long it took various species of animals to find the latch that would open its door, and permit the animal to eat the food placed just outside.

Because of the work begun by Thorndike and others, the turn of the 20th century saw the growth of systematic investigations of animal behavior. Before he began his studies of primates, Robert Yerkes examined intelligence (learning) in the turtle *(12)*. He devised a maze-like experimental environment, a rectangular box, in which, to find its nest, the turtle had to negotiate its way from one end to the other through a succession of portals cut through partitions and located in different positions along the walls. Yerkes found that, over successive days of testing, the time to pass through the correct route diminished, just as Thorndike's animals did in the puzzle box *(12)*. Thus, it seemed that all manner of infrahuman species were capable of learning by association, as had been observed in humans. Thorndike's Law of Effect, which would become one of the pillars of behaviorism, was articulated in 1910, as follows:

> Of several responses made to the same situation those which are accompanied or closely followed by satisfaction to the animal will, other things being equal, be more firmly connected with the situation, so that, when it recurs, they will be more likely to recur; those which are accompanied or closely followed by discomfort to the animal will, other things being equal, have their connections to the situation weakened, so that, when it recurs, they will be less likely to occur. The greater the satisfaction or discomfort, the greater the strengthening or weakening of the bond *(13)*.

H. S. Jennings *(14)* went further and argued that there was experimental evidence for the operation of many common psychological processes in humans and lower organisms. In 1906, in his *Behavior of the Lower Organism*, Jennings observed that, like single-cell organisms, humans reacted similarly to changes in temperature and other kinds of stimuli. In addition to responses to stimulation, humans also experienced subjective states of consciousness, e.g., pleasure, pain, anger, fear, and so on. Thus, he considered whether animals may also have subjective states like those found in humans. Jennings noted that, like humans, animals perceived objects, made discriminations and choices, showed evidence of attention, and exhibited characteristics of memory by what was then called habit. Based on these findings, Jennings concluded that the state of consciousness that we understand exists in humans must also be extant in infrahuman organisms, despite the fact that such an inference was neither verifiable nor falsifiable through experimentation and observation.

At approximately the same time, Sir Charles Sherrington *(15)* had written in *The Integrative Action of the Nervous System* that animal behavior—the "puppet-animal"—was more consistent with Descartes' automata than Jennings' conjectures, which were more liberal. That is, animals act without mind. Sherrington remarked that there was no evidence available of the thoughts, feelings, perceptions, memories, and so on, of which Jennings spoke. Although one could enumerate long lists of reflexive behaviors engendered in infrahuman organisms, their bodies react to physical, not psychical, stimuli. A decade later, Jacques Loeb *(16)*, who had written extensively on forced movements and tropisms in plants as models for the objective study of behavior, also remarked that subjectivism and anthropomorphism had no place in a science of biology. In accord with the associationists, he stated that a mechanistic biology would explain human behavior.

On the other hand, Sherrington also stated that, as the life of the organism developed, conscious behavior would replace reflexes *(15)*. In concert with these changes, there would be an increased role for habit. Habits always surface during conscious action, reflexive behaviors never do. The organ that continually modified reflexes in humans was the cerebrum, so that humans could adapt to, and gain advantage over, the environment. Habits are acquired, whereas reflexes are innate and inherited. By stating that habit should never be confounded with innate, reflex action, Sherrington distanced himself from Darwin.

In 1911, in *The Evolution of Animal Intelligence*, S. J. Holmes *(17)* recapitulated arguments by Wundt and others, that, whereas basic sensations, such as fear, may parallel our own sensations, the explanation of the experiences demonstrated by these animals was subject to interpretation and could vary greatly, even among trained observers. Holmes referred to Lloyd Morgan's example of the chick pecking at small objects; that it was not possible to differentiate whether this was the behavior of a sentient creature or an unconscious automaton. However, when some objects, such as pebbles, are rejected, whereas other objects, such as breadcrumbs, are selected, one could infer that some aspect of consciousness or experience had directed its behavior.

Ivan Pavlov, who was awarded the Nobel Prize for his studies of the digestive physiology and glandular secretion, brought the concepts of physical and psychical stimuli closer together. He noted that saliva secreted from submaxillary glands was often elicited by stimuli other than the act of eating food itself. The sight of food, or the sight of someone bringing food, even at some distance from the animal, would also cause the animal to salivate. Contrary to Sherrington's earlier remarks, in 1928, Pavlov observed that psychological and physiological stimuli worked in concert *(18)*, at least insofar as salivary processes were concerned. Nonetheless, in studying psychological phenomena, Pavlov was not concerned about the animal's hypothetical state of mind, but focused his investigations on the observation and description of the effects of distant objects on salivary secretion. He differentiated the innate, physiological salivary reflex from the acquired, psychological reflex evoked by external stimulation as, respectively, the "unconditional" and "conditional" (mistranslated as "unconditioned" and "conditioned") reflexes.

Unlike Pavlov, Margaret Washburn thought that all infrahuman behavior must be interpreted through the lens of human experience, that the meaning of terms such as perception, fear, anger, and so on, can only be understood as processes that occur in our own minds *(19)*. She also noted that the many factions within animal psychology, ranging from those who believed in animal consciousness to those who did not, had now been reduced to these two extreme positions. Like Loeb *(16)*, however, Washburn thought that associative memory was a fundamental process in all psychological phenomena, despite the fact that she also thought that any animal capable of demonstrating learning also showed evidence of consciousness.

As a student of Loeb's, John B. Watson adopted the mechanistic approach of his mentor. According to Watson, psychology was the study of the prediction and control of behavior, in humans or animals, and, to that extent, there was no division between them *(20)*. Unless observations indicated otherwise, reasoning by analogy from human to animal consciousness was unproductive for the behaviorist. The task of psychology was to avoid the introspectionism advanced by Wundt and the notion of consciousness promulgated by Romanes and others. Mental states were not the proper object of study for a science of behavior. Consequently, behaviorism circumvented the mind–body problem raised by Descartes. For Watson, humans and animals responded to the environment according to their hereditary and physiological machinery.

Twenty years after publication of Morgan's Canon, the field of experimental animal psychology had divided into two separate and distinct camps: one followed the principles of contiguity and association; the other, although mindful of these principles, continued to incorporate concepts of mind and consciousness into animal behavior. This second group gained additional support from the investigations by Wolfgang Köhler *(21)*, who had studied primate behavior for many years and reported that apes were capable of more than associational thinking. Köhler devised experiments that required the animal to coordinate several components to produce a solution to the problem. The solution to one of the parts of the problem involved the application of "insight." Although the associationists would explain the phenomenon using the principle of contiguity, Köhler thought that certain problem-solving behaviors were attained otherwise, by establishing "inner" relationships between two things, "based on the properties of these things themselves" *(21)*. Moreover, in 1927, Köhler thought that this other class of mental processes must be considered in addition to the principle of association.

3. ANIMAL BEHAVIOR: LATER STUDIES

The work by Thorndike, Pavlov, Watson, and, later, from the 1930s through the 1960s, B. F. Skinner, provided the foundations for a systematic study of animal behavior from a mechanistic perspective, as articulated in Thorndike's Law of Effect. Skinner, however, not only furnished animal experimental psychology with a standardized chamber into which infrahuman subjects could be isolated, systematically studied, and could have their responses measured; he also presented formal arguments to contrast his operant conditioning from Pavlovian respondent conditioning, and erected a philosophical and conceptual framework for experimental psychology modeled on the operationism embodied in the work of the physicist, Percy Bridgman *(22)*.

By the mid-1930s, nearly half a century of animal experimental research had been published. At this point, Maier and Schneirla *(23)* amassed a large fraction of these studies to better understand comparative psychology and the principles by which organisms adapt to their surroundings. In their encyclopedic text, *Principles of Animal Psychology*, Maier and Schneirla presented the results of many studies of how organisms acquired and maintained new behaviors, what neural mechanisms of behavior could be identified, and how to evaluate the "higher" mental processes.

These researchers wanted to formulate a grand theory of animal behavior from an evolutionary perspective, establishing a hierarchy of responses from inframammalian animals to mammals, and studying the range of activities from simple movements in single-cell organisms, such as the amoeba, to complex behaviors exhibited by primates *(23)*. Their emphasis shifted away from differentiating between innate and acquired behaviors to the methods used to examine the behavior of organisms in relation to their environment. Maier and Schneirla noted the importance of experimental methodology, but stated that the method used to study any one species should be contingent on the animal under investigation and the theory tested. They underscored the differences in the paradigms used to examine acquired behavior, contrasting the associationistic methods of Thorndike, Pavlov, Watson, and Skinner, with the problem-solving techniques developed by Köhler. Like Köhler, Maier and Schneirla thought there was a qualitative difference between the ways in which behaviors were acquired under conditioning paradigms compared with problem-solving activities. According to these authors, the formation of associations (learning) occurs through repetition, and the stereotypical responding, which could also be a behavioral marker for cortical lesions, was, therefore, a lower form of learning. Thought processes, e.g., problem solving, were regarded as higher and different from associative processes. The question was whether such higher processes could be demonstrated in infrahuman animals.

One procedure thought to identify the existence of higher mental processes in animals was known as the delayed reaction method. The technique, developed by Hunter *(24)*, was considered a valid measure of symbolic behavior, the results of which were frequently replicated, albeit with some variation in outcome across species. However, early results indicating that rodents were only able to maintain very short delayed reaction intervals compared with primates were not borne out by later experiments, and the method fell into disrepute. Later, it was thought that abstraction, in general, and concept formation, in particular, could be used to identify the existence of higher mental processes in infrahuman animals. Tasks involving discriminations of triangles—as abstract figures—of different sizes and colors, and what aspects constitute equivalence, were used in studies of various species. However, Maier and Scheirla found that such experiments could not differentiate abstraction from the processes of association and discrimination.

Indeed, one problem central to comparative psychology, as Hinde *(25)* noted, was that it may not be possible to find standard tests suitable to contrast abilities across species. Few procedures used to study learning were appropriate for a wide range of animals, and rates of acquisition may simply reflect sensorimotor differences. One of the few methods that were able to demonstrate cross-species similarities were the experimental boxes developed initially by Thorndike, then radically altered by Skinner to examine operant conditioning. Classical conditioning paradigms were also adapted for many different species. Although acquisition rates of conditioning vary little among vertebrates, rates of discrimination and extinction do *(26)*. Another paradigm, maze learning, also seems to vary among species. That is, mammals learn mazes more rapidly than invertebrates. Another set of procedures, learning sets, has been used to compare how rapidly animals of different species make discriminations. In the learning set, pairs of objects are presented over a fixed number of trials, then replaced with another pair of objects. Several hundred pairs are presented and the percent correct responses recorded for the second trial in the series is used as a measure of learning across species. Results show that primates learn to make discriminations more rapidly than other infrahuman animals *(27)*.

4. EVALUATING ANIMAL BEHAVIOR: OBSERVATIONAL METHODS

Although most animal psychologists continued to emphasize experimental methods to assess the acquisition and maintenance of behaviors, some chose to examine animals in their natural habitats, i.e., their innate ethological behaviors. A century earlier, Darwin's studies of animals, as well as human infants, were entirely observational and nonexperimental. Animal ethology grew out of this noninterventional tradition. The earliest ethologists were trained originally as zoologists, and, as such, developed behavioral taxonomies to demonstrate within-species similarities comparable to the morphological taxonomies of Darwin's time. Therefore, instead of developing experimental procedures, types of nest building among certain bird species and stereotyped behaviors, such as fixed action patterns of fighting and courting observed in fish and water fowl, imprinting and song learning, were all carefully described. These behaviors were thought to be different from both the processes of conditioning described by Pavlov, Watson, and Skinner, as well as the problem-solving skills described by Köhler *(25)*. By establishing phylogenetic relationships among taxonomic groups, researchers, such as Tinbergen *(28)* and Lorenz *(29)*, hoped to reconstruct their advance through the course of evolution. Once similar fixed action patterns could be established across species and their differences noted, it was thought that it should be possible to determine which of these variants were genetically determined, and which behaviors were evolutionarily more primitive. After the evolutionary origin of a specific action pattern was resolved, the transformations in its evolutionary course could be depicted.

Shettleworth examined animal behavior from both an ethological and operant conditioning perspective and observed that some activities initially portrayed as innate, e.g., imprinting and song learning, and thought to fall outside the purview of traditional associationism could, in fact, be examined as forms of associative learning *(30)*. For example, imprinting in goslings was thought to occur at full strength instantaneously, but is apparently a function of both the stimulus object and exposure time to the stimulus *(31)*. Imprinting was also supposed to occur for only one stimulus object, but deVos and van Kampen have shown that, under some conditions, a bird can imprint to three distinct stimuli *(32)*. Like imprinting, song learning was also thought to be acquired during an early critical period. More recent studies, however, have used operant conditioning paradigms to examine how different species of birds acquire songs of their own species as well as of other species, discriminate birdsongs of their own families compared with others within the species, and remember different birdsongs *(33,34)*.

Another problem facing ethologists in formulating the development of behavior from a phylogenetic perspective was that Darwin's theory of evolution did not entail a rank ordering of species, a notion that predated Darwin, as Mackintosh observed *(35)*. The premise that animals adapt to the environment is not equivalent to, or necessarily a function of, increasing complexity or intelligence. However, in his descriptions of animal intelligence, Romanes framed his arguments to show that only humans were capable of the highest forms of thought, i.e., abstraction, and that this ability allowed humans to adapt more quickly to novel situations. On the other hand, as Blackman noted, in applying the principle of biological continuity among animal species in a consistent fashion, one must consider the possibility of moving in either direction along the phylogenetic continuum *(36)*. For example, despite differences in primates' performances on learning sets shown by Warren *(27)*, other researchers have found out that rats can be as efficient *(37)*, and that putative differences in learning may be related to sensory processes *(38)*. In addition, increasing accuracy in learning sets may be related to physiological processes in vertebrates and, therefore, associative learning *(39)*. One solution to these paradoxical findings may be to disentangle some of the issues involving theories of discrimination learning and taxonomic categorization. As Pearce noted, there are many commonalities among the several competing theories that cover human and animal discrimination and categorization, and there are data to suggest that some animals are better able to make relational discriminations than others *(40)*. If supported by additional research, this could provide a means by which to compare learning ability among infrahumans.

Differences between human and infrahuman behavior have been debated at other levels of analysis. For example, Bluton-Jones has argued that cultural adaptation is not associated with modern evolutionary theory in any obvious way *(41)*. In particular, optimization in foraging, as articulated by Marvin Harris and others, is not an evolutionary process, and is not inherited genetically. Humans learn most of their behavior, although learning mechanisms would not have evolved if they had not attained reproductive advantage. Bluton-Jones also stated that humans are not ordinarily concerned about how many descendants they will spawn, nor is it likely that animals consider this an issue either. If mental processes were inherited as Darwin and his many successors suggested, and govern adaptive behavior to the environment and culture, how shall we account for the differences in the speeds at which these processes advance, the different directions taken by individuals, and in different cultures, unless interaction with environment is considered as part of the equation? Describing an event as an adaptation to the environment does not explain how the adaptation transpires.

In summary, the development of comparative and animal experimental psychology in the 100 yr between the mid-19th to the mid-20th century could be described as a radical reformation of thought regarding behavior in human and infrahuman organisms. Before Darwin, animals were considered automata; different and lesser creatures than humans. After Darwin, the evolutionists were determined to demonstrate the continuity of animals and humans, in structure and function. In terms of function, some comparative psychologists were keen to demonstrate the similarities between humans and animals, going so far as to link learning and behavior, and emotional responses, across species through the conduit of consciousness and mind. Others, such as the ethologists, preferred to regard much behavior as innate and triggered by stimuli. The associationists—Thorndike, Pavlov, Watson, Skinner—chose a simpler path, mindful of Morgan's Canon and Occam's Razor, examining behavior and identifying principles of learning and memory that eschewed references to models that embraced imprecise concepts or lacked operational definitions. Attempts to quantify unobservable mental processes, such as consciousness, problem solving, or abstract reasoning were primarily hypothetical and eschewed by behaviorists for their lack of rigor. The camps remained divided until cognitive psychology emerged as a discipline many experimental psychologists from both sides could accept.

Although it was a powerful influence in studying animal behavior, behaviorism never dominated the field of psychology. Alternative theories of behavior, such as those proposed by Köhler, or the ethologists, or by Freud and his followers during the early part of the 20th century and by applied psychologists after World War II, also found sizeable audiences. To these factions, associationism seemed limited in what could be achieved in understanding human behavior by using bar pressing, key pecking, and maze running as measures.

5. COGNITIVE PSYCHOLOGY AND NEUROSCIENCE

As enthusiasm for behaviorism began to wane, cognitive psychology offered accounts that reinstated the concepts of consciousness and mentalism to psychology, primarily through the writings by Hebb *(42)* and the earlier works of Lashley *(43)*. Although precursors can be found in the late-19th century, cognitive psychology was indebted to computer science and statistical decision theory for its modern foundations *(44)*. As Watkins noted *(45)*, cognitive psychology saw intelligence and problem solving as its principal concerns. To investigate these activities, Neisser *(46)* proposed that psychologists, and especially cognitive psychologists, examine the neurodynamics of sensorimotor processes. Two of cognitive psychology's other early proponents in the 1960s, Alan Newall and Herbert Simon, examined problem-solving behavior by synthesizing methodologies from several disciplines: mathematical logic, game theory, and computer science *(47)*. Individuals were no longer stimulus–response mechanisms, as the behaviorists argued, they were information processors containing dynamically organized, goal-seeking programs. From these latter theories and models, cognitive processes would emerge. Stimuli became signals to a processing system that coded and stored information.

At approximately the same time that cognitive psychology began to come into its own, neuroscience surfaced as a field unifying several brain-related disciplines. The early sources of neuroscience can be found in the late-19th century investigations by

Ramon and Cajal, but the origins of modern neuroscience are generally attributed to the studies of neural impulses in the 1930s by Cole and Curtis and the neurobiology of synaptic transmission delineated in the 1950s by Fatt and Katz (*see* ref. *48* for a comprehensive overview). A multidisciplinary approach of neuroscience emerged as investigators focused on problems pertaining to brain function: how signals were transmitted within and between neurons, how genetic variants presaged normal and abnormal behavioral development, and how learning and memory were influenced by these factors.

The commingling of molecular neurobiology, genetics, and learning and memory brought about the field of cognitive neuroscience. The amalgamation of neuroscience with cognitive psychology can be found in Eccles' *The Neurophysiological Basis of Mind (49)*. Research during this period of the mid-20th century centered on sensory processes, but, within the last 20 yr, cognitive neuroscience has concentrated its efforts on the "higher" mental processes: sensorimotor integration, learning and memory, and consciousness.

Storage and information processing became a hub of activity for cognitive neuroscience, but the experimental analysis of higher mental processes began with the work by Evarts *(50)*. Evarts examined the neural correlates of movement, and led the way for other researchers to investigate motor and perceptual tasks in infrahuman primate behavior (*see* ref. *48*). The development of noninvasive neuroimaging technologies, e.g., functional magnetic resonance imaging and positron emission tomography scans, made possible studies of real-time neurological activity in humans and infrahuman animals as they performed tasks associated with various cognitive processes.

6. BEHAVIOR AND GENETICS

The inheritance of traits by humans and infrahuman organisms proposed by Darwin evolved into the study of behavioral genetics, the systematic study of which is best attributed to Darwin's cousin, Francis Galton. Among other things, Galton was interested in the inheritance of human intelligence and, from his study of twins, remarked on the "preponderating effect of nature over nurture" *(51)*.

If intelligence could be inherited, it would only be a short inductive leap to infer that intellectual disabilities and mental disorders could be inherited as well. Langdon Down was the first physician to note organic forms of mental retardation (MR) and, along with other prominent physicians of the mid-19th century, began to classify MR according to its extent, clinical features, and possible causes. Down proposed three possible etiologies, one of which was congenital *(52)*. Many forms of MR that we now know to be brought about genetically were identified clinically during his time, e.g., neurofibromatosis 1 and tuberous sclerosis *(52)*. The role genetics played in producing MR, however, would have to wait for the biological, cytogenetic, and molecular genetic revolutions of the mid- to late-20th century, beginning with the discovery of deoxyribonucleic acid (DNA) in 1953. Using cytogenetic methods shortly after the discovery of the structure of DNA, trisomy 21 (Down syndrome) was identified by Lejeune and his associates *(53)*. A decade later, also using cytogenetic techniques, Lubs identified the fragile X mutation in a family in which MR was segregating as an X-linked disorder *(54)*.

Although the earliest studies of mental disorders in the United States are usually associated with the 18th-century physician, Benjamin Rush (1745–1813), perhaps the greatest influence in 20th-century psychiatry was Emil Kraepelin (1856–1926). Kraepelin was a keen observer and the first to organize the many details of psychiatric information into a cohesive set of procedures. He maintained copious records of his patients during the course of their illness, diagnoses, and treatment. As a result, he was able to group disorders along with their natural histories, which provided an archive for making future diagnoses, setting treatments, and determining prognoses. The English edition of his classic text, *Compendium der Psychiatrie*, set the stage for classification of psychopathology in the United States. His descriptions of dementia praecox (schizophrenia) and manic-depression (bipolar disorder) made possible later taxonomic groupings of these disorders *(55)*.

As with the development of a suitable taxonomy, research in psychopathology is a relatively recent phenomenon. Some of the earliest efforts can be traced to the late-19th century, when Blackburn attempted to establish a relationship between brain and mental dysfunction. In the 20th century, however, Adolph Meyer (1866–1950) brought a psychobiological approach to the field. Meyer eventually became the director of the Psychiatric Institute in New York City, where research in psychopathology had begun in the late-19th century, before his directorship. At approximately the same time, the psychophysiological work of Sherrington, Cannon, and Pavlov began to appear and become known. The relationship between psychological and physiological systems was essential to investigations of the association between physiology, biology, and psychopathology.

In the early mid-20th century, the role of genetics as an etiological factor in psychopathology was explored using Galton's twin-study techniques to determine whether bipolar disorder and schizophrenia could be inherited. Prevalence rates of bipolar disorder and schizophrenia were about 1% of the general population, but concordance rates among monozygotic and dizygotic twins ran more than an order of magnitude higher *(56)*. However, the genetic component, although strong *(57)*, is not sufficient to produce the disorder by itself; therefore, other developmental and environmental factors are involved *(58)*.

7. ANIMAL MODELS OF NEUROBEHAVIORAL DISORDERS

The ascendance of cognitive psychology and cognitive neuroscience provided the epistemological framework in which animal models of cognitive impairments could be developed and studied. If the Darwinian theory of continuity of species was correct, and if the processes of learning and memory in infrahuman organisms and humans developed in parallel, one could maintain that cognitive impairment and psychological dysfunction in humans could also be replicated in animals. Recently, Anderson *(59)* argued that, for an animal model to demonstrate cognitive deficits, it must exhibit deficiencies along two dimensions noted earlier in animal experimental research:

1. Learning, as defined by the associationists.
2. Insight, as promoted by Köhler.

Studies by Thompson and his colleagues *(60)*, in which lesion-induced learning deficits were assessed, produced one such model in rats. Despite the lack of surgical precision in lesioning selected portions of the brain, Thompson et al. were able to demonstrate cognitive deficits in white rats, using a battery of novel problem-solving devices designed to measure aversively motivated problems (visual discrimination), appetitively motivated problems (climbing detour), puzzle-box problems, and an inclined plane discrimination to examine sensorimotor difficulties. Using statistical factor analysis, these researchers obtained a psychometric measure of intelligence (g), from these six performance tasks and were able to discriminate rats whose brains had been lesioned from those that were sham-surgical controls *(60)*.

In addition to the recent work by Thompson, another animal behavioral procedure that advanced the study of cognitive impairment had been conceived a decade earlier. That was the development of the Morris water maze (MWM). Morris was originally concerned with spatial localization in the rat *(61)*. He designed a method that used a circular pool approx 1 m in diameter that could be filled with milky water. The tub could contain one of two types of platforms: one that was elevated above the surface of the water; the other was receded just beneath water level and not visible. When the platform was visible, the animal could locate it using visual cues on the platform itself ("proximal" cues). When the platform was under water, the animal could locate it by using cues on the walls outside the tub ("distal" cues).

Morris and his colleagues later demonstrated that spatial learning was impaired in rats in which the hippocampus was lesioned *(62)*. Previously, it had been observed that humans with hippocampal defects exhibited severe deficits in learning and memory *(63)*. Using operant conditioning procedures, investigations by Rudy and Sutherland *(64)* demonstrated that rats with hippocampal lesions could acquire simple stimulus discriminations, e.g., light or tone alone, but were incapable of acquiring compound negative patterning discriminations, e.g., light plus tone, or maintaining compound discriminations that had been acquired previously. From their results, Rudy and Sutherland concluded that the hippocampus was essential in learning and remembering complex discriminations. Using the MWM, Gerlai and his colleagues *(65)* obtained comparable results in mice that had sustained bilateral ablations of the hippocampus. The additional discovery of hippocampal long-term potentiation (LTP) in the mid-1960s, along with many findings about its properties—its persistence, synaptic specificity, and associativity—made it the leading candidate for establishing a neurobiological basis for learning and memory in humans and infrahuman organisms *(66)*. Indeed, Morris et al. showed that impaired LTP was associated with deficits in learning and memory in the MWM. In Eric Kandel's lab, Grant demonstrated that *fyn* mutant mice also exhibit deficits in spatial learning in the MWM and show altered LTP compared with wild-type controls *(67)*.

In addition to his work on conditional salivary responses, Pavlov examined the effects of aversive stimulation on behavior, and how emotional responses and psychological dysfunction could be acquired *(68)*. By establishing the appropriate experimental conditions in animals, Pavlov was able to generate a chronic behavioral disturbance, what he called a neurosis. He also stated that human neuroses should be understood and explained by animal dysfunction *(68)*. One brain structure associated with emotional responses, specifically, anxiety and fear, has been the amygdala. In humans, Andy *(69)* successfully treated a patient with seizures and anxiety-related psychiatric disorders by performing a bilateral amygdalotomy. Previously, Goddard showed the effects of avoidance conditioning in rats after cholinergic stimulation of the amygdala *(70)*. Reduced fear responses had also been recorded in mice whose amygdalas were lesioned *(71)*. However, the role of the amygdala in fear-related responses has often been debated *(72)*. Wilensky and her colleagues *(73)* have shown that Pavlovian fear conditioning (as opposed to inhibitory avoidance learning) exhibit dissimilar effects after the N-methyl-D-aspartate receptor antagonist, AP5, is infused into the amygdala. Their results suggest that there are likely differences in neural pathways mediating memory in the two fear conditioning procedures.

Thus, cognitive deficits and psychopathology were, to some extent, demonstrated in infrahuman organisms.

8. ANIMAL MODELS, GENETICS, AND NEUROBEHAVIORAL PSYCHOPATHOLOGY

The revolution in animal model development, particularly mouse models, was brought about by the rapid growth in molecular genetics and neuroscience in the late-20th century. Using embryonic stem cell technology and other molecular techniques for modifying gene expression, e.g., knockout or transgenic mice, the creation of mutant mice accelerated the means by which altered mouse strains could be generated and bred, and obviated the need to use pharmacological or surgical techniques to produce neurological dysfunction. In conjunction with studies of LTP and its relationship to cognitive deficits observed in the MWM, knockout and transgenic mice have produced some of the most powerful means by which to examine a wide variety of behavioral disorders, from autism to MR to schizophrenia. Crawley *(74)* has noted that the recent advances in gene mutation technology have produced more than 100 genes expressed in the central nervous system that generate a variety of dysfunctional behavioral phenotypes. Indeed, murine models of human genetic disorders producing MR have been generated for the fragile X mutation *(75)*, Down syndrome *(76)*, non-syndromal X-linked MR *(77)*, Rett syndrome *(78)*, Prader-Willi syndrome *(79)*, velo-cardiofacial syndrome *(80)*, Williams syndrome *(81,82)*, and myotonic dystrophy *(83)*. Animal models of psychiatric disorders and pervasive developmental disabilities have also been created to study the effect of age on cognitive ability *(84)*, Alzheimer's disease *(85)*, phenylketonuria *(86)*, lissencephaly *(87)*, depression *(88)*, anxiety *(89,90)*, and schizophrenia *(80,91,92)*.

Unfortunately, the benefits of technology do not come without cost. Wahlsten has observed that several factors that may affect the behavioral phenotype are not directly related to the intended genetic modification *(93)*. For example, when knockout mice are created, flanking alleles of the intended gene may also be knocked out, which may have additional unintended phenotypic consequences. Often, differing effects on learning and memory will be generated depending on the genetic strain into which the mutation is bred. Phillips et al. *(94)* summarized the strengths and weaknesses of genetic techniques used to investigate complex behavioral traits in mice. They noted that targeted gene mutations, whether obtained by knockout technology or transgenic overexpression, raise issues in two fundamentally important areas:

1. More often than not, the mutation is implemented in one or two genetic strains, when it should be implemented in several.
2. More often than not, only one or two tests to evaluate the phenotype are used instead of a battery of behavioral tasks.

Additionally, as Crabbe et al. found, despite attempts to standardize procedures, outcomes of behavioral studies may vary from laboratory to laboratory *(95)*.

In the end, the extent to which animal models, in general, and mouse models, in particular, emulate human dysfunction relate back to the underlying theory of Darwinian evolution and whether there is continuity of species and continuity at all levels of analysis. The problem with nearly all neuropsychiatric disorders, whether they refer to learning disabilities and MR; pervasive developmental disorders, such as autism; psychotic disorders, such as schizophrenia; or mood disorders, such as depression and bipolar disorder, is that they are not without difficulty transformed into operationalized measures at the infrahuman level. In addition to other aberrant behaviors, mood disorders are inherently associated with negative intrasubject verbal reports, accounts that cannot possibly be replicated at the infrahuman level. Those behaviors that can be evaluated, e.g., using the forced swim test, social interactions, and changes in physical activity and sleep patterns, may play a role in depression, but may not be sufficient to meet the criteria for face validity or causal validity. Human depression may not be directly correlated with mouse depression, assuming the latter exists. This is especially the case if one adopts the philosophical premises of mid-20th-century associationism and behaviorism, for which the notion of consciousness—not to mention depression or any other neuropsychiatric disorder—in infrahuman organisms is an unobservable phenomenon. Given the current limitations in our knowledge of the pathophysiology of neuropsychiatric disorders, can either construct or causative validity in murine models be evaluated in any meaningful way at the human level?

9. CONCLUSION

Nearly 150 yr after the publication of *The Origin of Species*, the Darwin-Wallace theory of evolution continues to dominate the way in which we think about the similarities and differences between human and infrahuman species, both in structure and function. The controversy swirling about the subjective notions of mind and consciousness, which began shortly afterward in the late-19th century, continues to this day. The rise of associationism during the 19th century, along with the parallel ascendance of empiricism, operationism, and behaviorism, brought about important developments in understanding animal and human behavior, but never triumphed over other theories of comparative animal psychology. The emergence of cognitive psychology as alternative paradigm to the study of behavior, as well as subsequent advances in cognitive neuroscience and molecular genetics, has taught us much about important relationships between genes, brain, and behavior. With the discovery of the structure of DNA and the revolution in molecular biology, molecular genetics provided the bases from which various neurobehavioral and neuropsychiatric dysfunctions, from learning disabilities to schizophrenia, may arise. However, it remains an open question whether abstract thought, problem solving, intelligence, and emotions—not to mention the disorders associated with each of these—can be adequately modeled in infrahuman organisms.

REFERENCES

1. Royce JE. Man and His Nature. New York: McGraw-Hill, 1961, p. 3.
2. McDougall W. An experiment for testing the hypothesis by Lamarck. Br J Psychol 1927;17:267–304.
3. Darwin CR. On the Origin of Species by Means of Natural Selection, or the Preservation of Favoured Races in the Struggle for Life. London: John Murray, 1859.
4. Darwin C. The Expression of the Emotions in Man and Animals. London: John Murray, 1872.
5. Darwin C. The Descent of Man. 2 Vols. London: John Murray, 1871.
6. Spalding DA. Instinct: with original observations oNYoung animals. MacMillan's Magazine 1873;27:282–293.
7. Lubbock J. Ants, Bees, and Wasps. New York: D. Appleton, 1884.
8. Romanes GS. Animal Intelligence. London: Kegan Paul, 1882.
9. Lloyd Morgan C. An Introduction to Comparative Psychology. London: Walter Scott, 1894.
10. Wundt W. Lectures on Human and Animal Psychology. Creighton JE, Titchener EB, trans. London: Swan Sonnenshein & Sons, 1894.
11. Thorndike EL. Animal intelligence: an experimental study of the associative processes in animals. The Psychological Review Monograph Supplements, 1898; 8.
12. Yerkes RM. The formation of habits in the turtle. The Popular Science Monthly 1901;58:519–529.
13. Thorndike EL. The Elements of Psychology. New York: Seiler, 1905, p. 203.
14. Jennings HS. Behavior of Lower Organisms. New York: Columbia University Press, 1906.
15. Sherrington C. The Integrative Action of the Nervous System. Reprinted 1947. New Haven: Yale University Press, 1906/1947.
16. Loeb J. Forced Movements, Tropisms, and Animal Conduct. Reprinted 1973. New York: Dover Publications, 1918/1973.
17. Holmes SJ. The Evolution of Animal Intelligence. New York: Henry Holt, 1911.
18. Pavlov IP. Lectures on the Conditioned Reflexes. Gantt WH, trans. New York: Liveright Publishing, 1928.
19. Washburn MF. The Animal Mind. 3rd ed. New York: MacMillan, 1930.
20. Watson JB. Psychology as the behaviorist views it. Psychological Review 1913;20:158–177.

21. Köhler W. The Mentality of Apes. Winter E, trans. New York: Harcourt Brace, 1927.
22. Skinner BF. The Behavior of Organisms. Reprinted 1966. Englewood Cliffs: Prentice-Hall, 1938.
23. Maier NRF, Schneirla TC. Principles of Animal Psychology. Reprinted 1964, New York: Dover Publications, 1935.
24. Hunter WS. The delayed reaction in animals and children. Behavior Monographs 1913;2:86.
25. Hinde RA. Animal Behaviour: A Synthesis of Ethology and Comparative Psychology. 2nd ed. New York: McGraw-Hill, 1970.
26. Voronin LG. Some results of comparative-physiological investigations of higher nervous activity. Psychol Bull 1962;59:161–195.
27. Warren JM. Learning in Vertebrates. In: Dewsbury DA, Rethlingshafer DA, eds. Comparative Psychology: A Modern Survey. New York: McGraw-Hill, 1973.
28. Tinbergen N. The Study of Instinct. Oxford: Clarendon, 1951.
29. Lorenz KL. The comparative method in studying innate behaviour patterns. Symp Soc Exp Biol 1950;4:221–268.
30. Shettleworth SJ. Biological Approaches to the Study of Learning. In: Mackintosh NJ, ed. Animal Learning and Cognition. San Diego: Academic, 1994, pp. 185–219.
31. Ten Cate C. Behavioral development: towards understanding processes. Perspectives in Ethology 1989;8:243–269.
32. de Vos GJ, van Kampen HS. Effects of primary imprinting on the subsequent development of secondary filial attachments in the chick. Behaviour 1993;125:245–263.
33. Ten Cate C, Los L, Schilperood L. The influence of differences in social experience on the development of species recognition in zebra finch males. Anim Behav 1984;32:852–860.
34. Stoddard PK, Beecher MD, Loesche P, Campbell SE. Memory does not constrain individual recognition in a bird with song repertoires. Behaviour 1992;122:271–284.
35. Mackintosh NJ. Approaches to the study of animal intelligence. Br J Psychol 1988;79:509–525.
36. Blackman DE. On the Cognitive Theories of Animal Learning: Extrapolation from Humans to Animals? In: Davey GCL, ed. Animal Models of Human Behavior. London: John Wiley & Sons, 1983, pp. 37–50.
37. Zeldin RK, Olton DS. Rats acquire spatial learning sets. J Exp Psychol 1986;37B:295–311.
38. Devine JV. Stimulus attributes and training procedures in learning-set formation of rhesus and cebus monkey. J Comp Physiol Psychol 1970;73:62–67.
39. Macphail EM. Brain and Intelligence in Vertebrates. Oxford: Oxford University Press, 1982.
40. Pearce JM. Discrimination and Categorization. In: Mackintosh NJ, ed. Animal Learning and Cognition. San Diego: Academic, 1994, pp. 109–134.
41. Bluton-Jones NG. Two Investigations of Human Behavior Guided by Evolutionary Theory. In: Davey GCL, ed. Animal Models of Human Behavior, London: John Wiley & Sons Ltd, 1983, pp. 179–204.
42. Hebb DO. The Organization of Behavior. New York: Wiley, 1949.
43. Lashley KS. Brain Mechanisms and Intelligence. Chicago: University of Chicago Press, 1929.
44. Hamilton V, Vernon MD. The Development of Cognitive Processes. London: Academic, 1976.
45. Watkins MJ. An Experimental Psychologist's View of Cognitive Science. In: Lister RG, Weingartner HJ, eds. Perspectives in Cognitive Neuroscience. Oxford: Oxford University Press, 1991, pp. 132–144.
46. Neisser U. Cognitive Psychology. New York: Appleton Century Crofts, 1967.
47. Newall A, Simon HA. Human Problem Solving. Englewood Cliffs: Prentice-Hall, 1972.
48. Cowan WM, Harter DH, Kandel ER. The emergence of modern neuroscience: some implications for neurology and psychiatry. Annu Rev Neurosci 2000;23:343–391.
49. Eccles JC. The Neurophysiological Basis of Mind: The Principle of Neurophysiology. Oxford: Clarendon, 1953.
50. Evarts EV. Relation of pyramidal tract activity to force exerted during voluntary movement. J Neurophysiol 1968;31:14–27.
51. Galton F. Inquiries into Human Faculty and its Development. New York: MacMillan, 1883.
52. Scheerenberger RC. A History of Mental Retardation. Baltimore: Brookes, 1982.

53. Lejeune J, Gautier M, Turpin R. Etudes des chromosomes somatiques de neuf enfants mongoliens. CR Academie de Science 1959;248:1721.
54. Lubs HA. A marker X chromosome. Am J Hum Genet, 1969;21:231–244.
55. Kraepelin E. Compendium der Psychiatrie zum Gebrauche für Studirende und Aerzte. Leipzig: Abel Verlag, 1883.
56. Kallman FJ. The Genetics of Mental Illness. In: Arieti S, ed. American Handbook of Psychiatry. New York: Basic Books, 1959.
57. Gottesman II. Schizophrenia Genesis—The Origins of Madness. New York: WH Freeman, 1991.
58. Cannon TD, Thompson PM, van Erp TG, et al. Cortex mapping reveals regionally specific patterns of genetic and disease-specific gray-matter deficits in twins discordant for schizophrenia. Proc Natl Acad Sci USA 2002;99(5):3228–3323.
59. Anderson B. Role for animal research in the investigation of human mental retardation. Am J Ment Retard 1994;99:50–59.
60. Thompson R, Crinella FM, Yu J. Brain Mechanisms in Problem Solving and Intelligence. New York: Plenum, 1990.
61. Morris RGM. Spatial localization does not require the presence of local cues. Learn Motiv 1981;12:239–260.
62. Morris RGM, Garrud P, Rawlins JNP, O'Keefe J. Place navigation impaired in rats with hippocampal lesions. Nature (London) 1982;297:681–683.
63. Scoville WB, Milner BJ. Loss of recent memory after bilateral hippocampal lesions. J Neurol Neurosurg Psychiatry 1957;20:11–21. Reprinted in J Neuropsychiatry Clin Neurosci 2000;12:103–113.
64. Rudy JW, Sutherland RJ. The hippocampal formation is necessary for rats to learn and remember configural discriminations. Behav Brain Res 1989;34:97–109.
65. Gerlai RT, McNamara A, Williams S, Phillips HS. Hippocampal dysfunction and behavioral deficit in the water maze in mice: an unresolved issue? Brain Res Bull 2002;57:3–9.
66. Morris RGM. The Neural Basis of Learning with Particular Reference to the Role of Synaptic Plasticity. In: Mackintosh NJ, ed. Animal Learning and Cognition. San Diego: Academic, 1994.
67. Grant SG, O'Dell TJ, Karl KA, Stein PL, Soriano P, Kandel ER. Impaired long-term potentiation, spatial learning, and hippocampal development in fyn mutant mice. Science 1992;258(5090):1903–1910.
68. Pavlov IP. Lectures on Conditioned Reflexes Volume II. Conditioned Reflexes and Psychiatry. WH Gantt, trans. New York: International Publishing, 1941.
69. Andy OJ. Psychomotor-psychic seizures treated with bilateral amygdalotomy and orbitotomy. South Med J 1976;69:88–93.
70. Goddard GV. Analysis of avoidance conditioning following cholinergic stimulation of amygdala in rats. J Comp Physiol Psychol 1969;68:1–18.
71. Slotnick BM. Fear behavior and passive avoidance deficits in mice with amygdala lesions. Physiol Behav 1973;11:717–720.
72. Maren S. The amygdala, synaptic plasticity, and fear memory. Ann NY Acad Sci 2003;985:106–113.
73. Wilensky AE, Schafe GE, LeDoux JW. The amygdala modulates memory consolidation of fear-motivated inhibitory avoidance learning but not classical fear conditioning. J Neurosci 2000;20:7059–7066.
74. Crawley JN. Behavioral phenotyping of transgenic and knockout mice: experimental design and evaluation of general health, sensory functions, motor abilities, and specific behavioral tests. Brain Res 1999;835:18–26.
75. Dutch-Belgian Fragile X Consortium. Fmr1 knockout mice: a model to study fragile X mental retardation. Cell 1994;15:23–33.
76. Reeves RH, Irving NG, Moran TH, et al. A mouse model for Down syndrome exhibits learning and behaviour deficits. Nature Genet 1995;11:177–184.
77. D'Adamo P, Welzl H, Papadimitriou S, et al. Deletion of the mental retardation gene Gdi1 impairs associative memory and alters social behavior in mice. Hum Mol Genet 2002;11:2567–2580.

78. Shahbazian M, Young J, Yuva-Paylor L, et al. Mice with truncated MeCP2 recapitulate many Rett syndrome features and display hyperacetylation of histone H3. Neuron 2002;35:243–254.
79. Muscatelli F, Abrous DN, Massacrier A, et al. Disruption of the mouse Necdin gene results in hypothalamic and behavioral alterations reminiscent of the human Prader-Willi syndrome. Hum Mol Genet 2000;9:3101–3110.
80. Liu H, Abecasis GR, Heath SC, et al. Genetic variation in the 22q11 locus and susceptibility to schizophrenia. Proc Natl Acad Sci USA 2002;99:16,859–16,864.
81. DeSilva U, Elnitski L, Idol JR, et al. Generation and comparative analysis of approximately 3.3 Mb of mouse genomic sequence orthologous to the region of human chromosome 7q11.23 implicated in Williams syndrome. Genome Res 2002;12:3–15.
82. Bayarsaihan D, Dunai J, Greally JM, et al. Genomic organization of the genes Gtf2ird1, Gtf2i, and Ncf1 at the mouse chromosome 5 region syntenic to the human chromosome 7q11.23 Williams syndrome critical region. Genomics 2002;79:137–143.
83. Seznec H, Agbulut O, Sergeant N, et al. Mice transgenic for the human myotonic dystrophy region with expanded CTG repeats display muscular and brain abnormalities. Hum Mol Genet 2001;10:2717–2726.
84. Gallagher M, Rapp PR. The use of animal models to study the effects of aging on cognition. Annu Rev Psychol 1997;48:339–370.
85. Kordower JH, Gash DM. Animal models of age- and disease-related cognitive decline: perspectives on the models and therapeutic strategies. Neurobiol Aging 1988;9:685–689.
86. Strupp BJ, Levitsky DA, Blumstein L. PKU, learning, and models of mental retardation. Dev Psychobiol 1984;17:109–120.
87. Paylor R, Hirotsune S, Gambello MJ, Yuva-Paylor L, Crawley JN, Wynshaw-Boris A. Impaired learning and motor behavior in heterozygous Pafah1b1 (Lis1) mutant mice. Learn Mem 1999;6:521–537.
88. Dalvi A, Lucki I. Murine models of depression. Psychopharmacology (Berl) 1999;147:4–6.
89. Clement Y, Calatayud F, Belzung C. Genetic basis of anxiety-like behaviour: a critical review. Brain Res Bull 2002;57:57–71.
90. Flint J. Genetic effects on an animal model of anxiety. FEBS Lett 2002;529:131–134.
91. Tremolizzo L, Carboni G, Ruzicka WB, et al. An epigenetic mouse model for molecular and behavioral neuropathologies related to schizophrenia vulnerability. Proc Natl Acad Sci USA 2002;99:17,095–17,100.
92. Ellenbroek BA, Cools AR. Apomorphine susceptibility and animal models for psychopathology: genes and environment. Behav Genet 2002;32:349–361.
93. Wahlsten D. Standardized tests of mouse behavior: reasons, recommendations, and reality. Physiol Behav 2001;73:695–704.
94. Phillips TJ, Belknap JK, Hitzemann RJ, Buck KJ, Cunningham CL, Crabbe JC. Harnessing the mouse to unravel the genetics of human disease. Genes Brains Behav 2002;1:14–26.
95. Crabbe JC, Wahlsten D, Dudek BC. Genetics of mouse behavior: interactions with laboratory environment. Science 1999;284:1670–1672.

2
Transgenic Mouse Models and Human Psychiatric Disease

Jonathan Flint

Summary

Genetic susceptibility to common psychiatric disease arises from the complex interactions between a multitude of genes and an unknown number of relevant environments. However, a common method for investigating gene function involves the creation of a mouse knockout of a candidate gene. Although this approach seems inappropriate to model such complexity, genetic effects on behavior attributable to null mutants in the mouse are in fact subject to the same set of complications, the same gene by environment and epistatic interactions that characterize genetic effects in psychiatric illness. Consideration of the genetic architecture of behavior indicates that even when the molecular lesion is sufficient to inactivate the gene or in other ways alter its function substantially, the effect on the phenotype is typically very mild. Overall, the explanation for the behavior may not be as complex, but it is the product of the same factors. Consequently, it may be possible to take apart the pathway from gene to psychiatric illness.

Key Words: Transgene; psychiatry, QTL; behavior; genetics.

1. INTRODUCTION

The ability to create animal models of human disease by specifically inactivating genes known to play a part in the human condition has transformed the way we investigate the pathophysiology of illness. Obtaining and analyzing a mouse knockout is now a routine part of the functional investigation of genetic disease, and knockout animals are a standard tool in many areas of biology. Nevertheless, using results obtained in a genetically abnormal mouse to solve a problem in human pathology has not always been easy, as is evident in relatively well-understood disorders, such as the inherited anemias *(1,2)*. We can expect interpretation of knockout experiments to be much more problematic when we deal with the behavioral effects of mutations in mice, because we know little about the biology of their behavior. Although there are clearly behavioral domains that are conserved between species, fearfulness for example, others, such as speech disorders, are not. Thus, in some cases, it is not at all clear whether a transgenic mouse will tell us anything about the

From: *Contemporary Clinical Neuroscience: Transgenic and Knockout Models of Neuropsychiatric Disorders*
Edited by: G. S. Fisch and J. Flint © Humana Press Inc., Totowa, NJ

phenotype, a consideration that will not stop those who have identified a gene for dyslexia from making a mouse mutant. Because so much of the genome is conserved between mouse and human species, the expectation is that any knockout will deliver some useful information. Is this really true? In this chapter, the features that complicate the use of transgenic animals for behavioral scientists working with models of psychiatric illness are reviewed.

Presented first are factors that impact the interpretation of any gene knockout experiment: the side effects of the gene knockout technology and the importance of taking into account developmental and tissue-specific effects. However, we are concerned here with factors that more specifically confound investigation of psychiatric disease. These include the compensatory effects of other genes, the difficulties of using a single gene knockout to model a complex phenotype, and the unknown determinants of variation in behavior.

2. EFFECTS CAUSED BY TRANSGENESIS: COMPLICATIONS OF ENGINEERING

The indirect effects from the process used to create the null mutation can be confused with the effects of the induced mutation itself. There are several instances in which disruption of a gene by homologous recombination and the subsequent introduction of exogenous sequence into the genome result in a change in the expression pattern of a neighboring gene or genes. To select for the rare occasions in which homologous recombination produces a knockout, a selectable marker for neomycin resistance is targeted to a crucial portion of the coding region of the gene of interest. This process can itself alter gene expression. Thus, for example, disruption of the *Hoxd-10* homeobox gene (a gene involved in patterning the embryo) leads to altered expression of other *Hox* genes: in embryos, ectopic expression of the *Hox-9* gene was found in the spinal cord of embryos carrying the targeted gene *(3)*. Furthermore in this example, the neomycin gene exhibited a *Hox*-like expression pattern, indicating that its promoter was regulated as if it were a *Hox* promoter *(3)*. Similarly, disruption of two other multigene complexes, the *granzymes B* locus, and the β-*globin* locus control region, also resulted in altered expression of other genes as much as 100-kb distant in the locus downstream insertion *(4)*. Although examples have not been documented for behavioral phenotypes, there is no reason to suppose that they do not exist.

3. EFFECTS CAUSED BY TRANSGENESIS: CONSEQUENCES OF DEVELOPMENTAL CHANGE

Since the introduction of targeted mutagenesis, there has been a concern that the behavioral effects could be a result of the secondary developmental consequences of inactivating the gene, particularly because so many genes implicated in behavior also have a role in development *(5)*. In many cases, the concern has been theoretical, as the accumulated evidence on knockouts influencing tests of spatial memory have shown: the effects are indeed caused by the mutations and not secondary to other processes. However, there is now one case in which the effect of the mutation has been shown to depend on the time at which it occurs. It has been shown that serotonin (5-hydroxytryptophan [5-HT]) 1a receptors act during development to establish anxiety responses in the adult mouse *(6)*.

The role of serotonin in regulating mood is well established: an increase in the levels of serotonin has an antidepressant effect whereas depleting serotonin is depressogenic. The pharmacological consequences are also well known, as attested by the effectiveness of selective serotonin reuptake inhibitors (such as Prozac) in treating depressive illness. After the discovery in the 1980s that buspirone, a drug that acts at 5-HT1a receptors, is anxiolytic, serotonin has also been implicated in the regulation of anxiety.

From tests of anxiety, mice lacking 5-HT1a receptors seem to be more fearful than wild-type animals. Three groups have made knockouts of the receptor and all reported that the mutants behave in way consistent with a role of the receptors in modulating fear-related behavior *(7–9)*. The knockout mice are less active in tests of novelty (in which the animal is free to explore a novel, potentially threatening environment). All groups concluded that the mutants displayed increased anxiety. However, the data do not necessarily overall support the view that the 5-HT1a receptors are involved in modulating anxiety. Ramboz and colleagues found no significant changes in total serotonin or its metabolites in any region of the central nervous system they examined, suggesting that the behavioral effect was secondary to the genetic lesion *(7)*. Furthermore, the pharmacological data contradict the genetic findings: compounds that block 5-HT1a receptors do not cause anxiety in adult mice.

Rene Hen and colleagues resolved this problem by analyzing a transgenic mouse that they had engineered so that the receptor could be switched off (by feeding the antibiotic, doxycycline, to the animals) *(6)*. They had additionally designed the animal so that the receptor was only expressed in the forebrain, and they were able to show that restoration of receptor function in the forebrain would rescue the anxiety phenotype found in the constitutive knockout (in which the receptor is absent throughout the brain).

Administration of doxycycline to animals aged between 10 and 12 wk eliminated forebrain 5-HT1a receptors in the adult, but had no effect on measures of anxiety. In contrast, administering doxycycline during gestation resulted in animals that had increased levels of anxiety. In other words, the genetic effect depended on the developmental stage; forebrain 5-HT1a receptors are required to modulate anxiety during embryonic and fetal life, but they are not required for that task in adult animals.

This important experiment is still a solitary example, perhaps not surprisingly, given the difficulties of making and using mice with region-specific knockouts under antibiotic control. However, there is no reason to think that the example is unique. Many other behavioral systems may show the same dependence on the timing of the genetic influence on behavior.

4. COMPENSATION

Many of the genes that have attracted the attention of neuroscientists play fundamental roles in a variety of tissues, not just the brain, so it was a considerable surprise to discover that null mutants were not always lethal. In fact, in many cases, the problem has been to find any phenotype at all. One explanation for the relative lack of phenotypic effect is that other genes compensate for the effect of the mutation. It is preferable to call this phenomenon compensation rather than redundancy, because redundancy implies that the gene might have no specific function at all; however, complete absence of a phenotypic effect for a null mutation is unlikely. It is also very difficult to prove that a mutant has no phenotype, because that requires testing the knockout in an almost limitless number of different environments, at different developmental stages, and for different phenotypic effects.

Compensation is not easy to establish, as becomes clear when we examine cases in which the null phenotype in mice is mild, or apparently missing. Four examples of single-gene conditions in which the underlying pathophysiology is relatively well understood, but the behavioral phenotype in the animal model does not replicate the human condition are presented. These examples are gene knockout models for metachromatic leukodystrophy, Lowe syndrome, Tay-Sachs disease, and Lesch-Nyhan syndrome. In each case, the disorder arises from a deficiency in a metabolic pathway and part of the phenotype is mental retardation. However, the behavioral phenotype is not a nonspecific deficit; there is certainly good evidence in Lesch-Nyhan disease that the behavior is so specific that it can be used to make the diagnosis. Thus, it was hoped that mouse models would be able to explain how genetic mutations result in behavioral phenotypes.

4.1. Metachromatic Leukodystrophy

Metachromatic leukodystrophy is a lysosomal sphingolipid storage disorder caused by deficiency of arylsulfatase A (ASA), an enzyme that metabolizes the sphingolipid, cerebroside-3-sulfate (sulfatide), a major lipid component of myelin *(10)*. Deficiency of the enzyme leads to progressive demyelination in the central nervous system. By approx 18 mo, patients present with ataxia and gait disturbance; they go on to develop loss of speech, epileptic seizures, and a spastic quadriplegia. Symptoms are progressive, and children die in a decerebrate state. The behavioral phenotype has some of the features of a psychosis, although, admittedly, the data are limited *(11)*.

ASA-deficient mice, created by transgenesis, have a remarkably mild phenotype compared with humans; they have a normal life span and do not develop widespread demyelination *(12)*. It is not obvious why the animals are so mildly affected as compared with humans. The storage pattern of cerebroside-3-sulfate is comparable in the two species, but gross defects of white matter are not observed in the mouse up to the age of 2 yr. Animals display an astrogliosis and a decreased average axonal diameter and there are abnormalities in Purkinje cells and Bergmann glia of the cerebellum. Demyelination is seen in the acoustic ganglion, resulting in deafness by 1 yr *(12)*. There is also evidence for a behavioral deficit, but not one that has a clear homolog in humans. Motor coordination and equilibrium is impaired in 12-mo-old ASA-deficient mice, and there are mild impairments in spatial discrimination and fear conditioning (assessed by a Morris water maze and passive avoidance task, respectively), but these impairments are only seen in animals older than 1 yr *(13)*. The decline in neuromotor and cognitive functions in metachromatic leukodystrophy patients is much more severe.

The example of metachromatic leukodystrophy demonstrates that there are unexpected relationships between genotype and phenotype, even in what seem to be relatively well-understood genetic disorders, but the example does not explain how this complexity arises. The next two examples (Lowe syndrome and Tay Sachs disease) indicate the importance of compensatory mechanisms.

4.2. Lowe Syndrome

The oculocerebrorenal syndrome of Lowe (OCRL) is a multisystem disorder affecting the lens (cataracts), and the kidney (resulting in renal Fanconi syndrome). It affects the central nervous system, with some evidence that there is a behavioral phenotype consisting of temper tantrums, stereotypy, stubbornness, obsessions, and unusual preoccupations. Comparison between affected children and controls matched for sex, age, and visual impairment indicated that the phenotype could not be attributed solely to the visual, motor, and intellectual disabilities characteristic of Lowe syndrome, but could represent a specific effect of the mutation *(14)*.

OCRL is an X-linked disorder, and the gene, *Ocrl1*, encodes a phosphatidylinositol 4,5-bisphosphate 5-phosphatase in the Golgi complex *(15)*. Mice deficient in *Ocrl1* do not develop the cataracts, renal Fanconi syndrome, or neurological abnormalities seen in the human disorder *(16)*. One possibility for this surprising result is that *Ocrl1* deficiency is compensated in mice by inositol polyphosphate 5-phosphatase (*Inpp5b*), an autosomal gene that encodes a phosphatidylinositol bisphosphate 5-phosphatase highly homologous to *Ocrl1*. It was possible to test this hypothesis by creating mice deficient in *Inpp5b* and crossing them mice to mice deficient in *Ocrl1*. The double deficiency is lethal, which certainly suggests that the lack of phenotype in *Ocrl1*-deficient mice may be a result of Inpp5b function, but does not help further in providing an adequate mouse model of Lowe syndrome.

4.3. Tay-Sachs Disease

Tay-Sachs and Sandhoff disease are lysosomal storage diseases caused by mutations in, respectively, the *A* and *B* isoforms of β-*hexosaminidase* (*Hex A* and *Hex B*). Disease occurs because GM2 ganglioside accumulates in the central nervous system. The phenotype is variable, but, typically, onset is early and marked by rapidly progressive neurological and behavioral deterioration. The phenotypes of mice with mutations in *Hex A* and *Hex B* do not recapitulate this pattern. As expected, targeted mutagenesis produces *Hex A*-deficient mice that have undetectable levels of β-*hexosaminidase* and accumulate GM2 ganglioside in their central nervous system in an age-dependent manner and, as in a patient with Tay-Sachs disease, gangliosides accumulate in neurons as cytoplasmic bodies. However, by 3–5 mo of age, the mutant mice show no apparent defects in motor or memory function *(17)*.

Sango and colleagues investigated the relationship between *Hex A* and *Hex B* deficiencies *(18)*. They replicated the finding that *Hex A*-knockout animals are normal; by contrast, the *Hex B*-knockouts displayed progressive deterioration in motor function and gait, to the extent that they were incapacitated by the age of 5 mo. The authors went on to report the phenotype of mice that have both *Hex A* and *Hex B* genes disrupted. Double-knockout mice displayed a total deficiency of all forms of lysosomal β-*hexosaminidase*. Surprisingly, these mice showed the phenotypic, pathological, and biochemical features of the mucopolysaccharidoses, lysosomal storage diseases caused by the accumulation of glycosaminoglycans. The mucopolysaccharidosis phenotype is not seen in the Tay-Sachs or Sandhoff disease model mice, or in the corresponding human patients. They were, therefore, able to argue that the lack of storage of glycosaminoglycans in Tay-Sachs and Sandhoff diseases is a result of functional redundancy in the β-hexosaminidase enzyme system *(18)*.

4.4. Lesch-Nyhan Disease

Lesch-Nyhan disease takes the investigation of compensatory mechanisms one step further. Lesch-Nyhan disease arises from a lack, or very low levels, of hypoxanthine phosophoribosyltransferase (HPRT), an enzyme involved in pathways that resynthesize the components of nucleic acids from their breakdown products. Nucleic acids contain polymers of purine bases that are degraded to urate for excretion or are phosphorylated for further use. The three purine bases involved (adenine, guanine, and hypoxanthine) are served by different enzymes: adenine phosophoribosyltransferase (APRT) works on adenine and HPRT works on guanine and hypoxanthine.

Lesch-Nyhan disease has a remarkably specific behavioral phenotype, characterized by compulsive self-injury, which is present in more than 85% of cases *(19,20)*. The degree and extent of self-injury is remarkably severe: sufferers may require restraint, even teeth extraction, to control the behavior. Typically, injuries occur from biting of lips, fingers, and the inside of the mouth, but affected individuals seek to hurt themselves in other ways as well, even using their own wheelchairs to inflict injuries *(19)*.

Transgenic mice for Lesch-Nyhan disease were created by two groups in 1987 *(21,22)*. The mice seemed normal; there was no evidence of a behavioral deficit *(23)*. One possible explanation for the discrepancy between the mouse and human phenotypes is differential regulation of the nucleotide pool; the relative activities of HPRT and APRT are different in mice, thus, the activity of APRT could be compensating for the deficiency of HPRT.

A simple test of this hypothesis is to inhibit APRT in HPRT knockout mice and see whether a Lesch-Nyhan phenotype arose. Wu and Melton administered 9-

ethyladenine (9-EA), an APRT inhibitor, to HPRT-deficient mice and reported persistent self-injurious behavior in the animals *(24)*. They found that the reduction to 80% of APRT activity relative to the saline-treated control group induced the abnormal behavior. This story seems to provide a good example of how biochemical differences between mouse and human can be overcome and suggests a way forward for other similar discrepancies between mouse and human physiology. Unfortunately, the results are not as straightforward as they seem at first sight.

If the compensation explanation is true, then the same behavioral phenotype observed in the 9-EA-treated HPRT knockouts should arise in a double knockout of HPRT and APRT. Engle and colleagues generated APRT-knockout mice and found that these mice develop kidney stones and renal failure, just as APRT-deficient humans do, confirming that the enzyme has the same role in both species, at least in some tissues. They then crossed the null APRT allele onto an HPRT-deficient mouse background *(25)*. HPRT/APRT double-deficient mice did not exhibit any obvious behavioral abnormalities. They concluded that HPRT/APRT deficiency has nothing to do with self-injurious behavior and is not a good model for Lesch-Nyhan syndrome. A similar conclusion was reached by Edamura and Sasai, who failed to replicate the 9-EA result *(26)*.

In conclusion, although there is evidence in some cases for deficiencies being corrected by known mechanisms, in other cases, the cause of the discrepancy between mouse and human phenotypes is obscure, even when the disturbance occurs in well-characterized biochemical pathways, such as purine degradation and resynthesis. It is evident that we have to further examine why mouse and human phenotypes differ.

5. INTERACTIONS BETWEEN GENES AND ENVIRONMENT

Behavioral geneticists have long stressed the fact that genes and environment work in interaction with each other. Certain phenotypes only emerge when genetically susceptible individuals are placed in a particular environment. Interactions between genes and environment should be easier to examine using knockout mice, and are likely to be important in understanding pathogenesis, but demonstrations of the interactions are still rare.

In human behavioral genetics, a few examples have been reported recently: for instance, Caspi and coworkers reported that the allelic variation at the *serotonin transporter* (*5-HTT*) gene locus moderates the influence of stressful life events on depression. This would explain why some people are more likely than others to develop depression or anxiety after exposure to adversity. It was found that individuals with one particular variant in the promoter of the *5-HTT* gene (known as the short allele) who were exposed to stressful life events were more likely to develop depression than those with an alternative allele (the long allele). It was also reported that childhood adversity predicted adult depression only among individuals carrying a short allele but not among individuals homozygous for the long allele *(27)*.

To date, interactions between genes and environment have been investigated only rarely in the behavioral analysis of mutants. Nevertheless, there is good evidence that such interactions will be found, and the interactions must be taken into account if we are to understand how genetic effects impinge on behavior. Cabib and colleagues demonstrate that a simple environmental change, 12 d of food shortage, can dramatically reverse or abolish differences between inbred strains of mice in behavioral responses to the psychostimulant, amphetamine *(28)*. When food was available *ad libitum,* mice from strains C57BL/6J and DBA/2J exhibited opposite behavioral responses; C57BL/6J mice preferred the place where they previously received amphetamine injections, whereas DBA/2J animals avoided it. Furthermore, amphetamine injection increased locomotor response in C57BL/6J mice more than DBA/2J mice.

Strain differences, however, were altered after several days of food deprivation. DBA/2J mice that experienced food shortage exhibited an amphetamine-induced place preference, and not avoidance, whereas C57BL/6J mice remained unaffected by the food deprivation protocol and continued to show place preference. That is, the strain differences disappeared because of transient food deprivation. Similarly, locomotor responses were also modified by food shortage, and the changes were strain dependent. Whereas C57BL/6J mice were again unaffected by previous food deprivation, DBA/2J mice became highly responsive to amphetamine injection and showed a robust elevation of activity.

This is one example that demonstrates how important a more sophisticated investigation of behavior in mutants must become to dissect the pathway from genetic lesion to behavior. There are likely to be other complex interactions between the induced mutation and the environment that have yet to be discovered. However, this is still only part of the story. It is also necessary to take into account the genetic background of the animal carrying the induced mutation. The effect that other genetic variants have on the behavioral phenotype of a transgenic mouse will be discussed next.

6. BACKGROUND GENES

The importance of genetic background has been appreciated for some time. Threadgill and colleagues' report in 1995 showed that, on a 129/Sv background, homozygous mutants for epidermal growth factor receptor died at mid-gestation because of placental defects, whereas, on a different genetic background (CD-1), the mutants lived up to 3 wk and showed abnormalities in the skin, kidney, brain, liver, and gastrointestinal tract *(29)*. Simply put, genetic variants at loci distinct from the null allele modify the phenotype, potentially vitiating a comparison between knockout animals and controls, as pointed out by Gerlai *(30,31)*.

Targeted mutations in mice are commonly made in embryonic stem (ES) cells derived from the 129 mouse strain and introduced into a blastocyst to generate chimeric embryos. Offspring are mated to wild-type (nonmutated) mice in the hope of obtaining germline transmission of the mutation. Often, mating is carried out with a different inbred strain, normally C57BL/6; thus, the offspring inherit one chromosome homolog from C57BL/6 and the other from the 129 strain. Consequently, the offspring are not only heterozygous for the null mutant allele but are also heterozygous at all loci that are allelic between strains 129 and C57BL/6. There are also differences between substrains of 129 mice, therefore, the effects will vary depending on the origin of the ES cells *(32,33)*.

The strain effects on behavior are important because the 129 mouse has an unusual behavioral profile. Moreover, strain 129 mice suffer from dysgenesis of the corpus callosum and possess a number of other neuroanatomical abnormalities *(34)*. The animals are impaired in spatial learning tasks, which are frequently assessed in behavioral profiling of mutants, and the strain has a relatively high emotional reactivity (for example, they show relatively increased conditioned freezing, and low levels of exploratory activity in tests of novelty) *(35–38)*.

One specific example will serve to demonstrate the importance of appreciating the behavior of the 129 strain. Dockstader and van der Kooy examined the rewarding effects of psychoactive drugs in the 129/SvJ and C57BL/6 mouse strains using the place-preference paradigm mentioned in Section 5. *(39)*. After completing four conditioning trials, C57BL/6 mice preferred the chamber in which they previously received morphine injection to the nonrewarded chamber. Using the same paradigm, morphine was found to be rewarding for the 129/SvJ mouse strain, but only when the animals were under the influence of morphine during testing. Did this represent a difference in drug response, motivation, or learning between the strains? Further experiments revealed that the inability of 129/SvJ mice to exhibit morphine-rewarded place preference could be fully reversed by a pretest injection of anxiolytic agents (diazepam and pentobarbital), implying that the strain's behavioral phenotype was caused by anxiety.

Elevated anxiety in the 129 can certainly confound interpretation of gene-targeting experiments, as Holmes and colleagues report. They investigated anxiety-like behavior in mice with a null allele of the *5-HTT* gene. On a C57BL/6 background, null mutants exhibited increased anxiety-like behavior and reduced exploratory locomotion *(40)*. Comparison of *5-HTT* mutants on a C57BL/6 with 129S6 congenic background revealed that the mutation did not manifest as an alteration in fear-like behavior on the 129S6 background. The authors concluded that high baseline anxiety-like behavior in the 129S6 strain could indeed have precluded detection of the anxiety-like effects of the *5-HTT*-null mutation *(41)*. It should also be noted that there are differences in anxiety-like behavior between substrains of 129 mice *(33)*.

One accepted way of dealing with the problem of genetic background is to compare the mutant with its wild-type littermates, because, taken together, the same loci will be segregating in the mutants as in the littermates. Although, individually, no two animals are identical, by including enough animals in the two groups (those with and without the mutation) the effect of the mutation can be separated out from other, independently segregating loci. If the effect of the mutation is large (for instance large enough to account for 50% or more of the phenotypic variation), then a dozen animals in each group will be enough (assuming the background genetic effects are of the usual magnitude of a few percent variance attributable to each locus). However, when the mutation has a small effect, much larger groups will be needed.

An alternative way of dealing with background genetic effects is to backcross the mutant to one parental inbred strain for sufficient generations to purge the genome of all variants. Then, comparisons can be carried out between the knockout and the relevant pure inbred. This approach works well when the parental strain is the same as the strain from which the ES cells are derived (the 129 strain), but becomes complicated in the more usual situation when the knockout is crossed onto a different strain (C57BL/6). In this case, it is almost impossible to remove regions of the genome physically close to the mutation. To do so requires obtaining recombinants that occur precisely at either side of the site of the knockout, which is extremely unlikely to happen. As Bolivar and Flaherty point out, the knockout is actually a form of congenic, an animal that contains one small chromosomal segment from one strain, and the rest of the genome from another *(42)*. Obviously, if the introgressed segment contains no variants that influence the phenotype, the effect is irrelevant. Unfortunately, most behaviors in mice are influenced by many genes *(43)*, therefore, the probability that the segment contains an allele with an effect is not negligible. It may be unlikely, but the problem cannot be ignored.

There are now numerous examples in which behavior in knockouts depends on the strain background *(44–46)*. For instance, neuronal nitric oxide synthase knockout males were first tested for aggression in a mixed 129/SV and C57BL/6J background, and found to be highly aggressive as compared with wild-type littermates *(47)*. However, after five backcrosses into C57BL/6J, the neuronal nitric oxide synthase knockout male offspring were no more aggressive than wild-type littermates *(48)*. Similar findings have been reported across a range of behaviors, including spatial learning *(49)*. Differences observed between mutant and control mice could be caused by the genetic differences between the inbred strains used in the generation of null mutant animals and not by the null mutation itself. For instance, Kelly and colleagues analyzed a mouse with a knockout of the dopamine D2 receptor and found that wild-type strain 129 mice with unaltered functional D2 receptors had the same locomotor deficits as those attributable to the mutation *(45)*. Analysis of mutant congenic strains (backcrossed to either B6 or 129 parental strains) showed a significant interaction between background genes and the targeted mutation, the former exhibiting a greater effect on the behavioral phenotype.

Effects of 129 alleles have also been documented: for instance Errijgers and Kooy review evidence that variation in test results of the *fragile X* knockout mouse depends on the residual effect of 129 alleles in the C57BL/6 background *(50)*. Paradee and colleagues present evidence of the 129 genetic effects on tests of visuo–spatial orientation *(51)*.

Finally, in a few cases, investigators have crossed a mutant onto different strains to assess the effect of background genes. For example, the estrogen receptor (ER)-

α-knockout mouse was examined on C57BL/6J, DBA/2J, BALB/c, and A/J strains, and dramatic effects on male sexual behavior were observed. ER-α-knockout males in the DBA/2J and BALB/c backcrosses displayed more intromissions compared with males in the C57BL/6J and A/J mixed background. Many fewer ER-α-knockout females than males displayed masculine sexual behavior in any of the three hybrid crosses *(52)*.

Background effects are important in any phenotype: the example given in Section 6 of the epidermal growth factor receptor makes this point. However, there is evidence that the effects are of particular concern to behavioral genetics. Detecting the effect of the null allele using littermate controls depends on the effect size of the mutation; large effects are less likely to be obscured by background effects, which usually consist of many small-effect loci *(43)*. As a rule, null alleles have a small effect on behavioral phenotypes. It is possible to work out the effect size from the means and standard deviations reported for each genotype (wild-type, heterozygote, and homozygote mutant), data which are generally reported with the behavioral analyses. For example, using the published data on the effect attributable to the corticotropin-releasing hormone receptor-2 knockouts *(53)*, the knockout allele accounts for approx 10% of the total phenotypic variation, a figure consistent with the size of naturally occurring genetic effects that contribute to individual differences in behavior *(43)*.

Of course, it could be argued that the lack of large effects merely reflects the relatively small number of available knockouts. If more genes were inactivated, then we might find examples of mutations that have a large effect on behavior. However, this argument can be countered by the relative failure to detect segregating behavioral mutations in the mouse mutagenesis projects; that is, the analysis of single-gene mutations systematically produced in mice through the administration of a highly mutagenic compound *N*-ethyl-*N*-nitrosourea. *N*-ethyl-*N*-nitrosourea introduces single basepair mutations randomly at a rate sufficiently high to make it realistic to screen for the effects of mutations *(54,55)*.

Results from the mouse mutagenesis projects suggest that the yield of behavioral mutants is less than expected. Behavioral assays were included in four mutagenesis screens, covering learning and memory, motor activity, fear (or anxiety)-related behaviors, and a test of sensorimotor gating (prepulse inhibition, a model of one aspect of schizophrenia). Three screens have now discarded these behavioral assays because of the low yield of heritable mutants. More than 10,000 mice were screened at the Medical Research Council mammalian genetics unit in Harwell for prepulse inhibition deficits, however, no mutants were found. Similarly low yields are reported from other screens (http://www.gsf.de/ieg/groups/enu/behaviour.html, http://www.neuromice.org/, and http://www.mgu.har.mrc.ac.uk/).

How many functional mutations would we expect the mutagenesis projects to have detected? Sequence analysis of 370 kb of DNA of mutagenized mice detected six sequence changes, suggesting that a mutation will be found approximately every 60 kb *(56)*. Using a per locus functional mutation rate of 1.08×10^{-3}, the probability of obtaining at least one mutant is 0.66 if 1000 animals are screened, 0.86 for 2000 animals, and 0.97 for 3000 animals. Because the largest screen has processed 10,000 mice a year (http://www.neuromice.org/), the lack of inherited mutations with an effect on behavior cannot be explained by inadequate numbers of animals. The likely explanation is the relatively small size of the genetic effects.

If, as these observations suggest, genetic effects on behavior are small, it raises the possibility that we rarely, if ever, observe an effect that is independent of background genes. It may be the case that the outcomes we see in knockouts are more complex than we imagine, and that, in fact, knockout mice model the genetic susceptibility to psychiatric disease more closely than has been suspected. We are misled by the analysis of null alleles into thinking that the abnormal phenotype is the product of a single mutation; it may be nearer the truth to say that the effect is polygenic (caused by the induced mutation in combination with effects from many other loci). As discussed in Section 7, this model approximates much better the genetic susceptibility to psychiatric disease than a single gene knockout. Although it is true that detecting and controlling for background genetic effects is an important issue in the behavioral analysis of knockout animals, it may also be true that we need to look more closely at the interaction between modifier loci and the knockout if we are to understand fully how genetic lesions produce disease, a point made by Errijgers and Kooy in their review of the difficulties of modeling the fragile X syndrome *(50)*.

7. COMPLEX GENETICS

Apart from the difficulties of determining which parts of a psychiatric phenotype can be modeled in mice, we need to decide whether the genetic models in knockout mice are at all comparable to those in human psychiatric disease. Genetic susceptibility to psychiatric disorder arises from the conjoint effect of many loci, each contributing only a relatively small amount to the total genetic liability, and it is likely that part of that liability is caused by interaction between genetic loci (epistasis).

The full complexity of genetic architecture of human behavioral variation is still not clear, but there is very little evidence that major gene effects influence psychiatric disease. Despite considerable effort to find families, there are no convincing reports that psychiatric disorders are caused by mutations in a single gene. There are a few instances in which psychiatric illness arises in the context of a genetically determined syndrome. For example, patients with a deletion on chromosome 22q giving rise to velocardiofacial syndrome also have psychotic symptoms *(57)*, but no one has published a pedigree with a segregating recessive or dominant mutation that gives rise to the common psychiatric conditions of anxiety, affective, or psychotic disorders.

The number of genes involved in a common psychiatric illness is unknown; using linkage data collected on a large set of affected sibling pairs, Neil Risch and colleagues put a lower boundary on the number of susceptibility loci for autism at approximately a dozen loci *(58)*, but the results for this, and other data sets are also compatible with the presence of hundreds of loci. Additionally, the mode of action of each locus is unknown; it is suspected that there may be considerable, but so far undetected, amounts of interaction between loci in many complex phenotypes (not just psychiatric illness, in fact). Again, autism provides an example, in which there is a concordance between monozygotic twins (those that are genetically identical) of approx 90%, whereas the concordance in dizygotic twins (who share half their genes) is only approx 10% *(59)*. Under a simple additive model, in which the phenotype is the outcome of the independent action of each locus, we would expect the concordance in dizygotic twins to be approximately half that found in monozygotic twins. The discrepancy can be accounted for by gene interaction (although there are other explanations).

In addition to the number of loci involved and their mode of action, the complex relationship between genetic susceptibility and phenotype in psychiatric illness has to be taken into consideration when assessing the suitability of a genetic model. In some cases, susceptibility seems to have a quantitative nature; thus, for example, rates of depression are higher in first-degree relatives of a patient with recurrent major depression than in unrelated controls. A pattern of this sort is consistent with additive genetic action, in which a number of loci contribute to increase susceptibility to the disorder. However, psychiatric genetics also has examples of patterns that are not so easily explained. For example, studies of psychosis show that there is phenotypic spectrum in first-degree relatives.

Seymour Kety introduced the term *schizophrenia spectrum* to refer to all disorders that are "to some extent genetically transmitted" with schizophrenia *(60)*, and, since then, there have been efforts to develop operational definitions of the schizophrenia spectrum, resulting in the appearance of criteria for a number of personality disorders (schizotypal, schizoid, avoidant, and paranoid personality disorders). Subsequent work has set out to determine whether there is indeed a genetic relationship between schizophrenia and the spectrum disorders. Three studies have examined the risk of developing schizophrenia in offspring reared by biological parents with schizophrenia *(61–64)*. All report an increase in the genetic liability for schizophrenia-related illness. Similar observations arise from studies of first-degree relatives *(65)*. Together, these reports indicate that the genetic liability is not restricted to narrowly defined, typical schizophrenia but includes schizotypal and schizoid personality disorders and nonschizophrenic nonaffective psychoses.

Knockout animals obviously cannot capture the multilocus and "spectrum" pattern of susceptibility that characterizes psychiatric genetics, because, in a knockout, the genetic liability is caused by a single gene. However, as discussed in Section 6, the phenotypic consequences of the mutation do not arise simply; the effects are frequently small and often depend on background genetic variants for the appearance of the phenotype. This aspect of the behavioral outcome of transgenesis has received little attention, but it may provide new avenues of investigating the pathogenesis of psychiatric illness, allowing us to explore the issues of complexity in a new way. It may also explain the confusion and contradictory results in current findings.

Some years ago, Sanes and Lichtman, observed that approx 100 genes had been implicated in a cellular phenomenon called long-term potentiation thought to underlie memory processes in the hippocampus *(66)*. How, they asked, could so many apparently unrelated molecules be said to explain long-term potentiation? Clearly, this is a challenge to the causal model adopted from biochemical genetics, in which genes can be arranged in simple linear pathways.

An alternative model is to look at the interactions of the gene products, the proteins, as has been achieved in yeast genetics *(67–71)*. Analysis of the patterns of interactions showed that it was possible to determine structure in the network, leading Barabasi to propose that protein interactions involved in metabolism had the properties of a scale-free network *(72)*. In a scale-free network, the components are organized in the same way as the Internet, or as airlines that connect different destinations via a series of hubs. Because of this feature, scale-free networks are robust and error tolerant *(73)*.

Grant and colleagues have asked whether similar networks apply for proteins involved in behavior, using an excitatory neurotransmitter receptor (the *N*-methyl-D-aspartate receptor) as a model system for investigating the nature of protein (and gene) interactions *(74–76)*. They found that the interactions did indeed fit the pattern of a scale-free network; furthermore, when they mapped the effects of knockouts onto the network they found that mutations which affected a hub had, as expected, much more profound effects than mutations that affected the spokes of the network. Rather than manifesting with a large effect on behavior, knockouts that involved hub genes were likely to be lethal. These initial forays into a network analysis of behavior give some idea of how further analyses could proceed, providing us with a much richer, more sophisticated picture of the relationship between mutation and behavior.

8. CONCLUSION

Transgenic mouse models have undoubtedly made immense contributions to our understanding of behavioral disorders and to neurobiology in general *(77)*. Their value is seen as a tool for the investigation of gene function; by inactivating a gene we can observe the consequences and infer what that gene does in the intact animal. At a time when we are beginning to identify the molecular components of disorders such as schizophrenia, for which we have few clues about the biology, the availability of transgenic animals is a great advantage, an indispensable resource for working out what genes do *(78)*.

The difficulties of engineering mice so that they have mutations that affect only the gene of interest are well-known and have been discussed in this chapter. They include the direct effects of the induced mutation, such as its ability to influence expression of neighboring genes in addition to the target, and the importance of developmental and tissue-specific effects. The complications that influence interpretation of behavioral experiments are also well appreciated. Because we know so little about the biological origins of behavior, we rely substantially on behavioral measurements to assess the nature of the mutant phenotype. However, when we use an outcome measure that is so distant from the molecular lesion, it becomes extremely difficult to observe the pathway that leads from mutation to phenotype. A host of intermediate influences, ranging from other genes, to cellular systems, and, finally, to environmental effects, combine to obscure the relationship between gene and behavior.

As I have attempted to explain in this chapter, a strictly reductionist analysis of knockouts is not necessarily the best way to proceed when tackling models of psychiatric illness. No doubt, when there are specific hypotheses to test about gene function in a well-understood system, this model is appropriate; although, as we have seen in the discussion on compensation, even in these circumstances, the phenotype of a knockout experiment can confound expectations. It is sobering to realize that we still do not understand why the phenotype of one of the first genes ever to be knocked out in mice, HPRT, is not the same as its human equivalent (Lesch-Nyhan disease), even though we know much about the biochemical pathway that has been disrupted.

I have suggested that there are alternative ways of using knockout animals. A consideration of how genetic effects operate in psychiatric disease leads to the now undeniable conclusion that genes of large effect are rare and probably nonexistent for many common psychiatric conditions. Instead, we see that genetic susceptibility to common psychiatric disease arises from the complex interactions between a multitude of genes and an unknown number of relevant environments. On the face of it, using the single-gene model that we have available in the mouse knockout seems completely inappropriate to model such complexity. However, genetic effects on behavior attributable to null mutants in the mouse are in fact subject to the same set of complications, the same gene by environment, and epistatic interactions. Consideration of the genetic architecture of behavior indicates that even when the molecular lesion is sufficient to inactivate the gene, or in other ways alter its function substantially, the effect on the phenotype is typically very mild. Overall, the explanation for the behavior may not be as complex, but it is the product of the same factors. Consequently, it may be possible to take apart the pathway from gene to psychiatric illness. Using networks of interactions to understand how genes and then proteins work, it will be possible to begin to explain cellular processes, and eventually to explain how cellular processes give rise to behavioral phenotypes.

REFERENCES

1. Weatherall DJ. Molecular genetics: mice, men and sickle cells. Nature 1990;343:121.
2. Greaves DR, Fraser P, Vidal MA, et al. A transgenic mouse model of sickle cell disorder. Nature 1990;343:183–185.
3. Rijli FM, Dolle P, Fraulob V, LeMeur M, Chambon P. Insertion of a targeting construct in a Hoxd-10 allele can influence the control of Hoxd-9 expression. Dev Dyn 1994;201:366–377.
4. Pham CT, MacIvor DM, Hug BA, Heusel JW, Ley TJ. Long-range disruption of gene expression by a selectable marker cassette. Proc Natl Acad Sci USA 1996;93:13,090–13,095.

5. Hall JC. Pleiotropy of Behavioral Genes? In: Greenspan RJ, Kyriacou CP, eds. Flexibility and Constraint in Behavioral Systems. Berlin: Dahlem Konferenz, 1994, pp. 15–28.
6. Gross C, Zhuang X, Stark K, et al. Serotonin1A receptor acts during development to establish normal anxiety-like behaviour in the adult. Nature 2002;416:396–400.
7. Ramboz S, Oosting R, Amara DA, et al. Serotonin receptor 1A knockout: an animal model of anxiety-related disorder. Proc Natl Acad Sci USA 1998;95:14,476–14,481.
8. Heisler LK, Chu HM, Brennan TJ, et al. Elevated anxiety and antidepressant-like responses in serotonin 5-HT1A receptor mutant mice. Proc Natl Acad Sci USA 1998;95:15,049–15,054.
9. Parks CL, Robinson PS, Sibille E, Shenk T, Toth M. Increased anxiety of mice lacking the serotonin1a receptor. Proc Natl Acad Sci USA 1998;95:10,734–10,739.
10. Berger J, Moser HW, Forss-Petter S. Leukodystrophies: recent developments in genetics, molecular biology, pathogenesis and treatment. Curr Opin Neurol 2001;14:305–312.
11. Black DN, Taber KH, Hurley RA. Metachromatic leukodystrophy: a model for the study of psychosis. J Neuropsychiatry Clin Neurosci 2003;15:289–293.
12. Hess B, Saftig P, Hartmann D. Phenotype of arylsulfatase A-deficient mice: relationship to human metachromatic leukodystrophy. Proc Natl Acad Sci USA 1996;93:14,821–14,826.
13. D'Hooge R, Van Dam D, Franck F, Gieselmann V, De Deyn PP. Hyperactivity, neuromotor defects, and impaired learning and memory in a mouse model for metachromatic leukodystrophy. Brain Res 2001;907:35–43.
14. Kenworthy L, Charnas L. Evidence for a discrete behavioral phenotype in the oculocerebrorenal syndrome of Lowe. Am J Med Genet 1995;59:283–290.
15. Attree O, Olivos IM, Okabe I, et al. The Lowe's oculocerebrorenal syndrome gene encodes a protein highly homologous to inositol polyphosphate-5-phosphatase. Nature 1992;358:239–242.
16. Janne PA, Suchy SF, Bernard D, et al. Functional overlap between murine Inpp5b and Ocrl1 may explain why deficiency of the murine ortholog for OCRL1 does not cause Lowe syndrome in mice. J Clin Invest 1998;101:2042–2053.
17. Yamanaka S, Johnson MD, Grinberg A, et al. Targeted disruption of the Hexa gene results in mice with biochemical and pathologic features of Tay-Sachs disease. Proc Natl Acad Sci USA 1994;91:9975–9979.
18. Sango K, McDonald MP, Crawley JN, et al. Mice lacking both subunits of lysosomal beta-hexosaminidase display gangliosidosis and mucopolysaccharidosis. Nat Genet 1996;14:348–352.
19. Nyhan WL. Behavior in the Lesch-Nyhan syndrome. J Autism Child Schizophr 1976;6:235–252.
20. Christie R, Bay C, Kaufman IA, Bakay B, Borden M, Nyhan WL. Lesch-Nyhan disease: clinical experience with nineteen patients. Dev Med Child Neurol 1982;24:293–306.
21. Kuehn MR, Bradley A, Robertson EJ, Evans MJ. A potential animal model for Lesch-Nyhan syndrome through introduction of HPRT mutations into mice. Nature 1987;326:295–298.
22. Hooper ML, Hardy K, Handyside A, Hunter S, Monk M. HPRT-deficient (Lesch-Nyhan) mouse embryos derived from germline colonization by cultured cells. Nature 1987;326:292–295.
23. Finger S, Heavens RP, Sirinathsinhji DR, Kuehn MR, Dunnett SB. Behavioral and neurochemical evaluation of a transgenic mouse model of Lesch-Nyhan syndrome. J Neurol Sci 1988;86, 203–213.
24. Wu CL, Melton DW. Production of a model for Lesch-Nyhan syndrome in hypoxanthine phosphoribosyltransferase-deficient mice. Nature Genet 1993;3:235–240.
25. Engle SJ, Stockelman MG, Chen J, et al. Adenine phosphoribosyltransferase-deficient mice develop 2,8-dihydroxyadenine nephrolithiasis. Proc Natl Acad Sci USA 1996;93:5307–5312.
26. Edamura K, Sasai H. No self-injurious behavior was found in HPRT-deficient mice treated with 9-ethyladenine. Pharmacol Biochem Behav 1998;61:175–179.
27. Caspi A, Sugden K, Moffitt TE, et al. Influence of life stress on depression: moderation by a polymorphism in the 5-HTT gene. Science 2003;301:386–389.
28. Cabib S, Orsini C, Le Moal M, Piazza PV. Abolition and reversal of strain differences in behavioral responses to drugs of abuse after a brief experience. Science 2000;289:463–465.
29. Threadgill DW, Dlugosz AA, Hansen LA, et al. Targeted Disruption Of Mouse Egf Receptor— effect Of Genetic Background On Mutant Phenotype. Science 1995;269:230–234.

30. Gerlai R. Gene-targeting studies of mammalian behavior: is it the mutation or the background genotype? Trends Neurosci 1996;19:177–181.
31. Gerlai R. Targeting genes and proteins in the analysis of learning and memory: caveats and future directions. Rev Neurosci 2000;11:15–26.
32. Simpson EM, Linder CC, Sargent EE, Davisson MT, Mobraaten LE, Sharp JJ. Genetic variation among 129 substrains and its importance for targeted mutagenesis. Nature Genet 1997;16:19–27.
33. Rodgers RJ, Boullier E, Chatzimichalaki P, Cooper GD, Shorten A. Contrasting phenotypes of C57BL/6JOlaHsd, 129S2/SvHsd and 129/SvEv mice in two exploration-based tests of anxiety-related behaviour. Physiol Behav 2002;77:301–310.
34. Wahlsten D, Ozaki HS, Livy D. Deficient corpus callosum in hybrids between ddN and three other abnormal mouse strains. Neurosci Lett 1992;136:99–101.
35. Contet C, Rawlins JN, Deacon RM. A comparison of 129S2/SvHsd and C57BL/6JOlaHsd mice on a test battery assessing sensorimotor, affective and cognitive behaviours: implications for the study of genetically modified mice. Behav Brain Res 2001;124:33–46.
36. Holmes A, Wrenn CC, Harris AP, Thayer KE, Crawley JN. Behavioral profiles of inbred strains on novel olfactory, spatial and emotional tests for reference memory in mice. Genes Brain Behav 2002;1:55–69.
37. Contet C, Rawlins JN, Bannerman DM. Faster is not surer—a comparison of C57BL/6J and 129S2/Sv mouse strains in the watermaze. Behav Brain Res 2001;125:261–267.
38. Crawley JN, Belknap JK, Collins A, et al. Behavioral phenotypes of inbred mouse strains: implications and recommendations for molecular studies. Psychopharmacology (Berl) 1997;132:107–124.
39. Dockstader CL, van der Kooy D. Mouse strain differences in opiate reward learning are explained by differences in anxiety, not reward or learning. J Neurosci 2001;21:9077–9081.
40. Holmes A, Yang RJ, Lesch KP, Crawley JN, Murphy DL. Mice lacking the serotonin transporter exhibit 5-HT(1A) receptor-mediated abnormalities in tests for anxiety-like behavior. Neuropsychopharm 2003;28:2077–2088.
41. Holmes A, Lit Q, Murphy DL, Gold E, Crawley JN. Abnormal anxiety-related behavior in serotonin transporter null mutant mice: the influence of genetic background. Genes Brain Behav 2003;2:365–380.
42. Bolivar VJ, Cook MN, Flaherty L. Mapping of quantitative trait loci with knockout/congenic strains. Genome Res 2001;11:1549–1552.
43. Flint J. Analysis of quantitative trait loci that influence animal behavior. J Neurobiol 2003;54:46–77.
44. Thiele TE, Miura GI, Marsh DJ, Bernstein IL, Palmiter RD. Neurobiological responses to ethanol in mutant mice lacking neuropeptide Y or the Y5 receptor. Pharmacol Biochem Behav 2000;67:683–691.
45. Kelly MA, Rubinstein M, Phillips TJ, et al. Locomotor activity in D2 dopamine receptor-deficient mice is determined by gene dosage, genetic background, and developmental adaptations. J Neurosci 1998;18:3470–3479.
46. Bowers BJ, Owen EH, Collins AC, Abeliovich A, Tonegawa S, Wehner JM. Decreased ethanol sensitivity and tolerance development in gamma-protein kinase C null mutant mice is dependent on genetic background. Alcohol Clin Exp Res 1999;23:387–397.
47. Nelson RJ, Demas GE, Huang PL, et al. Behavioural abnormalities in male mice lacking neuronal nitric oxide synthase. Nature 1995;378:383–386.
48. Le Roy I, Pothion S, Mortaud S, et al. Loss of aggression, after transfer onto a C57BL/6J background, in mice carrying a targeted disruption of the neuronal nitric oxide synthase gene. Behav Genet 2000;30:367–373.
49. Wolfer DP, Muller U, Stagliar M, Lipp HP. Assessing the effects of the 129/SV genetic background on swimming navigation learning in transgenic mutants: a study using mice with a modified beta-amyloid precursor protein gene. Brain Res 1997;771:1–13.
50. Errijgers V, Kooy RF. Genetic modifiers in mice: the example of the fragile X mouse model. Cytogenet Genome Res 2004;105:448–454.

51. Paradee W, Melikian HE, Rasmussen DL, Kenneson A, Conn PJ, Warren ST. Fragile X mouse: strain effects of knockout phenotype and evidence suggesting deficient amygdala function. Neuroscience 1999;94:185–192.
52. Dominguez-Salazar E, Bateman HL, Rissman EF. Background matters: the effects of estrogen receptor alpha gene disruption on male sexual behavior are modified by background strain. Horm Behav 2004;46:482–490.
53. Kishimoto T, Radulovic J, Radulovic M, et al. Deletion of crhr2 reveals an anxiolytic role for corticotropin-releasing hormone receptor-2. Nat Genet 2000;24:415–419.
54. Hrabe de Angelis MH, Flaswinkel H, Fuchs H, et al. Genome-wide, large-scale production of mutant mice by ENU mutagenesis. Nat Genet 2000;25:444–447.
55. Nolan PM, Peters J, Strivens M, et al. A systematic, genome-wide, phenotype-driven mutagenesis programme for gene function studies in the mouse. Nat Genet 2000;25:440–443.
56. Beier DR. Sequence-based analysis of mutagenized mice. Mamm Genome 2000;11:594–597.
57. Karayiorgou M, Gogos JA. The molecular genetics of the 22q11-associated schizophrenia. Brain Res Mol Brain Res 2004;132:95–104.
58. Risch N, Spiker D, Lotspeich L, et al. A genomic screen of autism: evidence for a multilocus etiology. Am J Hum Genet 1999;65:493–507.
59. Bailey A, Lecouteur A, Gottesman I, et al. Autism as a strongly genetic disorder—evidence from a British twin study. Psychol Med 1995;25:63–77.
60. Kety SS, Rosenthal D, Wender PH, Schulsinger F. Mental illness in the biological and adoptive families of adopted schizophrenics. Am J Psychiatry 1971;128:302–306.
61. Parnas J, Cannon TD, Jacobsen B, Schulsinger H, Schulsinger F, Mednick SA. Lifetime DSM-III-R diagnostic outcomes in the offspring of schizophrenic mothers. Results from the Copenhagen High-Risk Study. Arch Gen Psychiatry 1993;50:707–714.
62. Erlenmeyer-Kimling L, Squires-Wheeler E, Adamo UH, et al. The New York High-Risk Project. Psychoses and cluster A personality disorders in offspring of schizophrenic parents at 23 years of follow-up. Arch Gen Psychiatry 1995;52:857–865.
63. Erlenmeyer-Kimling L, Adamo UH, Rock D, et al. The New York High-Risk Project. Prevalence and comorbidity of axis I disorders in offspring of schizophrenic parents at 25-year follow-up. Arch Gen Psychiatry 1997;54:1096–1102.
64. Tienari P, Wynne LC, Laksy K, et al. Genetic boundaries of the schizophrenia spectrum: evidence from the Finnish Adoptive Family Study of Schizophrenia. Am J Psychiatry 2003;160:1587–1594.
65. Kendler KS, Karkowski LM, Walsh D. The structure of psychosis: latent class analysis of probands from the Roscommon Family Study. Arch Gen Psychiatry 1998;55:492–499.
66. Sanes JR, Lichtman JW. Can molecules explain long-term potentiation? Nat Neurosci 1999;2:597–604.
67. Lee I, Date SV, Adai AT, Marcotte EM. A probabilistic functional network of yeast genes. Science 2004;306:1555–1558.
68. Tong AH, Lesage G, Bader GD, et al. Global mapping of the yeast genetic interaction network. Science 2004;303:808–813.
69. Wuchty S, Oltvai ZN, Barabasi AL. Evolutionary conservation of motif constituents in the yeast protein interaction network. Nat Genet 2003;35:176–179.
70. Ihmels J, Friedlander G, Bergmann S, Sarig O, Ziv Y, Barkai N. Revealing modular organization in the yeast transcriptional network. Nat Genet 2002;31:370–377.
71. Schwikowski B, Uetz P, Fields S. A network of protein-protein interactions in yeast. Nat Biotechnol 2000;18:1257–1261.
72. Jeong H, Tombor B, Albert R, Oltvai ZN, Barabasi AL. The large-scale organization of metabolic networks. Nature 2000;407:651–654.
73. Barabasi AL, Bonabeau E. Scale-free networks. Sci Am 2003;288:60–69.
74. Grant SG. Systems biology in neuroscience: bridging genes to cognition. Curr Opin Neurobiol 2003;13:577–582.

75. Grant SG. Synapse signalling complexes and networks: machines underlying cognition. Bioessays 2003;25:1229–1235.
76. Choudhary J, Grant SG. Proteomics in postgenomic neuroscience: the end of the beginning. Nat Neurosci 2004;7:440–445.
77. Albright TD, Jessell TM, Kandel ER, Posner MI. Neural science: a century of progress and the mysteries that remain. Neuron 2000;25(Suppl):S1–55.
78. Harrison PJ, Weinberger DR. Schizophrenia genes, gene expression, and neuropathology: on the matter of their convergence. Mol Psychiatry 2004;9:729–745.

3
Transgenic and Knockout Mouse Models
The Problem of Language

Linda J. Hayes and Diana Delgado

Summary

It is argued that the more removed the things investigated are from those about which knowledge is sought, the more susceptible to misinterpretation is the knowledge achieved. By this logic, scientific propositions pertaining to human psychiatric disorders derived from investigative contacts with mutant mice under contrived conditions deserve special scrutiny. Of particular importance, in this regard, is the adequacy with which characteristic features of the original phenomena are represented in the models under investigation. We contend that adequacy in this regard cannot be achieved for events of the psychological domain because human behavior has a unique characteristic that so profoundly affects the psychological experiences of human beings that it renders them incomparable to those of other species. We conclude that scientific propositions pertaining to the psychological components of human psychiatric disorders are not possible to construct on the basis of observational contacts with animals.

Key Words: Psychopathology; language; human behavior; murine models; model construction; nonverbal behavior.

1. INTRODUCTION

When human phenomena become subjects of scientific inquiry, they tend to be approached indirectly. This is to say, in a great many circumstances, the subjects exposed to experimental conditions aimed at the resolution of human problems are not humans, at least not initially. Although the investigation of human phenomena by indirect means usually reflects concerns of an ethical sort, scientific considerations may also be grounds for this approach. In the field of psychology, for example, animals may be substituted for human subjects to prevent the long and largely unknown pre-experimental histories of the latter from compromising experimental control *(1)*. Moreover, when the problems calling for investigation are unusually complex and the procedures deemed suitable for their investigation are especially invasive, concerns of both sorts may be raised.

Modeling strategies are commonly used for indirect investigations of phenomena that cannot be investigated directly, and a strategy of this sort has been adopted for the investigation of human psychopathologies. More specifically, the factors participating in human psychiatric disorders, their precipitating conditions, and their susceptibilities to treatment are being extrapolated from the results of direct investigations with respect to murine models of these disorders.

The value of a modeling strategy in any given investigative circumstance depends on the adequacy with which the events of interest are represented in the model under investigation. Human psychopathologies are complex conditions in which usual patterns of various types of events are either known or believed to be present, including those of the psychological, neurological, and genetic domains, among possibly many others; and not all of these event types may be equally well represented in the murine models under investigation. More to the point, it is our contention that although a murine modeling strategy may be of considerable value in our efforts to understand some aspects of human psychiatric disorders, its value for this understanding will not be realized in contributions of a psychological sort.

Our plan in this chapter is to offer support for this contention. We will begin by addressing the logic and operations of investigative practices in science, with the aim of identifying the specific configuration of these practices entailed in a modeling approach to investigation. Specifications for the construction of adequate models, as they pertain to psychological factors in particular, are then reviewed, and murine models of human psychiatric disorders are evaluated in terms of these criteria. On the basis of this review, we argue that the most significant features of the psychological events participating in human psychiatric disorders are not represented in murine models of these disorders. Their investigation, thereby, affords an understanding of only those features of human psychological events that are not uniquely human, features that, for the most part, are irrelevant to human psychopathology. Finally, we will return to the assumptions underlying a modeling approach to investigation and argue that, although it might be possible to view complex events as summations of simpler events in some scientific domains, this view cannot be sustained in the domain of psychology because of the historicity of its subject matter.

2. INVESTIGATIVE PRACTICES IN SCIENCE

The basic work of science consists of investigating things in such a manner that their constitutions and their interrelations with other things can be ascertained, whereupon they may become amenable to the additional operations of prediction and control *(2)*. Although all investigative procedures entail contacting things and events by way of observation, four varieties of investigative procedure may be distinguished on the basis of the type of observational contact involved. These include contact by direct observation; and three varieties of contact by indirect means, including remote observation, transformation, and manipulation *(3)*. Because procedures of all of these types are used to varying degrees in investigations of animal models of human psychiatric disorders, some discussion of their logic and implications seems warranted. These considerations follow.

2.1. Direct Observation

Investigation of things or events by means of direct observational contact, with or without the use of specialized instruments to enhance human observational capacities, implies that the things contacted by this procedure are the phenomena of original interest. This is to say, direct observation assumes that the constitutions of the original phenomena and their interrelations with other things have not been altered in any way before, and for the purpose of, their investigation.

In the context of murine modeling of human psychiatric disorders, this procedure, although not used during the primary investigational stage of the enterprise, makes an important contribution. Specifically, before the investigation of murine models of psychiatric disorders, the psychological and biological conditions of human beings presenting these disorders are subjected to direct observation for the purpose of selecting particular features of these classes of phenomena for representation in their models.

Following these preliminary investigations is the primary investigational phase of this type of research, wherein indirect observational contacts are pursued until scientific propositions regarding the constitutional and functional properties of particular disorders are formulated. It is only after these products of investigation have been achieved that direct observation is resumed. The role of direct observation at this time is the auxiliary role of verifying propositions and testing hypotheses regarding such issues as diagnosis and treatment.

2.2. Indirect Observation

Indirect observational procedures, including remote, transformative, and manipulative observations, are more central to the investigation of murine models of human psychiatric disorders than direct observational procedures. In fact, the primary investigative procedure in this enterprise is one in which all three varieties of indirect observational contact are concatenated.

As such, the validity and significance of human understandings constructed from the results of murine investigations depends on the extent to which the results were properly obtained *vis-à-vis* three sets of procedural guidelines, and the extent to which understandings were properly constructed in accordance with three sets of noncontradictory premises. Determining the value of the knowledge products of these investigations, thereby, requires familiarity with these sets of procedural guidelines, and familiarity with what each set of premises dictates concerning construct formation. Toward this aim, the logic of indirect observational contacts by remote observation, manipulation, and transformation are discussed next.

2.2.1. Remote Observation

Remote observational contact is distinguished from direct observational contact on the basis of the things contacted; direct contacts are made with respect to the original phenomenon; remote contacts are made with respect to things other than the original phenomenon. The nature of these other things and their relation to the original phenomenon constitute the bases on which remote observational contacts are distinguished from the other two varieties of indirect observation, transformation and manipulation. In the case of remote observation, the things contacted are best described as features of the original phenomenon and the relation between them and this phenomenon is one of part to whole, roughly speaking. Further, the features contacted by remote observation may be of either a formal or a functional sort, that is, they may be material constituents of the original phenomenon or characteristics of its interrelations with other things, respectively.

Remote observational contact is required when the phenomena of interest cannot be observed directly because of some impediment, such as the time of their occurrence, their distance from the observer, their unusually small or large size, or some other condition. When this is the case, their existence, presence in a particular field of interaction, or previous occurrence must be inferred on the basis of other things. For example, anthropological phenomena, having long since ceased to occur, are remotely observed in existing material conditions taken to be the results of their previous occurrences. The presence or occurrence of a particular phenomenon may also be remotely observed in what are taken to be their functional results. For example, a phenomenon resisting direct observation may be inferred on the basis of changes in other things, such as when disturbances in the pattern or organization of the event fields in which it is assumed to have participated are detected.

In summary, when the presence of a particular phenomenon, implicit in which are its constitutional properties and patterns of interrelation with other things, is inferred on the basis of having detected only some of its features, including what are taken to be the results of its presence or occurrence, the procedure is one of remote observation.

Certain premises pertaining to the relation of parts to wholes and of one whole to another underlie the pursuit of scientific knowledge by such means. Specifically, if wholes can be identified by their parts, it must be assumed that wholes are nothing more than, or other than, the parts that make them up. This is to say, wholes are taken to be sums of their parts. Further, as summative operations are applicable only to things that are not qualitatively changed by such operations, it must be assumed that the parts of wholes, whether collected as such or in isolation, are the same. It follows that the parts of one whole may also be parts of another. Hence, there is reason to be skeptical of inferences regarding the presence of particular wholes based on evidence pertaining to one or more of their parts.

2.2.2. Transformation

As previously mentioned, the things contacted by indirect observational procedures are not the phenomena of interest but, rather, are other things. In the case of remote observation, the things contacted were identified as parts or aspects of the original phenomena. By contrast, the things contacted by transformative procedures are synthetic reproductions of the constitutional properties of those phenomena, achieved by combinatorial and reconstructive operations of various sorts *(3)*, and the relation between them and the original phenomenon is one of partial similarity.

More precisely, because the things contacted in such investigations consist only of deliberately constructed features of the original phenomena, they are best conceptualized as analogs of those phenomena.

As Kantor *(2)* points out, procedures of this sort are common wherever interrelations among substances are at issue, such as in the science of organic chemistry, for instance. They are also central to investigations of murine models of human psychiatric disorders. In the latter case, the things analogous to humans presenting psychiatric disorders contacted by transformative observational procedures are transgenic or knockout mice, organisms in which specific genes have been inactivated or replaced with mutated versions, or in which additional gene functions have been established *(4)*. Analogs of a wide variety of human neurodegenerative diseases, psychiatric disorders, and conditions of mental retardation have been developed by such procedures, enabling a great deal of research that would not have been possible to conduct with human subjects.

Analogizing practices of two varieties are commonly used in science. The first of these is of the sort we have been discussing, wherein an analog of some phenomenon is deliberately constructed to permit it to stand in place of that phenomenon for the purpose of investigations that would not otherwise be possible. Analogizing practices of the second type are not purposive in this sense. They operate, instead, by taking advantage of found similarities among phenomena. More specifically, analogizing of this second sort may be described as the practice of orienting scientists to newly discovered phenomena by likening them to phenomena with which they are already familiar. Skinner's *(5)* likening of the action of reinforcement on a behavioral repertoire to the action of natural selection on the genetic composition of a species is an example of this strategy. In most cases, including this one, the likenesses to which the analogies point are likenesses of principle, not of substance. Hence, the orientation to a new phenomenon ordinarily achieved by such practices pertains to matters of its substantive structure only as far as such is reasonably implied by shared principle; the more dissimilar the substance of the known phenomenon to that for which orientation is sought, the more unreasonable is this implication.[1]

In summary, analogizing practices serve a useful purpose in science when particular phenomena cannot be investigated directly. Still, by definition, an analogy expresses a similarity between two sets of phenomena along particular dimensions or with respect to particular properties, and also implies dissimilarities along other, unspecified, dimensions or properties. The most common misstep in the handling of the knowledge products of analogizing practices is the tendency to assume that similarities of one sort imply similarities of another. Logical errors of this sort have the effect of misdirecting research efforts *(6)*.

---

[1]When conditions are such that the substantive properties of a newly discovered thing are well-enough known that the reasonability of such an implication is ample, the orientation of scientists to such a thing would not require an analogy in the first place.

We suggest, thereby, that knowledge of particular phenomena achieved by way of analogizing practices is valuable only when it focuses inquiry on observed similarities between those phenomena and their source observations. Even in this case, the scientific value of such information is short lived, because analogies are tolerated in science only as long as information achieved by direct study is unavailable *(7)*.

2.2.3. Manipulation

To reiterate, the things contacted by indirect observational procedures are not the phenomena of interest. Indirect observations of the remote type are made with respect to isolated constituents of the original phenomena, the relation between those things and the phenomena of interest being one of part to whole. In the case of indirect observational procedures of the transformative type, the things contacted are synthetic versions of the original phenomena, whereby the relation between those things and the phenomena of interest is one of analogy.

The things contacted by indirect observational procedures of the manipulative sort are also not the phenomena of interest. Instead, like those of the transformative case, they are synthetic versions of those phenomena. The difference between these two procedures has to do with the features of original phenomena selected by each for synthesis. Any given phenomenon is defined and identified by two sets of features: its constituent factors, or what might be called its formal properties; and its interrelations with other phenomena, otherwise known as its functional properties. The former are synthesized for observations of the transformative type, the later for manipulative observations.

More precisely, the things contacted by manipulative observational procedures are simplified versions of the original phenomenon's interrelations with other things. Further, as was the case of the synthetic things contacted by transformative procedures, the simplified things contacted by manipulative procedures are deliberately constructed to permit them to stand in place of the original phenomena for purposes of investigation.

The simplification of the original phenomenon's interrelations with other things is achieved by various operations that differ across disciplines. In a psychological context, in which the focus of inquiry is on an organism's responding with respect to stimulation from the multitudinous features of its environmental circumstance, simplification of the pattern of interrelations is achieved by inducing excessive focus on a highly restricted set of such features. For example, an organism deprived to 80% of its free-feeding body weight before investigation will be observed to interact almost exclusively with the food-related features of the investigative space. Such exclusivity is further exaggerated by minimizing the size of the investigative space and by impoverishing it with respect to features unrelated to food.

In summary, the things contacted by manipulative observational procedures are synthetic versions of particular features of the original phenomena, namely, their patterns of interrelation with other things. As such, they are appropriately conceptualized as analogs, the implication being that their unapparent dissimilarities are coupled with their apparent similarities to the original phenomena. This circumstance does not undercut the scientific value of the knowledge products of manipulative procedures. However, it does make the knowledge products vulnerable to misinterpretation, as discussed in the case of the products of transformative operations.

2.3. Concatenated Indirect Observational Contacts in Murine Modeling

To recapitulate: observational contacts in investigations of murine models of human psychiatric disorders include contacts of both direct and indirect types, although direct observation plays only an auxiliary role in such investigations. Direct observational contacts occur only before and after the primary stage of investigation in this enterprise, serving as the means by which features of the original phenomena are selected for representation in their models, and as the means by which specific propositions, hypotheses, and other products of the research enterprise are able to be validated, respectively.

During the primary stage of murine model investigations, observational contacts are exclusively of the indirect type. Moreover, the procedure is one in which all three varieties of indirect contact are concatenated, as follows. First, because observations are not made of original phenomena (i.e., human psychiatric disorders), but, rather, of other things (i.e., murine models of human psychiatric disorders), observational contacts are of the remote sort. Added to this, the materials investigated by remote observational contacts are synthetic versions of the constitutional properties of the original phenomena, these phenomena having been transformed from previously existing materials (i.e., normal mice), into new materials (i.e., genetic mutants). Finally, remote observations of these transformed materials are made under conditions in which their relations with other things are deliberately manipulated for the purpose of investigating particular relations (i.e., contrived environing conditions) in isolation from the larger relational configurations in which they ordinarily participate (i.e., natural environing conditions).

The knowledge products of this enterprise, thereby, are constructed from observational contacts with things and events many steps removed from the phenomena of original interest, and because some potential for logical as well as methodological error is present at every step in the accumulation of scientific knowledge, their validity and significance are far from certain. Validity is a matter of coherence. It is a measure of the extent to which the products of an investigative enterprise are formulated in accord with the systemic foundations of the enterprise *(3)*. Significance is a matter of compatibility. It is a measure of the extent to which the investigative products of a scientific enterprise, along with its underlying systemic foundations, are noncontradictory with these features of related sciences *(3)*. Both are threatened when factors not confronted in observational contacts with events, having their sources in cultural tradition, are incorporated in event descriptions. Adding to the description of an event of remembering, observed as a particular bit of verbal activity occurring in a particular set of environing conditions, a factor of access to a hypothetical internal storehouse of encoded copies of historical experiences, may serve as an example of this practice.

The propensity of scientists to engage in practices of this sort varies as a function of the foundational assumptions on which they are operating, as well as the adequacy with which those assumptions have been articulated and organized into a scientific system. Regarding the former, it is our assumption that sustaining a continuity of event descriptions with the events described is essential for their understanding within a particular scientific domain and is also required for their understanding to have significance in the context of other domains; the implication being that the practice of infusing event descriptions with factors not derived from observational contacts with those events is prohibited by premise.

Regardless of the premises adopted within a particular scientific system, the validity of the system depends on the coherence of its premises on one hand and its practices and products on the other hand; sustaining this coherence is greatly facilitated if its premises are explicitly articulated by way of deliberate system-building activities. Psychological systems are notoriously inadequate in this regard, however, and, as a result of their incomplete specification as scientific systems, construct validity in these systems is a serious problem. Specifically, descriptive constructs lacking continuity with the events from which they were derived are less likely to be challenged in the psychological domain than in any other science.

This problem is evident in the investigation of human psychiatric disorders by way of murine models. We will return to this issue after considering the criteria by which the adequacy of models of particular phenomena are evaluated, and the extent to which these criteria are met in murine models of human psychiatric disorders.

3. MODELING PRACTICES

Briefly described, modeling procedures involve four operations:

1. Identifying characteristic features of the phenomenon of interest.
2. Constructing a relatively simple representation of those features as a model of this phenomenon.
3. Subjecting the model to scientific investigations.
4. Extrapolating the findings of these investigations to the phenomenon of interest.

In the models of psychiatric disorders currently under investigation, combinations of factors of both the psychological and biological type are represented, whereby multiple avenues for extrapolation from animal findings to human phenomena are afforded. As such, this line of research would seem to hold significant promise of achieving its aim of identifying the formal and functional properties of these disorders as manifested in human beings.

Nonetheless, as previously mentioned, the promise of a modeling strategy toward this end depends on the adequacy of the models constructed to reach it. The adequacy of a model is ultimately a matter of the extent to which investigating it leads to verifiable propositions and hypotheses concerning the phenomenon modeled; successful outcomes of this sort have led to the formulation of certain principles or rules of effective model construction. These rules articulate criteria by which the adequacy of models may be measured, and among them are the following three:

1. The events from which an extrapolation is made must be constituted of the same sorts of phenomena as those to which it is applied.
2. The model must not contain significant factors that are absent in the phenomenon of interest.
3. The phenomenon of interest must not contain significant factors that are absent in the model.

When these conditions are met, a model of some phenomenon is deemed sufficiently representative of that phenomenon for extrapolative operations to be productive of valid knowledge concerning the model. We next consider the extent to which these conditions are met in murine models of human psychiatric disorders and, as mentioned, our considerations pertain exclusively to the representation of psychological factors in these models.

3.1. Model and Original Phenomenon Are Constituted of Similar Materials

The first rule of model construction, by which practices of extrapolation are likely to produce valuable knowledge products, specifies that the things known by way of extrapolation must share a likeness with those from which this knowledge was derived. Phenomenal likeness, as here required, may be a matter of shared material constituents or shared relational properties, although preferably both. In the case of murine models, this first condition of phenomenal likeness would seem to be met. That is to say, behaving with respect to a stimulus is the activity of a biological entity in interaction with a physical environment, be the organism engaged in such an interaction a human or a nonhuman. The model is, thereby, sufficiently representative of the original phenomenon for extrapolative operations to be productive of valid knowledge concerning it.[2]

3.2. Absence of Significant Factors in the Original Phenomenon

The second rule for effective model building specifies that the model must not contain significant factors that are absent in the original phenomenon. Again, from a psychological perspective, the relevant factors are those entailed in organisms' responding with respect to environmental stimulation. In as much as the respond-

---

[2]By way of contrast, knowledge of organisms' interactions with their environments extrapolated from the operations of intelligent machines must be viewed with caution. Biological entities and machines, regardless of the sophistication of the latter, are not constituted of the same sorts of phenomena, formally or functionally. Machine operations are, thereby, not sufficiently representative of the actions of biological entities to constitute a source of valid knowledge concerning them. Knowledge of organisms' interactions with their environments achieved by this means is thereby metaphorical, not representative in kind.

ing of organisms is a function (in the mathematical sense) of their biological structures, at issue here is whether the responding of the model, in this case, a mouse, has features not shared with that of a human being because of their anatomical and/ or physiological differences; and, if so, whether those features are of sufficient prominence or preeminence to overshadow whatever might be their shared features. Of particular concern, in this regard, are the proportions of what have been called "involuntary" (i.e., reflexive) and "voluntary" (i.e., operant) behaviors in the repertoires of these two species, the grounds for this concern having to do with the preeminence of the former.

The behavioral repertoires of species lower on the phylogenic scale display a greater preponderance of reflexive behavior than those higher on this scale, a disparity assumed to reflect differences in the rapidity and magnitude of change in their environing circumstances over evolutionary time. Additionally, the more substantially a species' environment changes from one generation to the next, the larger is its proportion of operant behavior likely to become *(8)*. If a disparity of this sort is evident in the repertoires of mice and human beings, and especially if it is of considerable magnitude, scientific propositions concerning the predominantly operant behavior of human beings, based on observations of murine models exhibiting a preponderance of reflexive behavior, are not likely to be verified. We turn, then, to the differences between these two types of behavior.

3.2.1. Reflexive Behavior

Reflexive behavior has features that operant behavior does not have, and one of those features makes its occurrence more probable than that of operant behavior under conditions in which both are possible. More specifically, because responding always occurs with respect to stimulating, particular responses of either type are possible only when their coordinated stimuli are present. However, because reflexive responses are elicited by particular stimuli and those stimuli exert near absolute control, a circumstance in which reflexive responding is possible is one in which it is virtually certain to occur. For example, stimulation in the form of a puff of air into a person's eye will elicit a blinking response, regardless of other circumstances, such as the setting in which this stimulus is encountered, the person's organismic condition at the time of its occurrence, or the person's history of contact with the stimulus.[3]

3.2.2. Operant Behavior

By contrast, operant behavior is multiply controlled, which to say, its occurrence is not solely a function of its coordinated stimulus. Also relevant to its occurrence are setting conditions of both an environmental and organismic sort, as well as its history of reinforcement under similar circumstances. In short, the occurrence of operant behavior is probabilistic *(1,8)*. For example, stimulation in the form of

---

[3]An exception is "habituation," wherein a reflexive response may not be elicited by its coordinated stimulus after repeated presentations of this stimulus over a relatively brief period.

an attractive woman may evoke a winking response on the part of a man, but whether this response occurs given the stimulus opportunity for its occurrence will depend on a combination of other factors. Among them may be the presence of the woman's husband, the appropriateness of the occasion for a flirtatious response, the consequences of having engaged in this behavior under similar circumstances in the past, and so on.

Consequently, in any circumstance in which the stimuli coordinated with both reflexive and operant behavior are present, and both are thereby possible of occurrence, reflexive behavior will always occur, whereas operant behavior will only sometimes occur.

Added to this, operant behavior tends to be disrupted or suppressed in the presence of precursors to reflexive behavior *(9)*. Taken together, these conditions suggest that reflexive behavior is preeminent over operant behavior.

In summary, an adequate model is one that does not contain significant factors that are absent in the phenomenon of interest. A preponderance of preeminent reflexive behavior in the mouse repertoire may constitute such a factor.

3.3. Absence of Significant Factors in the Model

Because extrapolative practices operate on the basis of particular, significant commonalities between the events investigated and those for which an understanding is sought, at least some of the defining features of the phenomenon of interest must be represented in the model of that phenomenon for such practices to be pursued with good result. Hence, the third rule for effective model construction specifies that the phenomenon of interest does not contain significant factors that are absent in the model.

To determine whether murine models of human psychiatric disorders are sufficiently representative of significant factors of the human psychological sort to satisfy this requirement, we must take the preliminary step of identifying the defining properties of human psychological events.

4. DEFINING PROPERTIES OF HUMAN PSYCHOLOGICAL EVENTS

Although the most often-cited difference between human behavior and the behavior of all other species is the linguistic character of the former, little consensus has been reached regarding the defining properties of linguistic events, even among psychologists of the behavioral persuasion (e.g., refs. *7,10–13*) much less the larger psychological community. For example, Skinner *(7)* differentiates verbal from nonverbal behavior by the manner in which its effect on the environment is achieved. Specifically, an act is held to be verbal if its effect is achieved indirectly by way of conventional action on the part of a listener. The action of the listener in this regard, although not itself necessarily verbal in type, is, nonetheless, of a conventional sort. As Skinner puts it, the listener's action is explicitly conditioned by the verbal community "especially to create a means of control" (ref. 7, p. 225). In other words, from Skinner's perspective, what permits the identification of an act as verbal in type is the provenance of another person's[4] response to it.

This is not the feature on which the linguistic designation depends for Kantor. Kantor *(11,12)* acknowledges the indirect nature of linguistic behavior, but argues that, because this property is also entailed in some forms of nonlinguistic activity (e.g., feeling reactions), it does not distinguish linguistic from nonlinguistic events in an unambiguous manner. Instead, linguistic interactions are held to be distinguished by the property of bi-stimulation. Kantor *(12)* argues that, unlike behavior segments of all other types in which responding occurs with respect to stimulation arising from a single source, segments of the linguistic type involve responding with respect to stimulation arising from two sources simultaneously.[5] One of these sources is the primary thing to which both the speaker and the listener are adjusting, called the adjustment stimulus or referent. The other source of stimulation for each of the participants, called the auxiliary stimulus, is the other participant (i.e., the listener in the case of the speaker, and the speaker in the listener's case). In essence, linguistic interactions entail simultaneous adjustments of both speaker and listener to a common source of stimulation as well as to each other.

In still another behavioral interpretation *(10)*, a verbal act is said to involve a special type of "relating" activity, defined as responding to one thing in terms of another. According to Hayes et al. *(10)*, a behavioral history in which specific things have served as sources of stimulation for relating acts eventually gives rise to an abstracted form of relating that is independent of specific things in relation. Relating acts of this sort are said to be "arbitrarily applicable" to any set of things, their particular varieties (e.g., relations of sameness, difference, and so on) being determined not by the things to which they are applied, but, rather, by features of the context in which they are invoked. When an act of responding to one thing in terms of another also entails responding to the other in terms of the one, that is to say, when an arbitrarily applicable relating act has the property of mutual entailment (as well as the derived property of combinatorial entailment),[6] it is characterized as verbal in type.

Our purpose in describing these divergent views has been to substantiate our claim that there is no consensus regarding the defining properties of linguistic action even among scientists operating on similar philosophical foundations.[7] Indeed,

---

[4]Skinner *(7)* differentiates the behavioral repertoire of the listener from the listener as a substantive entity whereby the actions of both speaker and listener may occur on the part of the same person in a given episode of verbal behavior.

[5]Kantor *(12)* discusses two types of linguistic behavior, symbolic and referential. Both are distinguished from nonlinguistic events by the property of bi-stimulation; however, the sources of this stimulation are different for the two types. Exemplified here is the case of referential behavior.

[6]Combinatorial entailment is held to be a stimulus relation derived from the combination of two or more mutually entailed stimulus relations *(10)*.

[7]Behaviorism is not one philosophical entity, but, rather, a class of such entities; the divergence of these views is attributable to their having been constructed in accord with different varieties of this philosophy.

there are almost as many definitions of linguistic phenomena as there are scientists concerned with such matters. Hence, although we agree that human behavior may be differentiated from the behavior of nonhumans by its linguistic character, our plan is not to make this case on the basis of yet another definition of linguistic phenomena. Instead, our plan is to evaluate the degree to which processes responsible for the expansion and elaboration of the human behavioral repertoire are participating factors in the behavior lives of nonhumans. Two such processes will be considered in making this evaluation, one involving substitute stimulation, the other, substitute responding. As previously discussed, substitute stimulation is not unique to the human species. However, because of a number of circumstances, this process is much more prevalent in the behavior lives of humans than nonhumans, and this difference is partially responsible for the size of the human repertoire. More importantly, substitute stimulation affords massive opportunity for the uniquely human process of substitutional responding, and it is by virtue of capacity for the latter that the ever-expanding enormity and ever more elaborate complexity of the human repertoire may be understood. A more detailed description of these two processes and their reportorial outcomes follows.

4.1. Substitute Stimulation

Both humans and nonhumans act with respect to stimulation arising from immediately present environing things, wherein the formal characteristics of their responses are determined by the natural properties of those environing things. For example, the salivating action of a dog is commensurate with the chemical properties of the food powder. Both humans and nonhumans also engage in responding having formal properties determined by the natural properties of environing things when those things are not immediately present. Thus, the salivating action of the dog, which is commensurate with the chemical properties of the food powder, may occur in the absence of the food powder.

However, this response cannot be said to occur in absence of stimulation of any kind, because responding implies stimulation and vice versa *(2)*. On the contrary, for the salivating response to occur in the absence of the food powder, the immediate circumstance must contain a stimulus bearing some relation to the food powder, by virtue of which, it may serve as a substitute for the absent food powder.[8] More specifically, the present circumstance must contain a stimulus that is similar to food powder or was present when a salivating response with respect to food powder occurred on a previous occasion.

Technically, substitute stimulation is the action or functional properties of one stimulus object occurring through or inhering in the object properties of another stimulus *(6)*. The operation of substitute stimuli is observed as the occurrence of a response having formal characteristics that are commensurate with the natural prop-

---

[8]Relations of similarity and spatial/temporal proximity exhaust the possibilities in this regard, and, given the impossibility of defining a functional response in the absence of a functional stimulus *(11)*, one or the other must be assumed under such conditions.

erties of a previously encountered stimulus object under conditions in which that object is absent.

The development of conditional stimulus–response relations by way of classical conditioning is illustrative of substitute stimulation, as herein defined. For instance, the formal properties of a pupil contraction response are determined by the natural properties of bright light as a stimulus, that is, a response of this form is specific to stimulation of this type. If pupil contraction occurs in the absence of a bright light, it must be assumed that a functional property of the light is operating through another stimulus present in the immediate circumstance; and this, in turn, implies a history in which pupil contracting responses with respect to stimulation from a bright light had previously occurred under conditions in which a proximal relationship between this other stimulus and the light was obtained. For instance, if a tone sounds in close temporal/spatial proximity to the appearance of the light (coupled with its coordinated pupil contraction response), it may acquire a functional property of the light that would enable the tone to serve as a source of substitute stimulation for pupil contraction in the absence of the light. In short, the history implicated in the occurrence of pupil contraction in the absence of its original source of stimulation is precisely what the process of classical conditioning entails.

As mentioned, historical conditions of temporal/spatial proximity between original and substitute stimulus objects, as in the case of classical conditioning, constitute only one of the conditions favorable to the development of substitute stimulation. Added to these are conditions of formal similarity (either in whole or in part) between stimulus objects.[9] In such cases, the functional properties of an original stimulus object may inhere in another object by virtue of its formal or physical similarity to the former.

For purposes of clarity, note that substitute stimulation is not restricted to responses of the reflexive or respondent sort. On the contrary, operant behavior, particularly of the perceptual variety, is just as likely to be involved *(13,14)*. For example, given a history of hearing a person's voice while also seeing her face, the act of seeing her face when only her voice is present, as when speaking on the telephone, illustrates the process of substitute stimulation on the basis of proximal relations among stimuli. Substitute stimulation for operant behavior based on formal similarities among stimuli is also a prevalent form of human activity. Kissing a photograph of a lover exemplifies an event of this type.

The capacity of an organism to engage in responding, having formal properties determined by the natural properties of environing things when only the functional properties of those things are immediately present, is by no means a trivial matter. On the contrary, it reflects an organism's psychological potential, that is, its capacity to learn. By way of substitute stimulation, an organism's historical interactions with its environment are afforded a place in the effective present, and the cumula-

---

[9]The development of substitute stimulus functions by virtue of relations of formal similarity among stimulus objects is conceptualized in other technical formulations as stimulus generalization or stimulus induction (*see*, e.g., ref. *1*).

tive outcome of this process is a complex, conditional, behavioral repertoire, related to an equally complex environing circumstance *(15)*.

By contrast, when organisms lack the capacity for substitute stimulation, their historical interactions with their environment have no bearing on their current interactions, whereby their responses to previously encountered stimulus objects are invariant from one occurrence to the next. In other words, when an organism lacks the capacity for substitute stimulation, its interactions are interpretable as coincidental operations of the functional properties of an organismic structure in physical contact with the physicochemical properties of environing things *(16)*. In short, the actions of such an organism are biological, not psychological, in type, and its behavior life is restricted to the exigencies of immediately present conditions.

Although responding with respect to substitute stimulation is not unique to the human species, it is especially pervasive in the repertoires of human beings, and is completely absent in the repertoires of some other species. As such, it may serve as the basis for a hierarchical organization of species in terms of their psychological attributes. From the outset, its seems reasonable to assume that the survival value of a process by which an organism's historically acquired modes of interaction with stimulus objects are implicated under current circumstances would vary as a function of species' life cycles. Hence, at the pole opposite to that occupied by the human species, would be those species with life cycles of durations too short for any such value to accrue. However, this factor does not account for the placement of the human species at the other end of this continuum; nor does it provide guidance regarding the placements of other species along the continuum. For these purposes, we must consider the relevance of two other factors, namely the relative preponderance of reflexive behavior in species' behavioral repertoires and the relative complexity of their stimulating environments.

4.1.1. Preponderance of Reflexive Behavior

As discussed in the previous section, the stimulus objects in which substitute stimuli inhere are also sources of stimulation for direct actions coordinated with their natural properties. If these direct actions are compatible with those coordinated with the substitute functions inhering in the object, both may be expected to occur in same encounter with the object. However, if they are incompatible, the direct actions will always take precedence. This is because the relations between substitute stimulation and responding, being conditional, are more readily disrupted than those involving stimulation arising from an object's natural properties. As such, the greater the prevalence of reflexive action in the repertoire of a given species, the less likely is that species to demonstrate responding with respect to substitute stimulation.

4.1.2. Complexity of Environmental Circumstances

Inasmuch as substitute stimulation depends on relationships among stimulus objects, the more complex an organism's environment in terms of the number and variety of stimulus objects with which it is capable of interacting, the more prevalent are events of the substitutional sort likely to be in its behavior life. The pervasiveness of substitutional phenomena in the behavior lives of human beings is readily accounted for by this logic.

To a very large extent, the human environment is a constructed environment, populated with an enormous and ever-increasing quantity of an endless variety of things, deposited as the material products of human action. All of these things are potential sources of stimulation for individual organisms and, more importantly, their proximal relations and partial similarities to one another are conditions favorable for the development and operation of substitutional functions among them. No other species modifies its natural environment in this manner and, in none, thereby, is the opportunity greater for the operation of substitute stimulation.

However, environing *things,* including substantive features of the natural environment as well as the material products of human action, do not exhaust the potential sources of stimulation in the human environment. In the environing conditions of any member of the human species are other members of its species, which is to say, humans live in groups. Hence, a major source of stimulation for the behavior of any individual member of a human group is the behavior of its other members. The significance of this circumstance has to do with the difference between animate and inanimate sources of stimulation with respect to the multiplicity of their coordinated responses *(13,17).* Specifically, persons are sources of a larger number of stimulational functions than are inanimate objects.[10] To put it another way, responding with respect to stimulation arising from animate sources (i.e., other persons) tends to be more varied than that arising from inanimate sources. Further, because of the process of stimulus substitution on the basis of formal similarities among stimulus objects, the stimulus functions inhering in another person as a stimulus object are not entirely dependent on previous encounters of the responding person with this particular object. This is also the case for inanimate objects.

However, animate objects are not merely sources of stimulation for responding on the parts of others; they are also responding entities. As such, their stimulus functions arise not only from their object properties but also from their actions, and because these are constantly changing, so too are the responses of other persons coordinated with them. Hence, functional relations involving animate objects evolve more rapidly than those involving inanimate objects, and this situation is further exacerbated when the interaction between two such objects (i.e., persons) is one of mutual stimulation and responding *(13,17).*

This circumstance is not unique to the human species, of course. However, a unique property of human behavior is manifested in these sorts of interactions and its participation so drastically changes them to render whatever commonalities there may be between human and nonhuman intraspecies interactions trivial by comparison. We turn now to this all-important feature of human behavior.

---

[10]This difference is not entirely a matter of animation, because some inanimate objects may have rapidly changing characteristics, and, as a result, may be endowed with an unusually large number of stimulational properties. In general, animate objects are sources of more stimulational functions than inanimate objects.

4.2. Substitute Responding

Up to this point, we have argued that all species engage in behaviors having formal features determined by the natural properties of immediately present stimulus objects; that only some species engage in such behavior when those stimuli are absent from the immediate situation, through the operation of substitute stimuli; and that, of the latter, the human species reigns supreme because of the relatively small proportion of its behavior that is reflexive in type and the exceedingly complex environment in which humans live.

We turn now to the matter of behaviors having formal properties that are not determined by the natural properties of stimulus objects, either when those objects are present in the immediate situation or when they are absent. Behavior of this sort is evident only in the repertoires of human beings. To illustrate the distinction we are making here, interactions between stimuli and responses exemplifying responses forms commensurate with the natural properties of stimuli are contrasted with those in which responding is incommensurate with those properties.

4.2.1. Commensurate Response Forms

In picking up a cup, not just any response will do. It must be one with particular formal characteristics, namely those suited to the natural properties of the cup, including its size, shape, weight, whether or not it has an intact handle, and so on. If the cup is filled to the brim with hot coffee, additional properties of responding will be entailed, among them constancy of force and steadiness of adjustment to the horizontal plane (*see* ref. *13* for further discussion).

The formal properties of such responses are not impacted by the settings or contextual circumstances in which they take place. This is because these properties of responses are determined by the natural properties of stimulus objects, and the latter remain intact regardless of the context in which they are encountered. However, the formal properties of such responses are not fixed. Rather, they become refined as variants that are more effective are selected over the course of multiple interactions with stimulus objects of a particular variety. For example, an infant may pick up a cup by inserting her fingers into the cup and grasping it between them and the heel of her hand. As a toddler, she may clasp the cup between both hands. When slightly older, she may grasp the handle with one hand while clasping the cup with the other. With more experience, she may pick up the cup with one hand grasping the handle and, eventually, she may do so without spillage.

Although behavior of this type develops over a series of encounters with environing things that differ from one person to the next, and, further, development of this type of behavior occurs in the absence of instruction from other members of a group,[11] a commonality of responding across organisms in contact with the same (similar) stimulus objects is assured by the dependence of response forms on the natural properties of stimulus objects.

---

[11]Although instruction may not be required, the acquisition of this type of behavior in human circumstances is usually supplemented or supplanted by way of instructions or rules *(8)*.

4.2.2. Incommensurate Response Forms

When the object properties of a stimulus are properly distinguished from its functional properties, it becomes obvious that multiple forms of responding may be coordinated with stimulation arising from the same object *(3)*. In other words, an object may be a source of stimulation for more than one type of response. For example, a cup may stimulate the act of picking it up as well as that of saying "cup." The latter exemplifies the type of response with which we are now concerned, namely responses having formal properties that bear no relation to the natural properties of the stimuli with which they are coordinated. To put it another way, the formal characteristics of the vocal response, "cup," have nothing whatsoever to do with the size, shape, weight, or any other natural property of the physical object, cup. Instead, the response form is arbitrary. Because a response of this sort may take the place of a response of commensurate form, and for lack of a better term, they are called substitute responses *(6)*.

Responses of arbitrary form are coordinated with stimuli, nonetheless. Only some of the properties of stimuli inhere in their bare qualities and conditions as physical objects. Others have been attributed to them under the auspices of particular groups, these properties thereby being characteristic of, or conventional within, those group circumstances *(16)*. It is properties of this latter sort with which responses of arbitrary form are coordinated. For example, in an encounter with the object, cup, a response of the form, "cup," will be suitable for one group circumstance, namely one in which English is spoken; whereas, a response of the form, "la tasse," will be suitable for one in which French is spoken.

The effectiveness of responses coordinated with the attributed properties of stimuli is thereby critically dependent on the context in which particular stimulus objects are encountered. Moreover, the only standard by which their effectiveness can be measured is their formal correspondence with the responses of other members of the group *(16)*. As such, responses of this sort do not become refined over the course of repeated interactions with particular stimulus objects.[12] On the contrary, they are conventional forms of responding, requiring explicit instruction from the group for their acquisition, and, once acquired, their formal properties are fixed.

4.2.3. Implications of Substitutional Responding

Conventional responses of this sort are prominent members of virtually every category of human activity, including those of the aesthetic, religious, political, scientific, and domestic varieties, to name a few *(16)*. Indeed, much of what humans do in each of these domains, which is to say, how they respond with respect to the stimulus objects encountered within them, depends not on the bare qualities and conditions of the encountered objects, but rather on the group circumstances under which their behavioral repertoires were established. By way of illustration, the

---

[12]However, an individual's repertoire may contain multiple forms of responding coordinated with the same object encountered under different contextual conditions (i.e., a multilingual repertoire).

beauty of a sunset, the righteousness of a political position, the validity of a scientific fact, and so on, are not natural properties of sunsets, positions, and facts as stimulus objects. Rather, they are properties that have been attributed to these objects under the auspices of particular groups. A massive amount of human behavior is coordinated with such attributed properties of stimuli.

This circumstance has a number of implications, the first of which bears on the complexity of the stimulating environments of human beings relative to those of other species and the potential therein for processes of substitute stimulation to be an especially pervasive feature of the behavior lives of human beings. In this regard, the occurrence of actions not determined by the natural properties of stimulus objects, a condition unique to the human case, means that human actions display much greater variety than that of other species *(13,17)*; and, because the actions of one organism constitute stimuli for another, the stimulation inhering in these arbitrary response forms renders the human environment even more complex than the enormous quantity of deposited material products of human action have already been acknowledged as having engendered. The potential for substitute stimulation is greatly enhanced by physical presence of these additional stimulus objects.

More importantly, responses having formal properties that are not determined by the natural properties of stimulus objects, even in their physical presence, are particularly well suited for occurrence in the absence of those objects *(13,18)*. In other words, engaging in the response of picking up a cup depends on the physical presence of a cup in a manner that engaging in the response of saying "cup" does not.[13]

In short, from a psychological perspective, the effective environment for a human being includes not only that which is physically present (as is the case for all species), as well as that which was present in this person's experience at a previous time (as is the case for some species), but also that which has never been present in this person's experience because it happened before the onset of this person's experience, or because it has yet to happen in anyone's experience, or because it has always been present in a distant location, or because it has never been present in any location by virtue of its nonexistence.[14]

In summary, the environment of the human species affords its members a means of engaging in arbitrary forms of activity with respect to the past, the future, the distantly removed, and the nonexistent; and it does so by virtue of the stimulation inhering in actions of the very same sort on the parts of other human beings, as well as of themselves. This is not the environment in which other species live, and their psychological experiences have none of the properties that living in such an envi-

---

[13]In our view, this circumstance does not imply that such actions are occurring in the absence of stimulation, as suggested by Skinner *(7,8)*, but, rather, that they are coordinated with currently operating substitutional functions of physically absent stimuli.

[14]Although activities of various types may be coordinated with stimulus functions of these types, including perceptual activities of all kinds, it is our position that nothing of this sort would be possible in the absence of linguistic action, which, we believe, is also the means by which the material environment of human beings is able to be so radically altered.

ronment affords. Indeed, the extraordinary expansion and elaboration of the human psychological experience by way of mutual intraspecies interactions of this nature is a distinctly human phenomenon. More importantly, these unique forms of activity are not mere additions to the collection of other features of the human repertoire that are shared with other species. On the contrary, the human repertoire is fundamentally altered by their presence, as a growing body of research continues to show (*see* ref. *10* for a summary of findings). In short, activities of arbitrary form, coordinated with attributed properties of stimuli, are insinuated in virtually every kind and variety of human interaction. Hence, to whatever extent an understanding of human psychopathology must take these sorts of interactions and their implications into account, that is the extent to which a murine model of the same will fall short of its aim.

4.3. Summary of Significant Features of Human Psychological Events

Elucidating the defining properties of human behavior was undertaken as a preliminary step toward the goal of determining whether murine models of human psychiatric conditions satisfied the criterion for effective model construction specified in the third rule pertaining to such matters; namely, that the phenomenon of interest does not contain significant factors that are absent in the model. Our preceding analysis revealed three defining properties of human psychological events:

1. A disproportionately small amount of reflexive action.
2. An inordinately large involvement of substitute stimulation.
3. An immense factor of conventional responding of arbitrary form.

Given that mice display substantially greater amounts of reflexive behavior than humans; that their environing circumstances afford much less potential for the operation of substitute stimulation than those of humans; and that their repertoires are completely devoid of conventional responses of arbitrary form, we must conclude that murine models of human psychiatric conditions do not satisfy this third criterion.

On the basis of this review, we may conclude that the most obvious shortcoming of murine models of human psychiatric disorders, as pertains to their representation of psychological factors, is the absence of linguistic behavior in the repertoires of the model organisms. In as much as the presence of such behavior in the repertoires of human beings is the basis on which the psychological events of humans are differentiated from those of other animals, this is not a trivial problem for this research enterprise. Moreover, this is not a problem that can be solved by making improvements to these models. Nonhuman organisms do not engage in substitutional responses. It is thereby not possible to construct an animal model of a human psychiatric disorder that is representative of its most significant psychological feature. It follows that extrapolative operations from animal findings will not be productive of valid knowledge regarding the most significant psychological features of these disorders.

5. THE PROBLEM OF ADDITIVE SUMMATION

Investigations involving remote observations of transformed phenomena under manipulated conditions, as is the case of murine modeling of human psychiatric disorders, are not uncommon in science. Still, because the products of investigation eventuate on the completion of a great many operations of a logical as well as a methodological sort, all of which present opportunities for error; the products of investigations in which the things actually observed are many steps removed from those about which knowledge is sought are especially vulnerable to misinterpretation. This being the case, it would not be prudent to adopt an understanding of human psychiatric disorders based on the products of investigating murine models of the same, in the absence of scrutiny regarding the practices on which such an understanding had been constructed.

More importantly, the value of a modeling approach depends on the type of event on which extrapolations to the human condition are made. It is only when events are such that an understanding of their participation in complex conditions is able to be achieved by an additive procedure with respect to facts gleaned in isolated studies of simpler conditions that a modeling procedure has value. Events of the neurological, genetic, and other relevant subdivisions of the biological domain may fit this pattern. However, it is not possible to achieve an understanding of complex psychological events by an additive procedure of this sort, because of the historical nature of psychological phenomena.

No quantity of facts pertaining to the nonverbal behaviors of animals will ever be enough to constitute an understanding of predominantly verbal actions of human beings. Further, even without the problem of animals and humans having different repertoires, the additive procedure suggested by advocates of this strategy is questionable. The issue in this case is a philosophical one, having to do with variant interpretations of the concept of causality. This is a complex issue. In the broadest sense, the concept of causality is invoked as an explanation for the regularities observed among things of nature. Typically, the things of nature are divided into the dichotomous classes of causes and effects. "Cause" things are held to have the power to produce "effect" things. In behavior science, for example, environmental causes are held to produce organismic effects *(1)*.

Causal logic of this sort does not rule out the possibility that a given effect may be a product of more than one cause, or that a given cause may produce more than one effect. Accordingly, a particular effect may be explained by appeal to the combined powers of multiple causes, and multiple effects may be attributed to the power of a single cause. Further, no limitations are placed on the potential for these relations to be combined. This interpretation of causality suggests that a complex event is made up of some combination of simple events. Accordingly, it is assumed that, with enough facts about the simple events, the complex event may be understood in its entirety. It is this logic that sustains the view that investigations of animal behavior under controlled conditions will eventually add up to an understanding of the more complex case of human behavior *(1,8)*.

A less mechanistic view of causality would argue otherwise. For example, causal knowledge is sometimes understood as knowledge of the factors participating in a given event, along with their interrelations and their organization *(19,20)*. Understood in this way, a psychological event is not made up of a stimulus cause and an organismic effect organized as a dependency relation, but is rather comprised of an interdependent relation of stimulating and responding, taking place in an integrated field involving many other factors. From this perspective, any change in the factors making up the event field change the entire field, including the interdependent relation of responding and stimulating occurring in that field. In other words, the factors comprising a psychological event are not additive. By this reasoning, no collection of facts about simple events will ever add up to an understanding of the complex event in which all are participating.

The point we are making in this regard is simply that scientific expectations are derived from philosophical premises. Ours are such that, if an understanding of some particular set of events is the scientific goal, then the strategy most likely to be successful in reaching that goal will be to investigate those events directly. We conclude, therefore, that an understanding of the psychological aspects of human psychopathology are more likely to be achieved over the long run if the subject of our investigations is the pathological behavior of humans in typical human circumstances than if a murine model of the same is substituted for that subject. The significant risk in doing the latter is that what one may eventually understand by doing so is little more than how a tormented mutant mouse behaves.

REFERENCES

1. Skinner BF. Science and Human Behavior. New York: Knopf, 1953.
2. Kantor JR. The Logic of Modern Science. Chicago: Principia, 1953.
3. Kantor JR. Interbehavioral Psychology. Chicago: Principia, 1959.
4. Nicholl ST. An Introduction to Genetic Engineering. Cambridge, UK: Cambridge University Press, 2002.
5. Skinner BF. Contingencies of Reinforcement. New York: Appleton-Century-Crofts, 1969.
6. Kantor JR. Psychology and Logic. Vol. 1. Chicago: Principia, 1945.
7. Skinner BF. Verbal Behavior. New York: Appleton-Century-Crofts, 1957.
8. Skinner BF. About Behaviorism. New York: Knopf, 1974.
9. Henton WW, Iverson IH. Classical Conditioning and Operant Conditioning: A Response Pattern Analysis. New York: Springer-Verlag, 1978.
10. Hayes SC, Barnes-Holmes D, Roche B. Relational Frame Theory. New York: Kluwer/Plenum, 2001.
11. Kantor JR. An Objective Psychology of Grammar. Chicago: Principia, 1938.
12. Kantor JR. Psychological Linguistics. Chicago: Principia, 1977.
13. Parrott LJ. Listening and understanding. Behav Anal 1984;7:29–40.
14. Hayes LJ. Equivalence as Process. In: Hayes SC, Hayes LJ, eds. Understanding Verbal Relations. Reno: Context, 1992.
15. Hayes LJ. The psychological present. Behav Anal 1992:15:139–148.
16. Kantor JR. Cultural Psychology. Chicago: Principia, 1982.
17. Parrott LJ. On the Difference Between Verbal and Social Behaviors. In: Chase PN, Parrott LJ, eds. Psychological Aspects of Language. Springfield, IL: Thomas, 1986.

18. Hayes LJ. Substitution and Reference. In: Hayes LJ, Chase PN, eds. Dialogues on Verbal Behavior. Reno: Context, 1991.
19. Kantor JR. Interbehavioral Philosophy. Chicago: Principia, 1981.
20. Kantor JR. Psychology and Logic. Vol. 2. Chicago: Principia, 1950.

4
If Only They Could Talk
Genetic Mouse Models for Psychiatric Disorders

Trevor Humby and Lawrence Wilkinson

Summary

The advent of advanced molecular genetics methods has revolutionized biology, both in terms of identifying gene candidates and, once identified, in terms of manipulating gene function in the experimental setting to address issues of causality and mechanism. Mice are currently the most genetically tractable animal model available and have contributed significantly to furthering our understanding of gene action across multiple areas of biology and medicine. However, using mice to model psychiatric conditions raises particular challenges because of the complex and perhaps unique phenotypes affected by disorders such as autism, schizophrenia, attention deficit hyperactivity disorder, depression, and personality disorder. In this chapter, we present a realistic view of what can be modeled in this area. We discuss the concept of endophenotypes (intermediate traits), and how endophenotypes can be used to address the complexity of the clinical conditions. We also discuss the practicalities of carrying out valid behavioral studies in mice, taking in the issues raised by the ethobiological vs artificial approach debate and emphasize the positive contribution made by the increasing use of operant behavioral paradigms in mice. We conclude that, although we are very much at the "work in progress" stage, it is likely that mouse models will be of major importance in ensuring that psychiatry gets its share of the genomic dividend.

Key Words: Psychiatric disorders; genetic mouse models; endophenotypes/intermediate traits; construct validity; ethobiological behavioral tests; operant methods.

> What a piece of work is a man! how noble in reason! how infinite in faculties! in form and moving, how express and admirable! in action, how like an angel! in apprehension, how like a god! the beauty of the world! the paragon of animals! *(1)*

These words spoken by Hamlet, in one of his frequent melancholic moods, seem to epitomize the challenge faced by any attempt to use nonhumans to model psychiatric disorders, those conditions that inflict damage on mental functions closest to what we fondly imagine makes us the paragon of animals. Namely, our reason, our cognitive and emotional faculties, our self-awareness, and our apprehension of the world filtered through our unique ability to abstract and use symbols in the form of language. At the outset, therefore, we have to ask ourselves, is it ever going to be

From: *Contemporary Clinical Neuroscience: Transgenic and Knockout Models of Neuropsychiatric Disorders*
Edited by: G. S. Fisch and J. Flint © Humana Press Inc., Totowa, NJ

possible to convincingly model phenomena such as guilt, religiosity, delusions, hallucinations, body image distortion, and multiple personality in the mouse (reviewed by Tecott and Nestler, ref. *2*)? Notwithstanding the many myths and legends that have conferred attributes such as cunning, sagacity, fortitude, or even nobility to animals *(3)*, the short answer to this has to be no. So, that's that out of the way; a complete account of complex psychiatric disorders cannot, obviously, be achieved in any animal model, not least because of the reporting problem. However, as we attempt to argue in this chapter, some headway can be made by modeling discrete components of the brain and behavioral symptoms.

The problem posed by psychiatric conditions can be contrasted to the quite stunning progress made in our understanding of organic neurological conditions, in which genetic mouse models have been in the forefront in exploiting the advances in human molecular genetics by permitting, once a gene candidate has been identified, questions of causality and mechanism to be addressed in the experimental setting (*see* review by Duff and Rao, ref. *4*). One specific example is the use of genetically modified mice expressing human transgenes with mutations that are fully penetrant for familial forms of dementia, especially familial Alzheimer's disease *(5)*, but also the explicit genetic tauopathies, such as frontotemporal dementia with parkinsonism linked to chromosome 17 *(6)*. Such transgenic mice develop brain pathology remarkably reminiscent to that seen in patients, and develop changes in behavior, as we have shown in our own work *(7)*, that are consistent with effects on discrete aspects of psychology, including memory and response inhibition. It would seem that, if there is a clear genetic etiology, and the disorder gives rise to a well-defined brain pathology, the species-gap can be overcome, and genetic mouse models can make a significant contribution. Indeed, so rapid has been the pace allowed by the genetic tractability of mouse models, extending across numerous neurological conditions *(8,9)*, that, in some cases, data from mice studies have been of immediate use in guiding clinical trials of putative therapies, e.g., the recent assessment of antibody-based therapies in Alzheimer's disease *(10,11)*.

However, psychiatric conditions, such as schizophrenia, autism, depression, and personality disorder, have a more complex etiology, involving multiple interactions between environment and genetic background. Certainly, for the common psychopathologies, fully penetrant, single-gene causes, as can occur in specific neurological diseases, have not been, and are unlikely ever to be, found (however, *see* the extremely rare Brunner syndrome, a genetic condition found in a single pedigree, ref. *12*). However, at the same time, it is clear from numerous twin and sibling studies that for virtually all psychiatric conditions, genetics plays a significant role (e.g., refs. *13–16*) opening up at least the possibility of investigating these genetic effects in mouse models. One way of charting a path out of the modeling impasse is not to attempt to model specific psychiatric conditions as defined in the various psychiatrist's bibles, such as the Diagnostic and Statistical Manual of Mental Disorders, 4th ed (DSM-IV) *(17)* but instead to examine so-called intermediate traits/endophenotypes. There have been several excellent reviews describing the concept of endophenotypes and how they may be used in the context of psychiatric research (for details, *see* refs. *18–21*). Basically, an endophenotype is a behavioral, physiological, or anatomical trait that is closely associated with a psychiatric disorder but that, in itself, does not completely define the disorder. In other words, intermediate traits are essentially risk factors that are, in many cases, more common than the full disorder, and that can, importantly, be closer to a discrete genetic etiology than the full spectrum. For example, in the case of schizophrenia, one might model gene effects influencing the cognitive symptom of perseveration (the failure in inhibitory control leading to continued inappropriate control of responding by a contingency that was correct but that is now incorrect) *(22–24)*; or one might examine an animal task purporting to assay attentional set shifting (a form of cognitive flexibility) assayed by the Wisconsin Card Sort Test in humans, which is also abnormal in schizophrenia *(25–27)*. In the case of attention deficit hyperactivity disorder (ADHD), appropriate behavioral endophenotypes would be hyperactivity and impulsivity *(28–30)*. An impressive listing of endophenotypes of potential relevance to psychiatric conditions can be found in the recent review by Seong et al. *(20)*.

It can be appreciated that a given endophenotype may span a number of conditions as defined by the DSM-IV, an observation with interesting implications, insofar as it may suggest common underlying neurobiological abnormalities. In the present context, this raises an interesting irony, in that, although the primary aim of using intermediate traits is to be in a position to begin modeling the complexities of psychiatric conditions in nonhumans, recognition of their redundancy points to a likely inadequacy of the clinical definitions provided by the DSM-IV. Clearly, what is needed are diagnostic criteria informed by neurobiological definitions. Such a state of affairs would not only be optimal for modeling purposes but would also help clear much of the diagnostic fog that surrounds psychiatry. The problem is, of course, that, apart from notable exceptions, such as depression (in which there have been interesting data on hippocampal shrinkage, *see* ref. *31*), the neurobiological bases of psychiatric disorders are, at best, ill-defined, or, at worst, a complete mystery. At present, therefore, the prospects of systematic psychiatric disease specification based on defined neurobiological abnormalities must remain an aspiration.

If we accept the argument that, in principle, modeling intermediate traits, in particular, behavioral intermediate traits, is a valid starting position to assess genetic effects on psychiatric disorders, then the next question is to what extent can we use mice in this endeavor? No one would dispute that, currently, mice constitute the most genetically tractable mammalian species and offer a large variety of useful genetic models for the researcher, encompassing isogenic (inbred) strains, natural mutants, single-gene manipulations, and partial or whole chromosome deletions and duplications (*see* Table 1 for listing). However, there are those, perhaps most likely to be found in the field of human neuropsychology, who may find the idea of using mice to model aspects of human psychology and behavior dubious, almost risible. Indeed, there may also exist a degree of rodent chauvinism, summed up by

Table 1
Genetic Methods for Modeling Psychiatric Disorders in Mice

Genetic manipulation	Example of use
Transgenes	Altered expression of exogenous gene
Knockout/knock-in/knock-down	Disruption of endogenous gene expression—embryonic stem cell methods
	• Transgene expression on knockout back ground
	• Reporter genes (e.g., GFP)
	• Single amino acid substitutions/deletions (e.g., $GABA_A$ receptor subunits)
	• Amino acid repeat sequences (e.g., CAG in Huntington's disease models)
Chromosome manipulations	• Duplications (e.g., trisomy h21/m16, UPDs)
	• Deletions (e.g., h22q11.2/m16A1)
"Natural" spontaneous mutations	http://www.jax.org/mmr/previous.html
Engineered random mutations (ENU)	http://www.mut.har.mrc.ac.uk/
Inbred strains	http://jaxmice.jax.org/info/inbred.html

GFP, green fluorescent protein; GABA, γ-aminobutyric acid; ENU, ethylnitrosourea; UPD, uniparental disomy.

our own experiences some years ago when asking advice about using mice to assess cognitive functioning, "mice are stupid and, anyway, rats are bigger" we were told. In fact, as shall become apparent, in our own laboratory and, increasingly, in other laboratories, there is ample evidence to challenge this assumption.

The use of mouse models in behavioral studies attracts all of the usual caveats attached to the use of nonhuman animals, especially when assessing behaviors purporting to index the operation of underlying cognitive processes. Here, as emphasized by Martin Sarter (32), it is all too easy to fall into the trap of taking a superficial behavioral similarity (sometimes termed face validity) as evidence of identical or similar psychological substrates operating across species. Unfortunately, this is a problem that seems to be of particular relevance to mouse work, in which there has been a tendency to use what has been described by Sarter (32) as "fast and dirty laboratory assays, often used with limited regard to the multiple variables that influence performance and hence determine the interpretation of data." These problems are often associated with the inappropriate use of the most commonly used rodent tasks to assess learning and memory, the Morris water maze and rapid, aversively motivated memory tasks, such as the passive avoidance task. The reasons for the popularity and frequent, less than optimal use of these tasks is probably caused by a combination of their apparent (but deceptive) simplicity and their relative rapidity, coupled with the fact that, in many cases, the main drivers of a behavioral genetics research program may be the molecular biologists responsible for the creation of the genetically modified mouse. Under such circumstances, it is easy to see how using "off the peg" behavioral tasks, with limited examination of the controlling variables of any genetic effect, can become routine. Sarter *(32)* argues that, instead of being seduced by the deceptive simplicity of face validity it is much more important to determine the construct validity of a behavioral task, the extent to which the controlling variables of performance can be systematically varied to determine, much more exactly, the precise psychological functions being taxed. Only such parametric analyses of performance, both within and across tasks, will provide data with a high likely predictive validity, that is, data from animal models that are of relevance to the clinical situation. The concepts of face, construct, and predictive validity (the latter usually related to drug discovery) have been reviewed extensively elsewhere (*see* refs. *33–35*).

In another angle on the "good genetics needs good phenotypes" issue, Tarantino and Bucan *(21)* discussed the study of Crabbe et al. *(36)*, which caused much debate and not a little consternation, in apparently revealing a degree of variation between laboratories when assessing the same mutations in "identical" murine behavioral tasks. It has to be said that we, and a good many of our colleagues, were baffled by the way in which these data were used as evidence that all work on mouse behavior was likely to be fatally flawed. Such findings are, of course, nothing new in behavioral research and Crabbe and colleagues' work merely (re)emphasized the need for rigorous confirmation of experimental findings, both within and between different laboratories (*see* ref. *37*).

Another issue, which for some reason seems to have been given an unusual and unfortunate prominence in the debate about mice and behavior, is the ethobiological vs artificial approach spat. Advocates of the ethobiological approach, such as Gerlai and Clayton *(38)*, believe that behavior in genetic mouse models will only be informative in the context of measuring what mice do in their everyday lives, insofar as, "only naturalistic studies can inform us about the cognitive processes that are sensitive to natural selection" and, therefore, that "artificial tasks are less sensitive to differences between mutant and wild-type animals." Gerlai and Clayton *(38)* also state that, "a learning paradigm that ignores the behavioral ecology of the species might be swamped by environmental error variation." This kind of thinking can be contrasted with those who think that sticking solely to the shibboleth of ethobiological orthodoxy is an unnecessary limitation and that behavioral methods able to use arbitrary combinations of stimuli and outcomes (such as in many operant situations) offer substantial advantages in terms of dissecting out the specific psychological functions indexed by a given change in behavior. Moreover, by tapping artificially into natural talents and stripping away species-specific behavioral predispositions, it has been argued that such artificial approaches increase the ability to generalize across species *(39)*.

This is a particularly sterile argument because both approaches are valid, depending on the question. For example, in our own work, we have exploited a specific behavioral function in mice, whereby information about the suitability of eating a novel foodstuff is transferred from one mouse to another via social contact

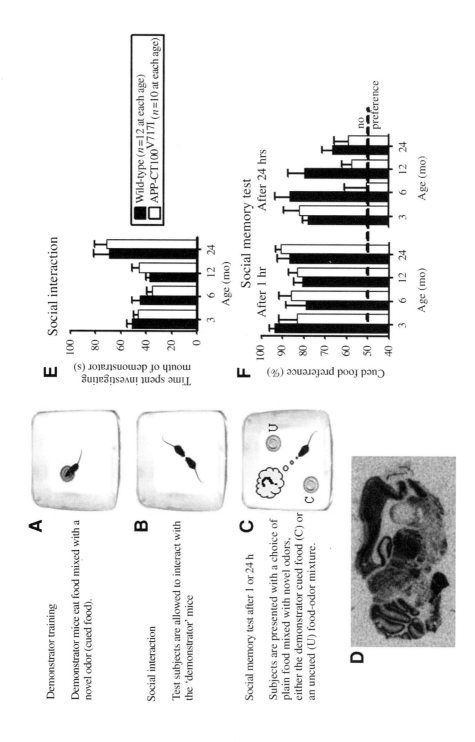

and then used at a later time and place to guide food choices in the naive subject (known as the social transfer of food preference paradigm) *(40,41)*. We have shown that the olfactory memory formed in the social interaction is sensitive to both normal and pathological aging effects, the latter associated with the transgenic expression of gene mutations implicated in the familial forms of Alzheimer's disease (*see* Fig. 1). It may also be valid to observe mouse behaviors based on social interactions in other contexts, in particular, with respect to modeling psychiatric conditions containing elements of social and emotional dysfunction. Here, there are interesting possibilities emerging from work illustrating the power of the oxytocin and vasopressin hormone systems in controlling response to social cues in situations such as social recognition and pair bonding *(42,43)*. Although it may be stretching it to try to use these specific data to model constructs such as affiliation, love, and fidelity in humans, these data do illustrate that many empathetic behaviors may share common neurobiological substrates across species. Indeed, there are exploratory findings, yet to be definitively confirmed, that genetic variants of the arginine vasopressin receptor 1A may constitute a risk factor in autism *(44,45)*. Social interactions are, of course, also important for emotional well-being. Another area of potential interest, which model affective disorders such as depression and its partner, anxiety, is the use of social defeat paradigms and the effects of social buffering in ameliorating the adverse effects of stress *(46–48)*. In general, the study of the social brain, to which mouse models can contribute, offers an important future area of relevance to increasing our understanding of a wide range of psychiatric conditions in which emotional and cognitive reactivity to social cues and situations goes awry.

On other occasions, the everyday behavioral repertoire of mice cannot be exploited easily, such as, for example, when the aim is to isolate and analyze complex psychological endophenotypes, such as attention and impulsivity. Abnormalities in these functions are well represented across multiple psychiatric condi-

---

Fig. 1. (*opposite page*) Social transfer of food preference paradigm; an ethobiological test of olfactory memory. (**A**) A demonstrator mouse consumes food mixed with a specific odor (cued food) before interacting with test or observer mice (**B**). During this time, the test mice smell the food odor on the demonstrator's breath. In a discrimination test session (**C**) some time later (in this case, 1 or 24 h after learning), the test mice are given the choice of food mixed with the demonstrator-cued odor or another novel uncued odor, and the preference for the cued food determined as the index of olfactory memory functioning. Data shows results from wild-type and transgenic mice (APP-CT100^{V717I}) that express human amyloid precursor protein with the London mutation (V717I) fully penetrant for familial Alzheimer's disease. In these mice, the transgene is expressed most strongly in cortical regions and the hippocampus (**D**). The behavioral data (**F**) shows highly specific effects of transgene expression on olfactory memory with evidence of accelerated memory loss in the transgenic mice (from 6 mo of age) in the 24-h but not in the 1-h retention test. The wild-type mice showed evidence of memory loss much later, at 24 mo of age. There were no confounds caused by differential learning of the olfactory information, as shown by equal amounts of time spent in close proximity to the mouth region of demonstrator mice (**E**).

tions, from schizophrenia to ADHD, but there have been few attempts to model them in mice. One reason for this may be the previously mentioned (and erroneous) assumption that mice are, in comparison with other animal models, somehow cognitively challenged. Another reason may be that, until comparatively recently, the behavioral tools to assay these functions in mice, such as appropriate operant methodologies, were not well established and/or available. Operant methods offer a number of advantages, including a high degree of stimulus control, the ability to measure response latencies accurately and, of particular relevance to the current discussion, the ability to make use of arbitrary stimuli and outcomes, which means that any observed effects may have greater cross-species generality. Of course, with even the most abstract contingency, it is important to be aware of the inherent behavioral predispositions that a mouse may possess (important, for example, during the initial shaping of an operant task). Similarly, there may be true qualitative differences between species, which ultimately place a limit on what can be modeled. However, it could be argued that, notwithstanding the additional caveats that stem from evolutionary arguments (i.e., the mouse brain has evolved to deal with mouse problems), mouse and man are linked, to a degree, by a common psychological tool kit. Operant methods reveal the operation of these core psychological functions, marshaled to solve novel, arbitrary problems, with minimal interfering ethobiological noise.

In the past few years we, and others *(29,30,49,50)*, have made use of operant methods to assay aspects of attentional and impulsive endophenotypes of relevance to conditions such as schizophrenia and ADHD using various configurations of the nine-hole box, in which mice respond to brief visual stimuli with a nose poke (*see* Figs. 2 and 3). These studies have confirmed the cognitive perspicacity of mice and have allowed (by virtue of being able to systematically isolate and manipulate the main controlling variables of performance) a much increased confidence in the interpretation of the behavioral data in terms of effects on specific underlying psy-

---

Fig. 2. (*opposite page*) Mouse five-choice serial reaction time task (5-CSRTT); a test of visuospatial stimulus detection and responding. The mouse 5-CSRTT allows assessment of dissociable aspects of attention and response control *(50,51)*, and uses operant boxes controlled by computers (Campden Instruments, UK). Mice are trained to nose poke, in response to a brief flash of light, in one of five stimulus locations in a curved rear wall (*see* ref. 52 for review). If they make a correct response, then a food reward is delivered to a food hopper at the front of the chamber, and a delay period is started before the next stimulus presentation. Data shows the effects of the muscarinic antagonist scopolamine on response accuracy in wild-type mice and neuronal nitric oxide synthase (nNos)-knockout mice. The nNos knockout mice are protected from the deleterious effects of scopolamine on five-choice task responding. When attentional load is reduced by presenting stimuli in a single location (one-choice) the effects of scopolamine in wild-type mice is reduced, and methyl-scopolamine, which does not cross the blood–brain barrier, is without significant effect, demonstrating the centrally mediated effects of scopolamine on five-choice performance. In addition to response accuracy, premature responding (responding in advance of the visual stimulus) can be used as an index of impulsive motor action *(51)*.

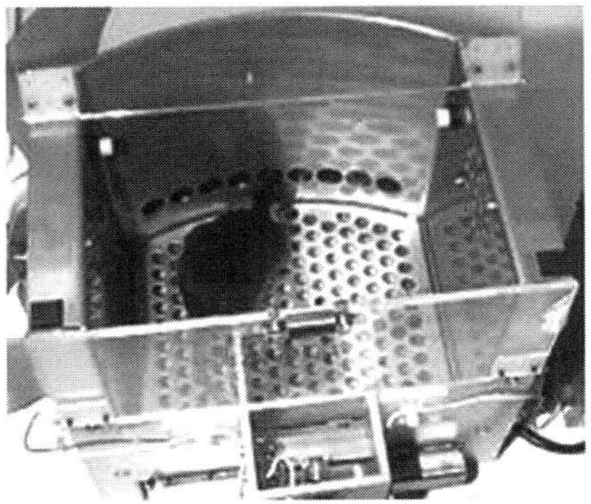

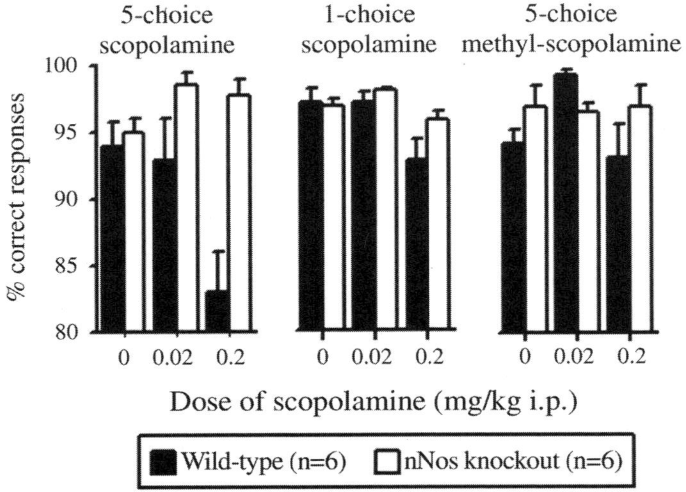
5-choice scopolamine
1-choice scopolamine
5-choice methyl-scopolamine
% correct responses
100
95
90
85
80
0
0.02
0.2
Dose of scopolamine (mg/kg i.p.)
Wild-type (n=6)
nNos knockout (n=6)

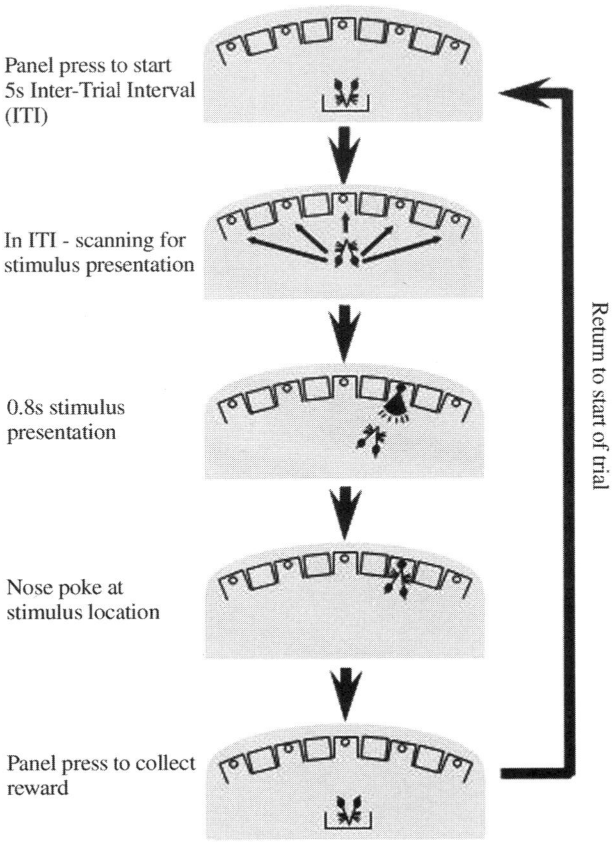
Panel press to start 5s Inter-Trial Interval (ITI)
In ITI - scanning for stimulus presentation
0.8s stimulus presentation
Nose poke at stimulus location
Panel press to collect reward
Return to start of trial

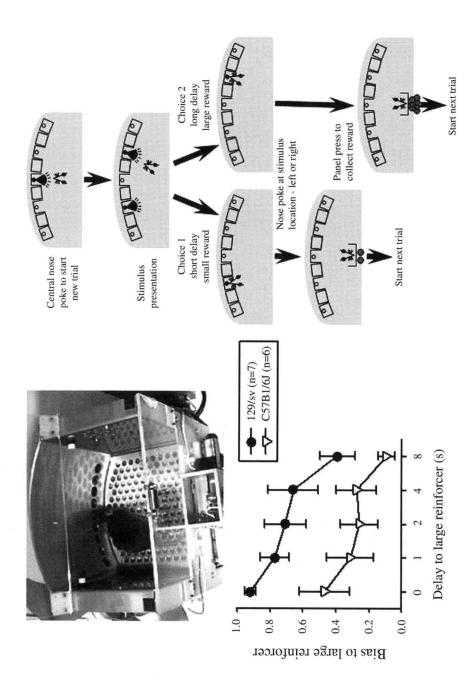

chological functions. Hitherto, much of the mouse behavioral literature has been dominated by maze paradigms, especially the Morris swim maze that, although when carried out properly can give rise to valid data (see, e.g., ref. 53), are often of ill-defined relevance to the behavioral endophenotypes exhibited in clinical conditions. We anticipate that the increasing use of operant methods, including some recent exciting work using a mouse touch screen (54), will add significantly to our ability to model intermediate traits of increased relevance to psychiatric disorders and provide behavioral data that can more easily cross the species gap. Moreover, the routine use of such methods will mean that we in the behavioral field can at least begin to approach the exquisite selectivity of the molecular genetics methods underpinning the revolution in functional genomics, and, in doing so, more precisely match a given gene effect to a discrete component of behavior.

There are many ways in which the genetic tractability of mice can be exploited in the field of psychiatric disease, encompassing both phenotype- and genotype-led approaches (also sometimes known as forward and reverse genetics). In the phenotype-led approach, gene candidates (or more often, in the case of behavioral phenotypes, relatively large chromosome intervals arising from quantitative trait loci analysis; ref. 55) are suggested either from the effects on behavior of systematic genome-wide breeding strategies, or, with luck, of discrete chromosome region deletions/duplications in which the choice of gene candidates is (helpfully) more limited (e.g., refs. 56,57). In reverse genetics, a gene candidate is manipulated in some way (often knocked out) and the effects on behavior noted. In both forward and reverse genetics, of course, good behavioral phenotypes are essential for success. One particularly imaginative angle on the reverse genetics approach has been the reverse pharmacology approach.

In reverse pharmacology, the properties of a known drug are further dissected to come up with new or more efficacious actions with fewer unwanted side effects (i.e., going from drug to mechanism rather than mechanism to drug). A good example is the well-known interaction between γ-aminobutyric acid $(GABA)_A$ receptors and the benzodiazepine class of drugs. Benzodiazepines are used extensively in medicine and form one of the main drug therapies for psychiatric conditions that present with abnormal levels of fear and anxiety. However, although being an excellent and relatively safe anxiolytic, benzodiazepines can have unwanted side

---

Fig. 3. (*opposite page*) Mouse delayed reinforcement; a test of impulsive choice. In this operant test, using a different configuration of the operant mouse box (Campden Instruments) mice must make a choice between a small amount of reinforcer (diluted condensed milk) provided immediately or larger amounts of reinforcer provided at increasing delays (*see* refs. 29 and 30). As shown in the schematic, mice are trained with either a left/immediate-right/long or left/long-right/immediate protocols, counter-balanced across test groups. Data shows the bias for the large reward/long delay of two inbred strains using this paradigm and demonstrates the change in impulsive choice that occurs as the reward delay increases. At the longest delays, the mice show a stronger tendency to choose the short delay/small reward option. These data demonstrate the sensitivity of the task in highlighting behavioral differences caused by subtle genetic differences.

effects on memory and other aspects of cognition and can cause sedation. Recent innovative work has used mice knock-ins with single amino acid substitutions at key sites of the various $GABA_A$ subunits ($GABA_A$ receptors are made up of five subunits and are variable, depending on the precise subunit composition, ref. *58*) to determine which subunit confers which functional attribute to benzodiazepine action *(59)*. In turn, this has allowed a generation of new drugs, targeted at specific GABA subunit components, with the potential for increased efficacy and reduced side effects *(58,60,61)*. Such elegant studies epitomize the benefit for the science of psychopharmacology in the molecular genetics era.

In coming to the end of this chapter, we wish to stress a number of points. First, mouse models can never completely recapitulate psychiatric conditions; what they can do is model intermediate traits or endophenotypes, elements of the human disorder that can be amenable to analysis across species. Freed from the unreasonable expectation of absolute homology, mouse models can make, and indeed are making, a significant contribution to unraveling the genetic contribution to variation in behavior and its underlying psychological and neurobiological substrates. Second, we have touched on the general danger, common to all animal models, of an over-reliance on the face validity of a behavioral task, and emphasized the importance of ensuring that a given change in behavior is, in fact, faithfully indexing the operation of a given underlying psychological function. We have also mentioned the strange polemics that the rapidly increasing and ever more varied use of mouse models in behavioral genetics seems to have generated, from the shocking observation that different laboratories can (sometimes) obtain different results, to the debate about the evolutionary validity of measuring naturalistic or artificial behaviors.

Although we have placed particular emphasis on the value of operant methods in the assessment of mouse behavior, observations of behaviors inherent to the species are also proving to be valuable. Ultimately, of course, the test of how useful mouse models are in furthering our understanding of psychopathology will be gauged by the extent to which the models engender a two-way traffic, not only from man to mouse but, equally importantly, from mouse to man, and the extent to which this traffic gives rise to real prospects for novel therapies. Currently, this is work in progress, but there are definite grounds for optimism that the appropriate use of mouse models will play a major role in ensuring that psychiatry gets its share of the genomic dividend.

ACKNOWLEDGMENTS

Our work is supported by the Biotechnology and Biological Sciences Research Council, United Kingdom. We thank Drs Anthony Isles and Sarah Lambourne of the Babraham Institute, Cambridge, United Kingdom, for access to data and their comments on the manuscript.

REFERENCES

1. Shakespeare W. Hamlet, Prince of Denmark. In: The Complete Works of William Shakespeare, Act II, Scene ii. London and Glasgow: Collins, 1960, p. 1141.
2. Tecott LH, Nestler EJ. Neurobehavioral assessment in the information age. Nat Neurosci 2004;7:462–466.
3. Aesop. Aesop's Fables. Translated by Handford SA. London: Penguin, 1996.
4. Duff K, Rao MV. Progress in the modeling of neurodegenerative diseases in transgenic mice. Curr Opin Neurol 2001;14:441–447.
5. St George-Hyslop PH. Molecular genetics of Alzheimer's disease. Biol Psychiatry 2000; 47:183–199.
6. Hutton M, Lewis J, Dickson D, Yen SH, McGowan E. Analysis of tauopathies with transgenic mice. Trends Mol Med 2001;7:467–470.
7. Lambourne SL, Sellers LA, Bush TG, et al. Increased tau phosphorylation on mitogen-activated protein kinase consensus sites and cognitive decline in transgenic models for Alzheimer's disease and FTDP-17: evidence for distinct molecular processes underlying tau abnormalities. Mol Cell Biol 2005;25:278–293.
8. Hafezparast M, Ahmad-Annuar A, Wood NW, Tabrizi SJ, Fisher EM. Mouse models for neurological disease. Lancet Neurol 2002;1:215–224.
9. Watase K, Zoghbi HY. Modelling brain diseases in mice: the challenges of design and analysis. Nat Rev Genet 2003;4:296–307.
10. Gelinas DS, DaSilva K, Fenili D, St George-Hyslop P, McLaurin J. Immunotherapy for Alzheimer's disease. Proc Natl Acad Sci USA 2004;101(Suppl 2):14,657–14,662.
11. Wilcock DM, Rojiani A, Rosenthal A, et al. Passive immunotherapy against Abeta in aged APP-transgenic mice reverses cognitive deficits and depletes parenchymal amyloid deposits in spite of increased vascular amyloid and microhemorrhage. J Neuroinflammation 2004;1:24.
12. Brunner HG, Nelen M, Breakefield XO, Ropers HH, van Oost BA. Abnormal behavior associated with a point mutation in the structural gene for monoamine oxidase A. Science 1993;262:578–580.
13. Faraone SV. Genetics of adult attention-deficit/hyperactivity disorder. Psychiatr Clin North Am 2004;27:303–321.
14. Gordon JA, Hen R. Genetic approaches to the study of anxiety. Annu Rev Neurosci 2004; 27:193–222.
15. McGuffin P, Tandon K, Corsico A. Linkage and association studies of schizophrenia. Curr Psychiatry Rep 2003;5:121–127.
16. Spence SJ. The genetics of autism. Semin Pediatr Neurol 2004;11:196–204.
17. American Psychiatric Association. DSM-IV. Diagnostic and Statistical Manual of Mental Disorders. 4th ed. Washington DC: American Psychiatric Association, 1994.
18. Gottesman II, Gould TD. The endophenotype concept in psychiatry: etymology and strategic intentions. Am J Psychiatry 2003;160:636–645.
19. Leboyer M, Bellivier F, Nosten-Bertrand M, Jouvent R, Pauls D, Mallet J. Psychiatric genetics: search for phenotypes. Trends Neurosci 1998;21:102–105.
20. Seong E, Seasholtz AF, Burmeister M. Mouse models for psychiatric disorders. Trends Genet 2002;18:643–650.
21. Tarantino LM, Bucan M. Dissection of behavior and psychiatric disorders using the mouse as a model. Hum Mol Genet 2000;9:953–965.
22. Crider A. Perseveration in schizophrenia. Schizophr Bull 1997;23:63–74.
23. Ridley RM. The psychology of perserverative and stereotyped behaviour. Prog Neurobiol 1994;44:221–231.
24. Robbins TW. The case of frontostriatal dysfunction in schizophrenia. Schizophr Bull 1990;16:391–402.

25. Clarke HF, Walker SC, Crofts HS, Dalley JW, Robbins TW, Roberts AC. Prefrontal serotonin depletion affects reversal learning but not attentional set shifting. J Neurosci 2005;25:532–538.
26. Dias R, Robbins TW, Roberts AC. Dissociable forms of inhibitory control within prefrontal cortex with an analog of the Wisconsin Card Sort Test: restriction to novel situations and independence from "on-line" processing. J Neurosci 1997;17:9285–9297.
27. Elliott R, McKenna PJ, Robbins TW, Sahakian BJ. Neuropsychological evidence for frontostriatal dysfunction in schizophrenia. Psychol Med 1995;25:619–630.
28. Heiser P, Friedel S, Dempfle A, et al. Molecular genetic aspects of attention-deficit/hyperactivity disorder. Neurosci Biobehav Rev 2004;28:625–641.
29. Isles AR, Humby T, Wilkinson LS. Measuring impulsivity in mice using a novel operant delayed reinforcement task: effects of behavioural manipulations and d-amphetamine. Psychopharmacology (Berl) 2003;170:376–382.
30. Isles AR, Humby T, Walters E, Wilkinson LS. Common genetic effects on variation in impulsivity and activity in mice. J Neurosci 2004;24:6733–6740.
31. Videbech P, Ravnkilde B. Hippocampal volume and depression: a meta-analysis of MRI studies. Am J Psychiatry 2004;161:1957–1966.
32. Sarter M. Animal cognition: defining the issues. Neurosci Biobehav Rev 2004;28:645–650.
33. Robbins TW. Homology in behavioural pharmacology: an approach to animal models of human cognition. Behav Pharmacol 1998;9:509–519.
34. Sarter M, Bruno JP. Animal models in biological psychiatry. In: D'haenen H, den Baer JA, Willner P, eds. Textbook of Biological Psychiatry. Chichester, UK: Wiley, 2002, pp. 37–44.
35. Willner P. Methods for assessing the validity of animal models of human psychopathology. In: Boulton AA, Baker GB, Martin-Iverson MT, eds. Animal Models in Psychiatry. Totowa: Humana, 1991, pp. 1–23.
36. Crabbe JC, Wahlsten D, Dudek BC. Genetics of mouse behavior: interactions with laboratory environment. Science 1999;284:1670–1672.
37. Dawson GR, Flint J, Wilkinson LS. Testing the genetics of behavior in mice. Science 1999;285:2068; author reply 2069–2070.
38. Gerlai R, Clayton NS. Analysing hippocampal function in transgenic mice: an ethological perspective. Trends Neurosci 1999;22:47–51.
39. Sarter M, Berntson GG. Tapping artificially into natural talents. Trends Neurosci 1999;22:300–302.
40. Galef B, Wigmore SW. Transfer of information concerning distant foods—a laboratory investigation of the information-centrer hypothesis. Animal Behaviour 1983;31:748–758.
41. Wrenn CC, Harris AP, Saavedra MC, Crawley JN. Social transmission of food preference in mice: methodology and application to galanin-overexpressing transgenic mice. Behav Neurosci 2003;117:21–31.
42. Bielsky IF, Young LJ. Oxytocin, vasopressin, and social recognition in mammals. Peptides 2004;25:1565–1574.
43. Keverne EB, Curley JP. Vasopressin, oxytocin and social behaviour. Curr Opin Neurobiol 2004;14:777–783.
44. Kim SJ, Young LJ, Gonen D, et al. Transmission disequilibrium testing of arginine vasopressin receptor 1A (AVPR1A) polymorphisms in autism. Mol Psychiatry 2002;7:503–507.
45. Wassink TH, Piven J, Vieland VJ, et al. Examination of AVPR1a as an autism susceptibility gene. Mol Psychiatry 2004;9:968–972.
46. Kaufman J, Yang BZ, Douglas-Palumberi H, et al. Social supports and serotonin transporter gene moderate depression in maltreated children. Proc Natl Acad Sci USA 2004;101:17,316–17,321.
47. Ruis MA, te Brake JH, Buwalda B, et al. Housing familiar male wildtype rats together reduces the long-term adverse behavioural and physiological effects of social defeat. Psychoneuroendocrinology 1999;24:285–300.
48. Wilson JH. A conspecific attenuates prolactin responses to open-field exposure in rats. Horm Behav 2000;38:39–43.

49. Bensadoun JC, Brooks SP, Dunnett SB. Free operant and discrete trial performance of mice in the nine-hole box apparatus: validation using amphetamine and scopolamine. Psychopharmacology (Berl) 2004;174:396–405.
50. Humby T, Laird FM, Davies W, Wilkinson LS. Visuospatial attentional functioning in mice: interactions between cholinergic manipulations and genotype. Eur J Neurosci 1999;11: 2813–2823.
51. Humby T, Wilkinson LS, Dawson GR. Assaying aspects of attention and impulse control in mice using the 5-choice serial reaction time task. Current Protocols in Neuroscience, Unit 8.5, 2005.
52. Robbins TW. The 5-choice serial reaction time task: behavioural pharmacology and functional neurochemistry. Psychopharmacology (Berl) 2002;163:362–380.
53. Micheau J, Riedel G, Roloff EL, Inglis J, Morris RG. Reversible hippocampal inactivation partially dissociates how and where to search in the water maze. Behav Neurosci 2004;118: 1022–1032.
54. Bussey TJ, Saksida LM, Rothblat LA. Discrimination of computer-graphic stimuli by mice: a method for the behavioral characterization of transgenic and gene-knockout models. Behav Neurosci 2001;115:957–960.
55. Flint J. Analysis of quantitative trait loci that influence animal behavior. J Neurobiol 2003; 54:46–77.
56. Kitsiou-Tzeli S, Kolialexi A, Fryssira H, et al. Detection of 22q11.2 deletion among 139 patients with Di George/Velocardiofacial syndrome features. In Vivo 2004;18:603–608.
57. Paylor R, McIlwain KL, McAninch R, et al. Mice deleted for the DiGeorge/velocardiofacial syndrome region show abnormal sensorimotor gating and learning and memory impairments. Hum Mol Genet 2001;10:2645–2650.
58. Reynolds DS, O'Meara GF, Newman RJ, et al. GABA(A) alpha 1 subunit knock-out mice do not show a hyperlocomotor response following amphetamine or cocaine treatment. Neuropharmacology 2003;44:190–198.
59. Wafford KA, Macaulay AJ, Fradley R, O'Meara GF, Reynolds DS, Rosahl TW. Differentiating the role of gamma-aminobutyric acid type A (GABAA) receptor subtypes. Biochem Soc Trans 2004;32:553–556.
60. Collinson N, Kuenzi FM, Jarolimek W, et al. Enhanced learning and memory and altered GABAergic synaptic transmission in mice lacking the alpha 5 subunit of the GABAA receptor. J Neurosci 2002;22:5572–5580.
61. McKernan RM, Rosahl TW, Reynolds DS, et al. Sedative but not anxiolytic properties of benzodiazepines are mediated by the GABA(A) receptor alpha1 subtype. Nat Neurosci 2000;3:587–592.

II Transgenic and Knockout Models of Neurocognitive Dysfunction

5
Spinocerebellar Ataxia Type 1
Lessons From Modeling a Movement Disease in Mice

Harry T. Orr

Summary

Expansion of a polyglutamine repeat within the spinocerebellar ataxia type 1 (SCA1)-encoded protein, ataxin-1, causes the neurodegenerative disease, SCA1. Animal models have been generated that recapitulate many of the aspects of SCA1 pathogenesis. These provide a good example of how animal models can be used to examine the pathogenesis of a human neurological disease. Studies using these animal models have led to numerous conclusions regarding the pathogenic potential of mutant ataxin-1. The data indicate that protein folding and clearance pathways are important in the development of disease. Aggregation of mutant ataxin-1 is not required for initiation of disease. In the case of SCA1, Purkinje cells are the last neurons to aggregate the mutant protein but the most susceptible to the toxic effects of mutant ataxin-1, suggesting that aggregation may be a protective event. Nuclear localization of mutant ataxin-1 is necessary but not sufficient to induce pathogenesis. A single amino acid, serine 776, within ataxin-1 was found to be important in disease progression. Serine 776 of both wild-type and mutant ataxin-1 is phosphorylated. Preventing phosphorylation of this residue by replacing it with an alanine results in a mutant protein found in the nucleus that is not pathogenic. Other modifiers of ataxin-1-induced neurodegeneration include components of RNA-processing and protein-processing pathways. Importantly, because wild-type ataxin-1 is found in the nucleus and is phosphorylated at serine 776, the disease pathway likely overlaps with the normal cellular pathway(s) in which ataxin-1 participates.

Key Words: Polyglutamate disorders; ataxin-1; transgenic mice; Purkinje cells; neuronal degeneration; knock-in mice.

1. INTRODUCTION

Advances in mouse genetics, along with the isolation of human genes in which mutations cause neurological disorders make it possible to model neurobehavioral diseases in the mouse. By creating a mouse carrying the human disease-associated mutation or mutant gene, the cellular and molecular basis underlying the disease phenotype can be probed. This approach can be used to identify the early processes that trigger the disease. It is hoped that a better understanding of a disease pathogenesis will lead to the development of better treatments. Application of this strat-

From: *Contemporary Clinical Neuroscience: Transgenic and Knockout Models of Neuropsychiatric Disorders*
Edited by: G. S. Fisch and J. Flint © Humana Press Inc., Totowa, NJ

egy is now routine for instances in which a human disorder is the result of a mutation in a single gene, e.g., Huntington's disease and the inherited forms of Alzheimer's disease, Parkinson's disease, and amyotrophic lateral sclerosis. Unfortunately, the more common human psychiatric/neurobehavioral diseases are very complex in their etiology and typically are uniquely human. However, because the application of mouse transgenic and knockout strategies are being considered for modeling for neuropsychiatric of disorders, it is prudent to review an instance in which the creation of a mouse model of a human single-gene neurological disorder has been profitable.

Spinocerebellar ataxia type 1 (SCA1) is an autosomal dominant, adult-onset neurological disorder in which individuals typically only survive 10–15 yr after the first appearance of symptoms *(1)*. Clinically, SCA1 patients exhibit ataxia, dysarthria, and bulbar dysfunction. Pathologically, there is Purkinje cell loss in the cerebellar cortex and loss of neurons in the inferior olivary nuclei, cerebellar dentate nuclei, and the red nuclei. SCA1 is caused by the expansion of a polyglutamine tract within the SCA1-encoded protein, ataxin-1 *(2)*. An inverse relationship exists between the length of the repeat tract and the age of onset and severity of disease *(3)*. That is, the longer the mutant CAG repeat, the earlier the age of onset and the more severe the clinical symptoms. In the general population, the number of CAG repeats is highly variable, ranging from 6 to 44 triplet repeats. Alleles with longer than 20 repeat tracts typically have between one and three CAT interruptions *(4)*. Mutant alleles have repeat tracts that contain 39–82 CAG repeats that lack CAT interruptions. Thus, the interruption might be an important factor in the intergenerational stability of the longer normal alleles.

SCA1 is a member of a group of neurodegenerative disorders known as the polyglutamine diseases. To date, this group includes spinobulbar muscular atrophy, dentatorubropallidoluysian atrophy, Huntington's disease, and the spinocerebellar ataxias (SCA1, SCA2, SCA3/Machado-Joseph disease, SCA6, SCA7, and SCA17) *(5,6)*. Other than the CAG repeat tract, the gene products associated with these disorders share no significant homology. A common feature among the polyglutamine diseases is the aberrant deposition of the mutant protein *(7)*. Most often, these accumulations are found within the nucleus of neurons. The role these deposits play in disease pathogenesis remains controversial. Of particular interest is the cell specificity of pathogenesis. Despite widespread expression of these genes in the brain and other tissues, only a certain subset of neurons is vulnerable to neurodegeneration in each of the disorders.

Ataxin-1 is a novel protein of unknown function. It has been shown to bind RNA and interact specifically with other cellular proteins *(8)*. The subcellular localization of ataxin-1 varies depending on cell type. For example, in lymphoblastoid cells, ataxin-1 is found primarily in the cytoplasm, whereas, in neurons, ataxin-1 is mostly nuclear *(9–11)*. Ataxin-1 shares a region of homology, termed the AXH domain, with the apparently unrelated transcription factor, HBP1 *(12)*. This region is postulated to be a protein–protein interaction and RNA-binding motif. Not surprisingly, the region of ataxin-1 that is involved in RNA binding and in its interactions with other proteins includes the AXH domain *(13)*.

The use of animal models has been crucial in understanding cellular and molecular aspects of disease pathogenesis in SCA1 and the other polyglutamine disorders. Numerous SCA1 models have implicated the importance of the subcellular localization of the mutant protein, the role of the aggregates, protein folding and clearance, and posttranslational modification in disease. In this chapter, we focus on the animal models used to investigate each of these aspects as they relate to SCA1 pathogenesis.

2. UNCOVERING PATHOGENIC ASPECTS OF SCA1 USING ANIMAL MODELS

2.1. Expanded Polyglutamine Mediates a Dominant Gain-of-Function

Both wild-type and mutant ataxin-1 are expressed in multiple tissues. To gain insight into the function of wild-type ataxin-1 and to determine whether SCA1 is caused by a loss or gain of function mutation, *Sca1*-null mice were generated. Mice lacking ataxin-1 failed to show any overt signs of ataxia or evidence of Purkinje cell pathology *(14)*. The *Sca1*-null mice displayed both motor and spatial learning deficits, suggesting that ataxin-1 has a role in both cerebellar and hippocampal-mediated learning. In humans, large deletions within 6p22-23 spanning the *SCA1* gene do not cause SCA1 *(15)*. These data indicate that SCA1 is not caused by loss of ataxin-1 function, supporting a toxic gain-of-function model.

In a further effort to test a gain-of-function hypothesis, SCA1 transgenic mice were made. The mice were generated to develop a model to allow the examination of pathogenesis in a single cell type that is a prominent site of pathology in patients, the cerebellar Purkinje cells. Studies on patient material have revealed reduced Purkinje cell dendritic arborization associated with a reduction in number of small spiny dendritic processes *(16,17)*. In addition, axonal torpedo bodies were observed in the Purkinje cells. To model Purkinje cell pathology seen in SCA1, the strong Purkinje cell-specific promoter, Pcp2/L7 *(18,19)*, was used to direct expression of a human *SCA1* complementary DNA (cDNA) to the cerebellar Purkinje cells *(20)*. Two transgenic lines were extensively characterized. The first, line A02, expressed a wild-type *SCA1* allele containing 30 glutamine repeats, ataxin-1[30Q], whereas the second, line B05, expressed a mutant *SCA1* allele containing 82 repeats, ataxin-1[82Q]. The B05 mice expressing mutant ataxin-1 developed severe ataxia and progressive Purkinje cell pathology. In contrast, A02 transgenic mice expressing wild-type ataxin-1 failed to develop any signs of ataxia or neurodegeneration. These studies indicated that neurodegeneration of Purkinje cells results from the expression of mutant ataxin-1 and reinforced the toxic gain-of-function model for polyglutamine-induced pathogenesis.

In B05 mice, mutant ataxin-1 localized to a single nuclear inclusion, similar to what has been seen in tissue samples from SCA1 patients *(10)*. The appearance of these inclusions, which were ubiquitinated, and contained components of the 26S proteasome and molecular chaperones, preceded the onset of ataxia by approx 6 wk *(11)*. The first sign of mild cerebellar impairment in the B05 mice occurred at 5 wk of age, as evidenced by a slight deficit in rotarod performance *(21)*. At this age, there were no signs of gait abnormalities or balance problems. By 12 wk, as assessed by home cage behavior, the motor impairment progressed to visible ataxia that worsened with age. The first pathological sign detected in the B05 mice was the appearance of cytoplasmic vacuoles in the Purkinje cells at postnatal d-25 *(21,22)*. By 6 wk of age, a loss of proximal dendritic branches and a decrease in the number of dendritic spines became apparent. Thus, Purkinje cells that overexpress mutant ataxin-1 develop pathological features that are characteristic of SCA1 in humans.

2.2. Importance of Subcellular Localization

In neurons, ataxin-1 is found primarily in the cell nucleus, although Purkinje cells also have a small amount of cytoplasmic ataxin-1 *(9)*. Cell culture studies found that mutant ataxin-1 confers nuclear alterations. Specifically, mutant ataxin-1 causes nuclear matrix-associated promyelocytic leukemia protein containing bodies to be redistributed, altering its normal nuclear localization *(10)*. Promyelocytic leukemia protein was often found sequestered with mutant ataxin-1 within the large nuclear inclusions. In both tissue culture cells and Purkinje cells of transgenic mice, ataxin-1 was found to be associated with the nuclear matrix. These results suggest that the primary site of pathogenesis is nuclear. The possible role of nuclear-mediated pathogenesis and predominant nuclear localization of ataxin-1 stimulated the identification of a functional nuclear localization signal (NLS) in the C-terminus of the protein *(23)*. To directly determine whether nuclear expression of mutant ataxin-1 is required to induce pathogenesis, transgenic mice expressing mutant ataxin-1 with a single point mutation in the NLS, K772T, preventing nuclear localization of the protein, were generated and characterized *(23)*. Similar to previous SCA1 mice, expression of this transgene was directed to the cerebellar Purkinje cells using the Pcp2/L7 promoter. In these ataxin-1[82Q]-K772T mice, the mutant protein was found predominantly in the cytoplasm of Purkinje cells. Ataxin-1[82Q]-K772T mice express a comparable amount of protein to the B05 transgenic mice; however, there was no indication of cytoplasmic aggregates in the ataxin-1[82Q]-K772T mice. Ataxin-1[82Q]-K772T mice do not develop ataxia or signs of Purkinje cell degeneration. Thus, the ataxin-1[82Q]-K772T mice do not develop the cerebellar dysfunction characteristic of the B05 SCA1 transgenic mice, despite expressing comparable levels of mutant ataxin-1 in Purkinje cells. This indicates that nuclear expression of mutant ataxin-1 is necessary for pathogenesis and that polyglutamine-mediated nuclear alterations are critical for disease in transgenic mice.

Demonstrating the importance of mutant ataxin-1 being in the nucleus raised the possibility that an alteration in gene expression is an early component of disease in the Purkinje cells. Initial studies on gene expression in the SCA1 mice used a subtractive cDNA cloning approach and found that, indeed, altered gene expression was apparent in the early phase of disease, before detectable morphological or neurological abnormalities in the mice *(24)*. More recently, DNA microarrays were used to assess the pattern of gene expression in SCA1 transgenic mice. Using three SCA1 transgenic mouse lines, each expressing a different form of ataxin-1, it was possible to identify a specific cohort of genes whose products could be placed within glutamate-signaling pathways in the Purkinje cell dendrite at the parallel fiber/ Purkinje cell synapse *(25)*. This suggests that SCA1 pathophysiology involves alterations in glutamate signaling.

2.3. Protein Folding, Clearance, and Solubility

A pathological hallmark in most of the polyglutamine disorders is the presence of nuclear protein aggregates containing the mutant protein *(5)*. The role that these aggregates have in disease pathogenesis remains controversial. In SCA1, ataxin-1 multimerization is mediated by intramolecular interactions via a self-association domain (SAD) *(26)*. To test the role of ataxin-1 self-association in pathogenesis, SCA1 transgenic mice were generated in which the self-association region was deleted *(23)*. Ataxin-1[77Q]-ΔSAD localized to the nucleus of Purkinje cells. However, despite nuclear localization, ataxin-1[77Q]-ΔSAD transgenic mice did not have nuclear aggregates, although they did develop ataxia and Purkinje cell pathology. A progressive Purkinje cell pathology similar to that observed in the B05 mice was apparent in these mice. This included the appearance of cytoplasmic vacuoles at 3 wk, shrinkage of the molecular layer by 8 wk, and the appearance of heterotopic Purkinje cells at 12 wk. The performance of the ataxin-1[77Q]ΔSAD mice on the accelerating rotarod at 12 wk of age was compromised compared with nontransgenic littermates. This deficit was similar to that observed for the B05 mice at 12 wk of age. Hence, despite the absence of detectable nuclear aggregates, ataxin-1[77Q]ΔSAD mice had histological and neurological alterations similar to B05 mice. Therefore, ataxin-1 aggregation is not required for initiation of SCA1 pathogenesis. Further characterization of the ataxin-1[77Q]ΔSAD mice found that the deleted protein–protein interaction region did not affect the initiation of disease, but did substantially suppress disease progression *(27)*. The protein–protein interaction region seems to have a role in disease progression. Moreover, mutant ataxin-1 induced initiation of disease and disease progression likely involve distinct events.

The nuclear polyglutamine aggregates were positive for ubiquitin as wells as chaperones and components of the proteasome *(28,29)*. Inhibition of the proteasome pathway increased ataxin-1 aggregation, which was accompanied by an increase in the amount of detergent-insoluble ataxin-1 *(30)*. Ubiquitin was found conjugated similarly to both wild-type and mutant ataxin-1. However, mutant ataxin-1 was more resistant to degradation in vitro, suggesting a role for the ubiquitin– proteasome pathway in SCA1 pathogenesis. To test this hypothesis, SCA1 transgenic mice were crossed with mice deficient in a component of the ubiquitin– proteasome pathway *(30)*. Ube3A is an E3 ubiquitin ligase responsible for the addition of a ubiquitin moiety to a substrate targeting it for 26S proteasome-mediated degradation. Ube3A has a unique expression pattern: the paternal allele is silenced in cerebellar Purkinje cells because of tissue-specific imprinting. Maternal deficiency of the Ube3A gene product (E6-AP) causes the human genetic disorder, Angelman syndrome. Severe mental retardation, hyperactivity, seizures, ataxia, and inappropriate laughter characterize Angelman syndrome. Ube3A deficiency results in no histological abnormalities of the cerebellum. Offspring from SCA1 B05 mice crossed with the Ube3A-deficient mice that carried both a mutant SCA1 allele and a targeted disruption of the maternal Ube3A allele were extensively characterized. The Purkinje cells of these bi-transgenic animals contained significantly fewer nuclear ataxin-1 aggregates. However, the Purkinje cell pathology in these mice was significantly worse than observed in the mice with a functional copy of Ube3A. Thus, E3 ligase deficiency greatly accelerates polyglutamine-induced pathology in SCA1 transgenic mice, indicating that aggregates are not necessary to induce pathogenesis and might instead serve a protective function.

Molecular chaperones work together with the ubiquitin-proteasome pathway to recognize and hydrolyze misfolded proteins. Nuclear inclusions from both SCA1 patient tissue and transgenic mice stained positive for the molecular chaperones, heat-shock proteins Hsp70 and Hsp40 *(11)*. This suggests that expansion of the polyglutamine repeat within ataxin-1 leads to a misfolded protein that is inefficiently cleared by the ubiquitin-proteasome pathway. In support of this, overexpression of an Hsp40 chaperone (HDJ-2) reduced mutant ataxin-1 aggregation in tissue culture cells. Studies using *Drosophila melanogaster* also found that overexpression of molecular chaperones suppressed polyglutamine-induced cell toxicity *(31–33)*.

To further investigate whether altered protein folding and deficient protein clearance are involved in polyglutamine-induced pathogenesis SCA1, ataxin-1[82Q] transgenic (B05) mice were crossed with mice overexpressing the molecular chaperone, HSP70 *(34)*. Overexpression of HSP70 did mitigate the phenotype of the B05 mice. SCA1 B05 transgenic mice overexpressing HSP70 recovered the ability to perform on the rotarod. This rescue was dose dependent, because B05/HSP70$^{+/+}$ animals performed better than B05/HSP70$^{+/-}$ animals on the rotarod. HSP70 also ameliorates Purkinje cell pathological changes. B05/HSP70$^{+/+}$ mice have Purkinje cells with thicker and more robust dendritic arborization and fewer heterotopic Purkinje cells compared with the neurons within B05 cerebella. Interestingly, overexpression of HSP70 did not alter the prevalence of nuclear inclusions. Suppression of the disease phenotype in the absence of any notable alterations in the properties of aggregated mutant ataxin-1 further indicates that inclusions are not a prominent aspect of disease pathogenesis. Molecular chaperones may modulate the folding and degradation of pathogenic nonaggregated mutant ataxin-1.

Purkinje cells are not the sole type of neuron affected in SCA1. Neurons within the brain stem and spinal cord also degenerate. In an attempt to faithfully recapitulate the full spectrum of pathological changes observed in SCA1, an expanded polyglutamine tract was introduced via homologous recombination into the endogenous murine *Sca1* locus. The endogenous murine *Sca1* locus contains two glutamines. Two knock-in lines have been generated, one with 78 CAG repeats, and the other with 154 CAG repeats. In the first line, murine *Sca1* with 78 repeats, mutant ataxin-1 was not pathogenic *(35)*. These mice were not ataxic by home cage behavior, nor did they have any of the neuropathological changes by 18 mo of age that are observed in other transgenic models of SCA1. Therefore, mutant ataxin-1 with 78 repeats expressed at endogenous levels was not sufficient to induce any detectable pathological changes. In contrast, overexpression of mutant ataxin-1 in the SCA1 B05 transgenic mice was pathogenic. Together, these data suggest that, in mice, disease is a function of polyglutamine length, protein level, and time of exposure. To circumvent these variables, a second knock-in was generated that harbored 154 repeats within the murine *Sca1* locus *(36)*. These mice did develop a neurological disorder reminiscent of the human disease. The mice had motor coordination deficits and Purkinje cell pathology with cell loss that worsened progressively with increasing age. Growth retardation, cognitive defects, and premature death were also observed. Of note, reduced Purkinje cell dendritic arborization was observed early in the disease process, whereas cell loss was not evident until the end stage of the disease. This provides further support that Purkinje cell dysfunction and not cell loss is responsible for the early stages of the neurological phenotype.

The 154 repeat knock-in Sca1 mice also revealed an interesting feature regarding the role of nuclear inclusions in disease. In the $Sca1^{154Q/2Q}$ mice, intranuclear inclusions were found within numerous neuronal populations. Of importance, inclusions were less frequent in the most susceptible groups of neurons compared with cortical or hippocampal neurons, which are only mildly affected in SCA1. For example, by 20 wk of age, less than 0.5% of the Purkinje cells contained inclusions, whereas more than 80% of the neurons in other regions had inclusions. This is consistent with previous reports that the nuclear inclusions do not initiate disease in *SCA1* transgenic animals. It also suggests that the most susceptive neurons are those unable to efficiently sequester mutant ataxin-1 into inclusions in a timely manner. Purkinje cells are the last cells in which mutant ataxin-1 containing aggregates form, but are the most susceptible cells in SCA1. Selective neuronal vulnerability in SCA1 and the other polyglutamine disorders most likely result from a combination of protein solubility and factors involved in a pathway that results in the aggregation of the mutant protein. As discussed earlier, these factors may include components of the ubiquitin–proteasome pathway and molecular chaperones. Such factors may be rate limiting in susceptible neurons, resulting in their increased vulnerability to polyglutamine-mediated toxicity.

3. GENETIC MODIFIERS OF DISEASE

Drosophila have proven to be a powerful model system that has been extremely useful in elucidating the genetic pathways that impact polyglutamine-induced neurodegeneration. In the case of SCA1, directed expression of mutant ataxin-1 to the retina produced a severe form of neurodegeneration with nuclear inclusions *(33)*. Interestingly, the wild-type form of the protein produced a somewhat milder form of neurodegeneration. Whether the intrinsic toxicity of normal ataxin-1 is relevant to the human disease is not known. The power of *Drosophila* genetics is evident in the ability to carry out genetic screens to identify genes that modify the SCA1 phenotype. Toward this end, a genetic screen of modifiers of the polyglutamine-induced pathogenesis in these SCA1 transgenic flies yielded several insightful genes. Modifier genes that suppress or enhance ataxin-1-induced neurodegeneration were identified. Mutations within numerous genes within the ubiquitin–proteasome pathway enhanced ataxin-1-induced neurodegeneration. These include ubiquitin, a ubiquitin carboxy-terminal hydrolyase, and genes involved in ubiquitin conjugation. Another group of enhancers included numerous transcriptional cofactors. Both enhancers and suppressors contained a RNA-binding domain were also found. A *Drosophila* protein (dDNAJ-1) homologous to the HSP40 chaperone suppressed neurodegeneration. The modifiers identified further cement the importance of protein folding and clearance pathways, RNA-processing pathways, and transcriptional homeostasis in the development of disease.

In SCA1, the cerebellar Purkinje cells are affected; in Huntington's disease, the striatal medium spiny neurons are affected; and, in spinobulbar muscular atrophy, the spinal cord motor neurons are most susceptible *(5)*. Although progress has been made in understanding the selective vulnerability of specific subsets of neurons in each of the polyglutamine disorders, the underlying mechanism remains a mystery. One hypothesized cellular response for mutant polyglutamine damage is the activation of apoptotic machinery, suggesting apoptotic cell loss as a central component of the disease *(37)*. However, in SCA1 transgenic mice, the ataxic phenotype is present well before any evidence of cell loss, arguing that neuronal dysfunction and not cell loss is central to SCA1 pathogenesis *(21)*. In contrast, numerous studies have implicated the activation of an apoptotic pathway in polyglutamine-mediated neurodegeneration *(31,38–40)*. In addition to being a central component of apoptotic pathways, p53 has been linked to neuronal cell death, including ties to polyglutamine-mediated pathogenesis *(41,42)*. To further investigate the role of p53 in SCA1, B05 transgenic mice were crossed with mice lacking p53 *(43)*. In these mice, a reduction in the severity of later features of the disease was observed. These included fewer heterotopic Purkinje cells, slowed dendritic thinning, and amelioration of the shrinkage of the molecular layer. Interestingly, no signs of apoptotic cell death were observed in the SCA1 mice, suggesting that the mitigating effects of p53 are independent of neuronal cell death. Thus, p53 acts after the initiation of neuropathological alterations to promote the progression of SCA1 pathogenesis.

Neuronal susceptibility and disease progression are not solely dependent on the length of the mutant polyglutamine tract. Numerous reports have found that a mutant polyglutamine peptide isolated from the intact sequence is much more toxic than when it is imbedded within the entire protein *(44,45)*. This argues that the context of the protein surrounding the mutant tract may lessen the pathogenic consequences of a mutant polyglutamine tract. Sequences outside of the polyglutamine tract may also be involved in cell-specific neurodegeneration. Animal models have been crucial in understanding the processes that modify SCA1 disease progression.

Nuclear localization of mutant ataxin-1 is necessary for pathogenesis, because SCA1 transgenic mice in which the NLS was mutated (to prevent nuclear import) did not develop ataxia *(23)*. In the absence of the E3 ubiquitin ligase, Ube3A, aggregate formation and pathogenesis were accelerated *(30)*. These results implicate a role for protein folding, clearance, and subcellular localization in the pathogenesis. Protein phosphorylation is a prominent mechanism to regulate protein degradation and subcellular localization. To investigate this possibility, further purified ataxin-1 was analyzed by mass spectroscopy to identify potential sites of phosphorylation. Toward this end, the serine at residue 776 of ataxin-1 was found to be phosphorylated *(46)*. To further investigate the role of this phosphorylation event in mutant ataxin-1 subcellular localization and neuronal dysfunction, the serine residue was mutated to alanine, thus preventing its phosphorylation. Mutation of this residue to an alanine in transgenic mice (A776) resulted in a significant dampening of the pathogenic effects of mutant ataxin-1. By both behavioral analyses and by assessing Purkinje cell morphology, mutation of serine 776 to alanine rendered mutant ataxin-1, in effect, innocuous. Mutation of this residue did not affect the subcellular localization of mutant ataxin-1, indicating that phosphorylation of this residue is not required for nuclear transport.

In tissue culture cells, mutation of serine 776 to alanine drastically reduced that ability of mutant ataxin-1 to aggregate. In cells expressing ataxin-1[82Q]-A776, less than 1 in 1000 cells contained nuclear aggregates. In contrast, in cells expressing ataxin-1[82Q]-S776, at least 60% of the cells contained aggregates. As expected, this reduction in aggregate formation coincided with an increase in extractable soluble mutant ataxin-1. In ataxin-1[82Q]-A776 transgenic mice, the ability of the mutant protein to aggregate was similarly diminished. The reduction of ataxin-1[82Q]-A776 aggregates suggests that the cell is able to more efficiently handle the mutant protein. The increased solubility of the ataxin-1[82Q]-A776 protein is consistent with this hypothesis. Polyglutamine tract expansion and nuclear localization of mutant ataxin-1 within Purkinje cells are not sufficient to induce disease SCA1 pathogenesis. A single amino acid within mutant ataxin-1, serine 776, plays a critical role in pathogenesis.

4. RECOVERY FROM DISEASE

A very useful strategy that can be applied to a mouse model of disease is the ability to generate a conditional model, a transgenic mouse line in which expression of the disease-causing gene can be regulated by the administration of a small molecule drug to the animals. An often-used approach involves an adaptation of the tetracycline-regulated system *(47,48)*. This system takes advantage of the ability of the tet activator (tTA) to differently regulate the expression of a gene in the presence and absence of tetracycline or its derivative, doxycycline, which crosses the blood–brain barrier. In the absence of tetracycline, tTA binds to tet operon elements, inducing expression of the respective downstream gene. In the presence of tetracycline or its derivative doxycycline, the ability of the tTA protein to bind tet operon sequences is abolished because of a conformational change in tTA brought about by the interaction of tetracycline with the tTA. Thus, expression of the respective gene is suppressed. To generate a transgenic mouse expressing full-length ataxin-1[82Q] in a tet-regulated manner, two transgenic lines were required. The first line was constructed to express tTA in a Purkinje cell-specific manner using the *Pcp2/L7* regulatory region. The second line was constructed to contain the target gene, a *SCA1* cDNA encoding a mutant polyglutamine tract with 82 repeats, joined to a tet-responsive element. The two lines were crossed, to place both genes in a single mouse *(49)*.

With cessation of *SCA1[82Q]* transgene expression after doxycycline administration; mutant ataxin-1, including that in nuclear inclusions, was cleared rapidly from Purkinje cells. At an early stage of disease, Purkinje cell pathology and motor dysfunction were completely reversible. After halting *SCA1* expression at later stages of disease, only a partial recovery was seen. Interestingly, restoration of the ability to perform a complex motor task, the accelerating rotarod, correlated with localization of the metabotropic glutamate receptor 1α to the Purkinje cell/parallel fiber synapse. These results show that the progression of SCA1 pathogenesis is dependent on the continuous expression of mutant ataxin-1. Of note, even at a late-stage of disease, Purkinje cells retain at least some ability to repair the damage caused by mutant ataxin-1.

5. CONCLUSION

Animal models of the polyglutamine diseases have been extremely useful in gaining insights into the pathogenic mechanism(s) of these disorders. In the case of SCA1, numerous transgenic mouse and *Drosophila* models have been used to uncover many important aspects of the disease. At this time, the data indicate that nuclear localization of the mutant protein is required for disease, that nuclear inclusions are not necessary to initiate disease, and that protein folding and clearance pathways are vital to the disease process. In fact, aggregation of the mutant protein may be a cellular defense mechanism against polyglutamine-mediated toxicity. Neurons that are the last to aggregate the mutant protein are the most vulnerable to neurodegeneration. Hence, perhaps neuron specific, rate-limiting components of protein folding and clearance pathways are responsible for the cell-specific neurodegeneration observed in each of the polyglutamine disorders. Finally, events outside of the CAG repeat in *cis* can modulate pathogenesis. Mutation of serine 776 of ataxin-1 to alanine dampened the pathogenicity of mutant ataxin-1. Moreover, this mutation rendered the mutant ataxin-1 substantially more soluble, perhaps allowing for more prompt disposal of the mutant protein by the neuron.

An important and relevant conclusion has come out of the numerous SCA1 transgenic mouse studies: the disease pathway seems to be acting through the normal cellular pathway of ataxin-1. Aspects involved in the normal cellular role of ataxin-1 seem to require nuclear localization and phosphorylation of the protein on serine 776. Both the wild-type and the mutant ataxin-1 are primarily found in the nucleus of neurons. Preventing the mutant protein from entering the nucleus by mutating the NLS signal prevents disease. Second, both wild-type and mutant ataxin-1 are phosphorylated. Mutation of serine 776 in the mutant protein mitigates disease. Thus, defining the role of phosphorylated ataxin-1 in the nucleus may provide further insights into SCA1 pathogenesis.

The progress made to date in developing an understanding of the disease process in SCA1 has uncovered numerous potential sites for therapeutic intervention. These include upregulation of molecular chaperone or protein degradation pathways, regulation of subcellular localization, and altering the solubility of the mutant protein. In addition, modulation of ataxin-1 phosphorylation and mutant-specific protein–protein interactions are promising avenues for further investigation.

REFERENCES

1. Zoghbi HY, Ballabio A. Spinocerebellar Ataxia Type 1. In: Scriver CR, Beaudet AL, Sly WS, Valle D, eds. The Metabolic and Molecular Basis of Inherited Disease. New York: McGraw Hill, 1995, pp. 4559–4568.
2. Orr HT, Chung MY, Banfi S, et al. Expansion of an unstable trinucleotide CAG repeat in spinocerebellar ataxia type 1. Nat Genet 1993;4:221–226.
3. Ranum LPW, Chung M-y, Banfi S, et al. Molecular and clinical correlations in spinocerebellar ataxia type 1: evidence for familial effects on the age at onset. Am J Hum Genet 1994;55: 244–252.
4. Chung MY, Ranum LP, Duvick LA, Servadio A, Zoghbi HY, Orr HT. Evidence for a mechanism predisposing to intergenerational CAG repeat instability in spinocerebellar ataxia type I. Nat. Genet. 1993;5:254–258.
5. Zoghbi HY, Orr HT. Glutamine repeats and neurodegeneration. Ann Rev Neurosci 2000;23: 217–247.
6. Nakamura K, Jeong S-Y, Uchihara T, et al. SCA17, a novel autosomal dominant cerebellar ataxia caused by an expanded polyglutamine in TATA-binding protein. Hum Mol Genet 2001;10: 1441–1448.
7. Kaytor MD, Warren ST. Aberrant protein deposition and neurological disease. J Biol Chem 1999;274:37,507–37,510.
8. Orr HT, Zoghbi HY. SCA1 molecular genetics: a history of a 13 year collaboration against glutamines. Hum Mol Genet 2001;10:2307–2311.
9. Servadio A, Koshy B, Armstrong D, Antalffy B, Orr HT, Zoghbi HY. Expression analysis of the ataxin-1 protein in tissues from normal and spinocerebellar ataxia type 1 individuals [see comments]. Nature Genetics 1995;10:94–98.
10. Skinner PJ, Koshy B, Klement IA, et al. Ataxin-1 with an expanded glutamine tract alters nuclear matrix-associated structures. Nature 1997;389:971–974.
11. Cummings CJ, Mancini MA, Antalffy B, DeFranco DB, Orr HT, Zoghbi HY. Chaperone suppression of aggregation and altered subcellular proteasome localization imply protein misfolding in SCA1. Nat Genet 1998;19:148–154.
12. de Chiara C, Gianninia C, Adinolfia S, et al. The AXH module: an independently folded domain common to ataxin-1 and HBP1. FEBS Lett 2003;551:107–112.
13. Yue S, Serra H, Zoghbi HY, Orr HT. The spinocerebellar ataxia type 1 protein, ataxin-1, has RNA-binding activity that is inversely affected by the length of its polyglutamine tract. Hum Mol Genet 2001;10:25–30.
14. Matilla A, Roberson ED, Banfi S, et al. Mice lacking ataxin-1 display learning deficits and decreased hippocampal paired-pulse facilitation. J Neurosci 1998;18:5508–5516.
15. Davies AF, Mirza G, Sekhon G, et al. Delineation of two distinct 6p deletion syndromes. Hum Genet 1999;104:64–72.

16. Koeppen AH. The Purkinje cell and its afferents in human hereditary ataxia. J Neuropath Expt Neurol 1991;50:505–514.
17. Ferrer I, Genis D, Davalos A, Bernado L, Sant F, Serrano T. The Purkinje cell in olivopontocerebellar atrophy. A golgi and immunohistochemical study. Neuropath Applied Neurobiol 1994;20:38–46.
18. Vandaele S, Nordquist DT, Feddersen RM, Tretjakoff I, Peterson AC, Orr HT. Purkinje-cell-protein-2 regulatory regions and transgene expression in cerebellar compartments. Genes Devel 1991;5:1136–1148.
19. Oberdick J, Smeyne RJ, Mann JR, Zackson S, Morgan JI. A promoter that drives transgene expression in cerebellar Purkinje and retinal bipolar neurons. Science 1990;248:223–226.
20. Burright EN, Clark HB, Servadio A, et al. SCA1 transgenic mice: a model for neurodegeneration caused by an expanded CAG trinucleotide repeat. Cell 1995;82:937–948.
21. Clark HB, Burright EN, Yunis WS, et al. Purkinje cell expression of a mutant allele of SCA1 in transgenic mice leads to disparate effects on motor behaviors, followed by a progressive cerebellar dysfunction and histological alterations. J Neurosci 1997;17:7385–7395.
22. Skinner PJ, Vierra-Green CA, Clark HB, Zoghbi HY, Orr HT. Altered trafficking of membrane proteins in purkinje cells of SCA1 transgenic mice. Am J Pathol 2001;159:905–913.
23. Klement IA, Skinner PJ, Kaytor MD, et al. Ataxin-1 nuclear localization and aggregation: role in polyglutamine-induced disease in SCA1 transgenic mice. Cell 1998;95:41–53.
24. Lin X, Antalffy B, Kang D, Orr HT, Zoghbi HY. Polyglutamine expansion downregulates specific neuronal genes before pathologic changes in SCA1. Nat Neurosci 2000;3:157–163.
25. Serra HG, Byam CE, Lande JD, Tousey SK, Zoghbi HY, Orr HT. Gene profiling links SCA1 pathophysiology to glutamate signaling in Purkinje cells of transgenic mice. Hum Mol Genet 2004;13:2535–2543.
26. Burright EN, Davidson JD, Duvick LA, Koshy B, Zoghbi HY, Orr HT. Identification of a self-association region within the SCA1 gene product, ataxin-1. Hum Mol Genet 1997;6:513–518.
27. Skinner PJ, Vierrra-Green CA, Emamian E, Zoghbi HY, Orr HT. Amino acids in a region of ataxin-1 outside of the polyglutamine tract influence the course of disease in SCA1 transgenic mice. Neuromolecular Med 2002;1:33–42.
28. Cummings CJ, Zoghbi HY. () Fourteen and counting: unraveling trinucleotide repeat diseases. Hum Mol Genet 2000;9:909–916.
29. Berke SJS, Paulson HL. Protein aggregation and the ubiquitin proteasome pathway: gaining the UPPer hand on neurodegeneration. Curr Opin Genet Dev 2003;13:253–261.
30. Cummings CJ, Reinstein E, SuNY, et al. Mutation of the E6-AP ubiquitin ligase reduces nuclear inclusion frequency while accelerating polyglutamine-induced pathology in Sca1 mice. Neuron 1999;24:879–892.
31. Warrick JM, Paulson HL, Gray-Board, et al. Expanded polyglutamine protein forms nuclear inclusions and causes neural degeneration in Drosophila. Cell 1998;93:939–949.
32. Kazemi-Esfarjani P, Benzer S. Genetic suppression of polyglutamine toxicity in Drosophila. Science 2000;287:1837–1840.
33. Fernandez-Funez P, Rosales MLN, de Gouyon B, et al. Identification of genes that modify ataxin-1-induced neurodegeneration. Nature 2000;408:101–106.
34. Cummings CJ, SuNY, Opal P, et al. Over-expression of inducible HSP70 chaperone suppresses neuropathology and improves motor function in SCA1 mice. Hum Mol Genet 2001;10:1511–1518.
35. Lorenzetti D, Watase K, Xu B, et al. Repeat instability and motor incoordination in mice with a targeted expanded CAG repeat in the Sca1 locus. Hum Mol Genet 2000;9:779–785.
36. Watase K, Weeber EJ, Xu B, et al. A long CAG repeat in the mouse Sca1 locus replicates SCA1 features and reveals the impact of protein solubility on selective neurodegeneration. Neuron 2002;34:905–919.
37. Paulson HL, Bonini NM, Roth KA. Polyglutamine disease and neuronal cell death. Proc Natl Acad Sci USA 2000;97:12,957–12,958.

38. Sanchez I, Xu C-J, Juo P, Kakizaka A, Blenis J, Yuan J. Caspase-8 is required for cell death induced by expanded polyglutamine repeats. Neuron 1999;22:623–633.
39. Ona VO, Li M, Vonsattel JPG, et al. Inhibition of caspase-1 slows disease progression in a mouse model of Huntington's disease. Nature 1999;399:263–267.
40. Li S-H, Lam S, Cheng AL, Li X-J. Intranuclear huntingtin increases the expression of caspase-1 and induces apoptosis. Hum Mol Genet 2000;9:2859–2867.
41. Trettel F, Rigamonti D, Hilditch-Maguire P, et al. Dominant phenotypes produced by the HD mutation in STHdh(Q111) striatal cells. Hum Mol Genet 2000;9:2799–2809.
42. Jana NR, Zemskov EA, Wang G-h, Nukina N. Altered proteasomal function due to the expression of polyglutamine-expanded truncated N-terminal huntingtin induces apoptosis by caspase activation through mitochondrial cytochrome c release. Hum Mol Genet 2001;10:1049–1059.
43. Shahbazian MD, Orr HT, Zoghbi HY. Reduction of Purkinje cell pathology in SCA1 transgenic mice by p53 deletion. Neurobiol Dis 2001;8:974–981.
44. Wellington CL, Hayden MR. Caspases and neurodegeneration: on the cutting edge of new therapeutic approaches. Clin Genet 2000;57:1–10.
45. Zoghbi HY, Botas J. Mouse and fly models of neurodegeneration. Trends Genet 2002;18:463–471.
46. Emamian ES, Kaytor MD, Duvick LA, et al. Serine 776 of ataxin-1 is critical for polyglutamine-induced disease in SCA1 transgenic mice. Neuron 2003;38:375–387.
47. Gossen M, Bujard H. Tight control of gene expression in mammalian cells by tetracycline-responsive promoters. Proc Natl Acad Sci USA 1992;89:5547–5551.
48. Yamamoto A, Hen R, Dauer WT. The ons and offs of inducible transgenic technology: a review. Neurobiol Dis 2001;8:923–932.
49. Zu T, Duvick LA, Kaytor MD, et al. Recovery from polyglutamine-induced neurodegeneration in conditional SCA1 transgenic mice. J Neurosci 2004;24:8853–8861.

6
Mouse Models of Hereditary Mental Retardation

Hans Welzl, Patrizia D'Adamo, David P. Wolfer, and Hans-Peter Lipp

Summary

This chapter describes a number of genetic mouse models of syndromic and nonsyndromic mental retardation (MR), focusing primarily on X-linked retardation models: the fragile X model, involving the fragile site mental retardation 1 gene (*FMR1*) the FRAXE model, involving the fragile site mental retardation 2 gene (*FMR2*); the Coffin-Lowry syndrome model, involving ribosomal S6 kinase 2 (RSK2); models involving GDP dissociation inhibitor (GDI)-1 mutations; the Rett syndrome model, involving the methyl-CpG-binding protein 2 (MECP2); the lacking angiotensin receptor 2 (AGTR2) model; the corpus callosum hypoplasia, mental retardation, adducted thumbs, spastic paraplegia, and hydrocephalus (CRASH) syndrome model, involving mutations of the cell adhesion molecule, L1; and models involving mutations of rho guanine nucleotide exchange factor 6 (ARHGEF6). Autosomal dominant models include neurofibromatosis type 2 (NF1) and phenylketonuria (PAH). The phenotypes of experimentally altered mouse genes mostly include relatively moderate pleiotropic changes in neuroanatomy, electrophysiology, and behavioral test scores, the latter rarely matching the severity of the human phenotype. Interpretation is hampered by a general lack of understanding the causation of mental variation, and by neglecting species-specific peculiarities of mouse neuroanatomy and cognitive abilities. On the other hand, these mouse models provide unique and invaluable tools for an empirical analytical approach deciphering the complex pathways between genotype and mental phenotype, chiefly because the developmental end point is, at least for nonsyndromic human MR, always severely impaired cognition. This is not the case for mouse models generated on the basis of theoretical expectations for memory and learning.

Key Words: Mental retardation; mouse models; genetic disorders; X-linked MR.

1. INTRODUCTION

1.1. Definition of Mental Retardation and Dementia

Individuals with mental retardation (MR) form the greater part of the developmentally disabled. MR is characterized by significant limitations both in intellectual functioning and in adaptive behavior as expressed in conceptual, social, and practical skills. Symptoms of the disability are first noticed at birth or early in child-

From: *Contemporary Clinical Neuroscience: Transgenic and Knockout Models of Neuropsychiatric Disorders*
Edited by: G. S. Fisch and J. Flint © Humana Press Inc., Totowa, NJ

hood, before 18 yr of age *(1)*. A later onset of this state of functioning is referred to as dementia. Similar definitions of MR can be found in the classification systems of mental disabilities, the International Statistical Classification of Diseases and Health Related Problems, 10th Revision *(2)* and the Diagnostic and Statistical Manual of Mental Disorders, 4th Edition *(3)*. Based on the severity of intellectual limitations, MR is categorized as mild, moderate, severe, or profound *(1)*. Individuals with MR make up approx 1% of the general population in the United States, with men being affected more commonly than women (1.5:1). Approximately 85% of mentally retarded individuals have intelligence quotients (IQs) in the mild range (IQ of approx 55–70).

MR is not a single disorder but a behavioral syndrome describing the level of an individual's functioning. The different manifestations of MR do not share a uniform cause, mechanism, course, or prognosis *(1)*. Many behavioral traits such as impulsivity, passivity, aggression, self-injury, and low frustration tolerance are associated with MR but are not considered to be diagnostic criteria for MR. Individuals with MR may also display specific physical traits in, for instance, facial features and malformations of skeleton and organs. This condition is known as syndromic MR, and is often associated with neurological symptoms, such as mood disorders, sensory and motor impairments, or seizures *(4)*. Nonspecific (nonsyndromic) MR is sometimes referred to as MR without other clinical manifestations.

1.2. Etiology of MR

Known biological causes of MR are numerous and can be roughly categorized into genetic/chromosomal abnormalities and nongenetic biological causes. The latter group of causes can affect an individual's brain development prenatally (maternal substance use, *in utero* infections, toxins, fetal malnutrition), perinatally (premature birth of several weeks, hypoxia, birth trauma), or postnatally (medical conditions in infancy or childhood caused by infection, trauma, or toxins). Further, psychological factors can cause MR (gross deprivation), or MR may be present in mental disorders (e.g., early-onset schizophrenia).

1.3. Genetic Causes of MR

That MR occurs in association with specific genetic conditions was suspected several decades ago *(5,6)*. As early as 1959, researchers discovered that trisomy of chromosome 21 causes Down syndrome, the most common genetic form of MR *(7)*. Fragile X syndrome, the most common hereditary and second most common genetic form of MR, has been linked to X chromosomal aberrations that were later attributed to an alteration in the fragile X gene, *FMR1 (8,9)*. Phenylketonuria, long known to be hereditary and causing disturbances in phenylalanine metabolism, was finally found to be caused by a defect in the gene coding for phenylalanine hydroxylase (for review, *see* ref. *10*). With the advent of modern molecular techniques, the list of known, genetically determined causes of MR is rapidly expanding, and specific locations in the genome can be pinpointed. Using the National

Center for Biotechnology Information (NCBI) gene-centered resource, LocusLink, 177 entries can be found for genetic loci associated with MR *(11)*. For many more forms of MR, a Mendelian inheritance is evident or has been suspected, although the underlying gene or genes have not yet been identified. A comprehensive list of all of the genes found to be involved in X-linked MR can be found at www.ggc.org/xlmr.html.

Syndromic MR is more easily discovered because of the accompanying clinical symptoms, which also facilitate the search for underlying genetic causes. The search for genetic causes of nonsyndromic MR is conceptually more interesting, because it may identify molecular mechanisms specific to cognitive processes. However, the border between normal variation of cognitive abilities *(12)* and nonsyndromic MR is typically much less evident. Likewise, the distinction between syndromic and nonsyndromic MR is often based on missing recognition of more subtle pleiotropic consequences of mutations. Genetic mutations causing nonsyndromic MR seem to most frequently affect genes on the X-chromosome *(13)*. This also explains why nonsyndromic MR is more common in men than in women.

2. MOUSE MODELS

Animal models are usually developed to assist better understanding of the etiology of a disease, that is, to recognize what metabolic processes are affected and are responsible for the pathological phenotype. Hopefully, this knowledge leads to techniques for early detection, prevention, and successful therapy of the disease, all of which can be first tested in the model. Animal models of MR are also providing insights into development of cognitive functions in normal individuals.

Knowledge of the exact mutations causing MR in humans as well as of techniques to generate genetically manipulated lines of mice allow researchers to produce mutant mouse models of specific forms of human MR. The goal of this chapter is to provide an overview of currently available animal models for hereditary forms of MR generated by molecular manipulation.

2.1. Fragile X Syndrome

2.1.1. Human Genetics and Disease

Fragile X syndrome is an X-linked moderate to severe form of MR associated with mutations in the *FMR1* gene. Physical signs such as large testicles, large ears, prominent jaw, and high-pitched jocular speech also characterize the phenotype. Behavioral signs other than MR include hyperkinesia and, occasionally, autistic features (for review, *see* ref. *14*). It is the most common hereditary form of MR, affecting approx 1 in 4000 men and a smaller percentage of women *(15)*.

Fragile X syndrome is most often caused by an unstable expansion of the trinucleotide repeat sequence CGG within the 5' untranslated region in the *FMR1* gene (Xq27.3) in more than 95% of all patients. The methylation of this sequence prevents the synthesis of the encoded fragile X MR protein (FMRP) (for review, *see* ref. *14*). FMRP belongs to a family of RNA-binding proteins associated with polyribosomes, and as part of a larger messenger ribonucleoprotein, it modulates translation *(16)*.

2.1.2. Animal Model

This mouse model will be described in more detail than the other mouse models because it has been investigated the most thoroughly and, therefore, also serves to illustrate the inherent problems of phenotype analysis in this field. Attempts to model fragile X syndrome in the mouse are promising, because of a high nucleotide (95%) and amino acid (97%) identity between human *FMR1* and mouse *Fmr1* on the X-chromosome *(17)*. Furthermore, tissue-specific expression patterns of *Fmr1* in the mouse resemble human expression patterns *(18)*. Transgenic or knock-in mice carrying a mouse or human expanded CGG repeat showed no or only moderate CGG repeat instability *(19)*. *Fmr1*-knockout mice more successfully modeled the fragile X syndrome phenotype, displaying several mild physical and behavioral symptoms consistent with the human fragile X syndrome phenotype. Characteristic changes affecting facial features and extremities present in human patients were missing. However, the most prominent feature, enlarged testes, could be found in more than 90% of all adult mice *(20–22)*. The mild behavioral phenotype of these mice seemed to depend on the genetic background.

Electron microscopic analysis of brain tissue revealed abnormal dendritic spines in patients as well as knockout mice, implicating the FMRP in dendritic spine formation (for review, *see* ref. *23*). The morphological spine abnormality might be caused by a deficit in synaptic activation-induced local translation of specific proteins *(24)*. Dendritic transport of messenger RNA seems intact in *Fmr1*-knockout mice *(25)*. The infrapyramidal mossy fiber projection in the hippocampus of *Fmr1*-knockout mice was found to be increased on a FVB background *(26)* and decreased on a C57BL/6 background *(27)*. Long-term depression and long-term potentiation are two forms of protein synthesis-dependent synaptic plasticity that can be tested in vivo as well as in vitro. Long-term depression triggered by activation of metabotropic glutamate receptors in the hippocampus was enhanced in mutant mice. Long-term potentiation was intact in the hippocampus *(28–30)* but reduced in the cortex of *Fmr1*-knockout mice *(28–30)*. The selective reduction of long-term potentiation was paralleled by a selective reduction of the glutamate receptor subunit, GluR1, in the cortex but not the hippocampus of knockout mice. These findings indicate that FMRP plays an important functional role in regulating activity-dependent synaptic plasticity in the brain. The involvement of metabotropic glutamate receptors in long-term depression suggests new therapeutic approaches for fragile X syndrome *(31)*.

As cited in the Chen and Toth study *(32)*, increased responsiveness of fragile X patients to sensory stimulation has been reported, and epilepsy appears in approx 20% of fragile X patients *(33)*. Similar patterns of sensory responsiveness are seen in the mouse model. *Fmr1*-knockout mice differed from wild-types in their response to auditory stimuli. Knockouts on two different genetic backgrounds showed greater startle responses than wild-types to low intensity auditory stimuli but decreased responses to high-intensity stimuli *(34)*. Prepulse inhibition of the acoustic startle response was significantly greater in *Fmr1*-knockouts than in wild-types *(35)*. The increased prepulse inhibition of *Fmr1*-knockout mice was strain-dependent and possibly also influenced by the intensity of prepulse stimuli *(34)*. Recently, a study investigating prepulse inhibition in children with fragile X as well as in Fmr1-knockout mice found prepulse inhibition to be reduced in patients but increased in *Fmr1*-knockout mice *(36)*. These data suggest that FMRP plays a role in sensory processing in humans and in mice; but they also caution us that results obtained with mouse models cannot always be transferred to human patients on a one-to-one level.

A loud sound could induce audiogenic seizures in knockouts but not in wild-types. Manifestation of such seizures was absent in 6- to 8-wk-old mice and appeared first in 10- to 12-wk-old knockouts. In contrast, sensitivity to chemical convulsants was equal in knockouts and wild-types *(35)*. In an earlier study, Musumeci and coworkers *(37)* also could induce audiogenic seizures in younger knockouts but only in a significantly smaller number of wild-types. The difference between the two studies was very likely caused by differences in the intensity and quality of the stimuli.

In the context of these studies, it is interesting to note that, in response to somatosensory stimulation, the production of FMRP in the cortex increased in intact rats *(38)*. Thus, the protein might be involved in controlling the mediation of sensory stimuli to brainstem reflex centers. This provides a nice example of how animal models can mirror human symptomatology and provide information about the substrates underlying those symptoms.

Lack of FMRP in mice affected activity- and anxiety-related responses only mildly or not at all. Compared with wild-types, *Fmr1*-knockouts were hyperactive in an open field *(27)* and showed reduced anxiety-related responses and increased exploratory behavior *(20,39,40)*. In another study, however, no increase in activity and no signs of anxious behavior in the open field and elevated plus maze was evident *(34)*. Further, knockouts tended to improve their performance in the rotarod task slightly less than wild-types *(39)*.

When submitted to a passive avoidance task, *Fmr1*-knockouts and their littermates performed equally well in one study *(20)* and worse than wild-type controls in another study *(40)*. However, the results of the latter study are difficult to interpret because knockouts might have entered the aversive dark chamber sooner because of their hyperactivity.

According to several studies, *Fmr1*-knockout and wild-type mice did not differ in fear conditioning *(39,41,42)*. Comparable performance of wild-type and knockout mice was observed on different genetic backgrounds. In contrast to these results, Paradee and coworkers *(29)* found that their knockouts froze less than their wild-type littermates on a more than 99.99% pure C57BL/6 background. The studies are difficult to compare because of the many differences in the experimental set up. One possibility could be that training was less intense in the latter study.

In a spatial task, the Morris water maze, *Fmr1*-knockout mice were only slightly impaired during the initial phases of reversal learning *(20,29,43)*. During a probe test, spatially guided searching for the escape platform was the same in knockout mice and wild-types. The subtle deficits restricted to reversal learning seemed to depend on genetic background. Two other groups found similar results using a cross-shaped water maze *(41,42)*. Knockouts performed worse than wild-type littermates, and the effect was again dependent on genetic background. Radial maze learning was moderately impaired in *Fmr1*-knockout mice *(27)*.

A surprising result was the superior performance of *Fmr1*-knockout mice over controls in a complex discrimination task *(44)*. The small number of mice used and the different genetic background of knockouts and controls make an interpretation of these results difficult. Studies that are more detailed, with littermates as controls, will be necessary. Along the same lines are data in support of improved instrumental learning in *Fmr1*-knockout mice *(36)*. Knockout and wild-type mice performed equally well during acquisition of instrumental responding, but knockout mice reacted more strongly to reward devaluation and to reward omission. Although alterations in cognitive processes could lead to these results, the specificity of the improvement might also be caused by changes in motivational or emotional mechanisms of *Fmr1*-knockout mice.

Peier and coworkers *(39)* attempted to correct the *Fmr1*-knockout phenotype by crossing knockout mice with transgenic mice overexpressing FMRP. Transgenic mice were generated by inserting a yeast artificial chromosome carrying the full length of the human gene. Presence of the transgene resulted in a cell- and tissue-specific expression of the human protein at 10–15 times higher levels than that of the endogenous protein. Knockout mice carrying the transgene alongside the knockout did not develop enlarged testes. On the behavioral level, presence of the transgene in wild-types as well as knockouts produced a phenotype opposite to that seen in knockouts, i.e., activity in the open field was reduced, an indicator of anxiety was increased, and the startle response was heightened. The data suggest that expression levels of FMRP have to be fine-tuned because levels that are too high result in a phenotype of their own.

To summarize, the rather subtle behavioral phenotype will make future evaluation of new therapies in *Fmr1* mice difficult. Attempts to correct the phenotype by inserting a transgene *(39)* clearly demonstrated that the success of a therapy in mouse models has to be evaluated with a test battery, including several behavioral, anatomical, and biochemical parameters. The great variability in the genetic background of *Fmr1*-knockout mice investigated hinders the comparison of data. However, a careful investigation of how background strains affect the manifestation of deficits in knockouts might, by itself, be an interesting topic that might help to explain the substantial variability of the syndrome observed in patients.

Two autosomal gene homologs of *FMR1* have been identified, fragile X MR gene 1 on chromosome 11 and fragile X MR gene 2 on chromosome 17 (*FXR1* and *FXR2*). The proteins encoded by the homologs and by FMRP show high sequence similarity. Their tissue distribution overlaps, with a high expression of all three proteins in the brain *(16)*. To gain insight into the function of the two homologs of FMRP, knockout mice for *Fxr1* and *Fxr2* were generated. Homozygous *Fxr1*-

knockout mice die shortly after birth, most likely resulting from cardiac or respiratory failure *(45)*. Missing FXR1 protein (FXR1P) disrupted the cellular architecture in skeletal and cardiac muscles. In contrast to the two proteins encoded in *Fmr1* and *Fxr2*, FXR1P seems to be essential for survival of the newborn mouse.

Brains and testes of *Fxr2*-knockout mice were not different from those of wild-types *(46)*. When compared with controls, *Fxr2*-knockouts on a mixed background including FVB were hyperactive and moved faster in the open field, performed significantly worse on the rotarod, and showed lower prepulse inhibition *(46)*. When tested on the hotplate, knockouts were less sensitive than wild-types *(46)*. Knockout mice displayed reduced contextual fear conditioning and were impaired during the acquisition phase of the water maze task. However, during the probe trial, a critical test to evaluate spatial navigation abilities, knockouts performed as well as wild-types *(46)*. Anxiety-related responses were not different from wild-types *(46)*. In summary, the behavioral phenotype of *Fxr2*-knockouts partly resembled that of *Fmr1*-knockouts; but *Fxr2*-knockouts also displayed behavioral characteristics of their own. The findings implicate a role for *Fxr2* in central nervous system development and/or function.

2.2. The FRAXE Syndrome

2.2.1. Human Genetics and Disease

After identifying the unusually large CGG repeat expansion in *FMR1* as the cause of fragile X syndrome, several cases of fragile X syndrome with MR were detected that lacked this critical CGG repeat expansion. Subsequently, another fragile X site, FRAXE, was detected in the vicinity of FRAXA *(47)*. Similar to FRAXA, an unstable expanded CCG repeat sequence in the *FMR2* gene (Xq28) was responsible for the folate-sensitive fragile X site, FRAXE (for review, *see* ref. *48*). Behavioral symptoms of FRAXE patients include very mild to borderline forms of MR. However, there are also individuals carrying the mutation without any signs of MR. This made it difficult to establish a causal relationship between the mutation and MR. It has also been speculated that the mutation is involved in other forms of behavioral deficits (for review, *see* ref. *49*). Furthermore, physical symptoms, including a long, narrow face have been inconsistently reported (for review, *see* ref. *48*). *FMR2* belongs to the *AF4/FMR2* gene family comprising four genes. The function of FMR2 protein (FMR2P) is not completely known, but present data suggest that it might possess transcriptional activation potential, a function that is conserved from yeast cells to mammalian cells *(50)*.

2.2.2. Animal Model

The mouse FMR2P has been isolated and shown to have 88% amino acid homology to the human FMR2P. As in humans, the protein is highly expressed in the mouse brain *(51)*. These similarities make the mouse a valuable model in which to investigate the expression and function of the *Fmr2* gene.

Fmr2-knockout mice backcrossed to C57BL/6 were generated using homologous recombination *(52)*. The mutation did not affect activity in the open field, startle response, prepulse inhibition, rotarod performance, or measures of anxiety in the light-dark box. In the conditioned-fear test, knockouts consistently froze less than wild-types in response to a context associated with shock. This difference appeared only at 24 h—but not at 30 min—after conditioning. Freezing in response to an auditory-conditioned stimulus was of equal duration in knockouts and wild-types. Aside from a small impairment at the beginning of training, knockouts and wild-types did not differ in their water maze performance. The enhanced long-term potentiation in hippocampal slices from knockout mice suggested a disturbance in synaptic functioning or plasticity *(48,52)*. Thus, knockout mouse data suggest that a lack of FMR2P results in selective impairments of brain function.

2.3. Coffin-Lowry Syndrome

2.3.1. Human Genetics and Disease

Coffin-Lowry syndrome is a syndromic form of X-linked MR characterized by MR, growth retardation (short stature), and various facial, hand, and skeletal abnormalities, which include a coarse face, prominent forehead, prominent supraorbital ridges, irregular or missing teeth, large and soft hands, and tapering fingers *(53,54)*. The incidence has been tentatively estimated to lie in the range of 1 in 50,000–100,000 men. The symptoms could be traced back to a large number of heterogenous mutations in the ribosomal S6 kinase (*RSK*) 2 gene (Xp22.2-p22.1) coding for RSK2. The mutations disrupt translation and/or interfere with the kinase activity *(55,56)*.

RSK2 is a member of a family of closely related 90-kDa RSKs that also includes RSK1, RSK3, and RSK4. The members of the RSK family are growth factor-regulated serine/threonine kinases implicated in the activation of the mitogen-activated kinase cascade and the stimulation of cell proliferation and differentiation, cellular stress response, and apoptosis. The kinases have a large variety of targets in the cytosol and nucleus and seem to play a major role in transcriptional control of gene expression (for review, *see* refs. *53* and *54*). Expression of *RSK* messenger RNAs can be found in all human tissues and brain regions tested, with tissue-specific variations in levels *(57)*.

2.3.2. Animal Model

The expression of *Rsk2* in the mouse brain resembled the expression of the gene in the human brain; in both cases, it was most abundantly expressed in the cerebellum, the occipital pole, and the frontal lobe *(57)*. To model Coffin-Lowry syndrome, Rsk2-knockout mice were generated that were viable and showed no obvious physical abnormalities. Knockouts weighed 10% less and were 14% shorter than wild-type littermates and revealed a role for RSK2 in body-weight regulation and glycogen metabolism in skeletal muscle *(58,59)*. Knockouts performed worse than littermates in a test for motor coordination with the rotarod apparatus, and they took longer to find a submerged platform in an abbreviated form of a water maze task *(58)*. In another series of experiments, wild-types and knockouts did not differ in emotional or exploratory behavior and motor coordination as measured in an open field, an elevated o-maze, a light–dark box and a rotarod apparatus (ref. *60*, and unpublished observations by Jacquot, Zeniou, Usiells, Dierich, Pannetier, Wolfer, Hanauer, and Lipp). However, in comparison to littermates, *Rsk2*-deficient mice were impaired in two tests for spatial working memory, the eight-arm radial maze and a matching-to-place version of the Morris water maze (unpublished observations, Jacquot et al.). Thus, this mutant mouse seems to be a promising model to study the function of RSK2 and how its lack might cause MR in Coffin-Lowry syndrome. It must be noted, however, that a spontaneous mutation in mice said to include complete deletion of the *Rsk2* gene results in intrauterine death of male mice between E10 and E14. On the other hand, analysis of Drosophila melanogaster mutant lines has shown that differential mutations of the *Rsk2* analog gene entail different deficits in operant and classical conditioning, thus underscoring the role of RSK2 in mammalian mental deficits *(61)*.

2.4. GDI1 MR

2.4.1. Human Genetics and Disease

A rare mutation causing nonsyndromic X-linked MR in humans has been identified as a mutation of *GDI1 (62)*. This gene encodes one of the proteins regulating the small GTPases of the Rab family. The Rab proteins are a group of small Ras-like GTPases, involved in vesicle fusion in the exocytic and endocytic pathways *(63)*. A detailed description of the human *GDI1* mutant phenotype is still pending, and no neuropathological information is available because the afflicted individuals are still alive.

2.4.2. Animal Model

Null mutations of the mouse *Gdi1* gene have been generated in two different laboratories *(64,65)*. One model has been investigated only in terms of electrophysiological properties, reporting that synaptic potentials in the CA1 region of the hippocampus revealed larger enhancement during repetitive stimulation *(64,65)*. D'Adamo and coworkers *(65)* investigated their model using an extensive behavioral test battery, which revealed an interesting set of alterations. Exploration, fear-related behavior, and motor abilities, as assessed by the dark-light box, elevated Null-maze, and open-field seemed normal in *Gdi1*-mutant mice. Mutant and wild-type mice also performed equally well in the water maze (acquisition, reversal learning, and probe trials) and in delayed fear conditioning (shock presentation during the last 2 s of the auditory cue). On the other hand, the *Gdi1*-mutant mice showed impaired radial maze learning, suggesting deficient spatial working memory. They also showed deficits in trace fear conditioning when the presentation of the shock followed 15 s after the termination of the auditory cue. This was apparent as an impaired development of a freezing response during the conditioning trials, and, subsequently, during retention tests after 3 or 14 d. Interestingly, they also showed severely reduced aggressive behavior in a resident–intruder paradigm, in which *Gdi1* mutants were often attacked by the intruder, apparently because of ethologically inappropriate contact behavior. This anomalous behavior was neither caused by olfactory disability nor lowered testosterone levels.

Neurophysiologically, in vitro slice studies using stimulation parameters other than those used by Ishizaki et al. *(64)* revealed no impairment of hippocampal or amygdalar long-term potentiation, the only impairment observed was a transiently reduced excitatory postsynaptic potential slope in hippocampal slices *(65)*. Likewise, a study of event-related auditory potentials did not reveal significant impairments in *Gdi1*-deficient mice *(66)*.

Inspection by light microscopy revealed a minor lamination anomaly of the mossy fiber projection only. However, electron microscopy showed a severe anomaly of synaptic vesicles, primarily in the hippocampus of *Gdi1*-deficient mice, and, to a lesser degree, also in the somatosensory cortex. Synaptic vesicles were sparse and lumped, strongly suggesting that at least part of the behavioral phenotype was caused by deficient recycling of synaptic vesicles (*see* ref. *67)*.

Although not yet investigated in detail, Gdi1 seems to interact with members of the Rab protein family, and, together, these proteins are involved in synaptic vesicle fusion and neurotransmitter release. Therefore, it was surprising to observe an incongruity of the rather subtle behavioral phenotype of *Rab3a*-knockout mice with that of *Gdi1*-knockout mice *(68)*.

2.5. Rett Syndrome

2.5.1. Human Genetics and Disease

Rett syndrome is an X-linked dominant progressive neurological disorder that is diagnosed predominantly in girls at an incidence of approx 1 in 15,000 to 20,000 (for review, *see* ref. *69*). Symptoms appear after a relatively normal development during the first 6–18 mo of life. Patients cease to acquire new skills, develop autistic symptoms, and eventually loose learned skills, such as speech and purposeful hand use. Other motor disturbances include breathing abnormalities during the waking state, ataxia and apraxia of trunk and gait, and stereotyped hand wringing. Seizures develop in approx 50% of girls with Rett syndrome. Motor deterioration continues, and a state of gross cognitive and motor impairment is usually reached. The clinical picture can appear in variants characterized by less or more severe manifestation of symptoms. Neuropathological investigations reveal a reduction in brain weight (by 14–34%) and impaired development of dendrites (for review, *see* ref. *69)*.

Genetic approaches could back trace the disease to mutations in the methyl-CpG-binding protein 2 (*MECP2*) gene (Xq28). Whereas more than 99.5% of all cases are caused by sporadic mutations, a small number of familial occurrences have been described. Many different mutation sites have been detected, and the exact locus and type of a mutation might also determine the severity of the symptoms. There are still a number of—especially familial—Rett syndrome cases for which no mutation in the coding region of the *MECP2* gene could be found. It is possible that, in such cases, mutations changed regulatory elements for the *MECP2* gene. Men can develop severe forms of Rett syndrome when they carry mutations that would lead to only mild forms of the disease in women (for review, see ref. 69).

The expression pattern of the *MECP2* gene is expressed in most tissues at varying levels. When the distribution of the MECP2 protein is analyzed, high expression is found in the brain in most neurons but not in glia. The expression correlates with the maturation of neurons *(69)*. The protein seems to inhibit gene transcription by binding to methylated CpG sites. Its binding domain resembles the binding domain of other methyl-CpG-binding proteins *(69)*.

2.5.2. Animal Model

The relatively normal development during the first few months of life raises the hope that an early presymptomatic therapeutic intervention in individuals carrying a *MECP2* mutation might prevent the manifestation of the disease. A suitable mouse model would help in investigating the effectiveness of therapeutic interventions. Thus, *Mecp2*-null mice, and mice with a brain-specific deletion of *Mecp2* beginning at E12 (Nestin-Cre) or beginning postnatally (CamK-Cre) were generated *(32,70)*. All mutant mice eventually developed neurological symptoms resembling those of Rett syndrome patients, although with a different time course. In *Mecp2*-null mice and Nestin-Cre mice, symptoms appeared at 5 wk of age and the disease led to death between 6 and 12 wk of age *(32,70)*. CamK-Cre mice with a postmitotic brain-specific *Mecp2* deletion did not develop symptoms before 3 mo of age, and they lived much longer than mice carrying one of the two other mutations. Symptoms included hindlimb clasping, tremor, reduced activity, irregular respiration, and jaw misalignments *(32)*. At autopsy, brain weight was reduced in mutants compared with wild-types, and a histological analysis revealed smaller neuronal cell bodies in the cortex and hippocampus of mutants *(32,70)*. Some, but not all, heterozygous *Mecp2*-mutant female mice showed mild behavioral symptoms beginning at 3 mo of age *(70)*. Interestingly, body weight of mutants depended on genetic background. Whereas a C57BL/6 background reduced body weight, a 129 strain background increased it *(70)*. This observation suggests an interaction of MECP2 with modifier genes in the control of body weight.

Shahbazian and coworkers *(71)* generated male mice expressing a truncated MECP2 protein that led to a milder phenotype than that observed in the null allele mice. Symptoms detected with an extensive test battery revealed motor deficits, including tremor, stereotypic forepaw movements, impaired clinging to a wire, loss of balance on a thin horizontal bar, and progressively lower activity during the 30 min in an open field arena. However, wild-types and mutants performed equally well in two learning tasks: the conditioned-fear test and the Morris water maze. A social-interaction test suggested that wild-types avoided mutants. Histological examination of brain tissue postmortem did not reveal any abnormalities in mutant mice.

In an attempt to prevent manifestation of Rett-like symptoms in *Mecp2*-null mutant mice, Luikenhuis and coworkers *(72)* generated mice expressing *Mecp2* under the control of a tau promoter. This line of mice specifically expressed *Mecp2* in postmitotic neurons and was crossed with *Mecp2*-null mutant mice. Heterozygous expression of the *Mecp2* transgene prevented early death of *Mecp2*-null mice at approx 10 wk of age and rescued them from loss in body weight, reduction of brain weight, and hypoactivity. Homozygous expression of the transgene in wild-types as well as *Mecp2*-null mutant mice, however, proved to be detrimental. Such mice suffered from profound motor dysfunction, including side-to-side swaying, tremors, and ataxia. These animals were of lower body weight than controls, and the disease symptoms intensified when they reached an age of 9 mo. The emergence of a specific *Mecp2* overexpression phenotype has been recently confirmed in mice overexpressing the wild-type human protein *(73)*. At 10 wk of age, these mice displayed enhanced motor (rotarod) and contextual (conditioned fear) learning, and enhanced plasticity in the hippocampus (long-term potentiation). Eventually, however, they developed seizures, became hypoactive, and 30% of them died by 1 yr of age.

In summary, mouse models of Rett syndrome showed that symptoms are primarily caused by the reduced or failing function of *Mecp2* in the brain, and that the protein is not only important for developing neurons but also for postmitotic neurons. Depending on the mutation, symptoms can be more or less severe. Recent research on female mice with a truncated *Mecp2* (*see* ref. *71*) revealed unbalanced X-chromosome inactivation patterns in favor of the wild-type allele for *Mecp2 (74)*. Rescue of at least some symptoms with transgenic expression is possible, but apparently needs exact titration of *Mecp2* gene expression.

2.6. Angiotensin II Receptor, Type 2 (AGTR2) Mutations With X-linked MR

2.6.1. Human Genetics and Disease

In very rare cases, a profound X-linked MR can be caused by disturbed expression of the *AGTR2* gene at gene map locus Xq22-q23 *(75–77)*. Patients with this type of MR also display other symptoms, such as epileptic seizures, restlessness, hyperactivity, and disturbed development of speech. However, Erdmann and coworkers could not confirm a link between a mutation in *AGTR2* and MR *(78)*.

2.6.2. Animal Model

As expected, *Agtr2*-null mutant mice had a number of physiological disturbances, including increased blood pressure, increased vasopressor response to injection of angiotensin II, impaired drinking response to water deprivation, and lowered body temperature *(79,80)*. In addition, mutants compared with wild-types showed attenuated exploratory behavior, anxiety-like behavior, and a lowered pain threshold *(79–82)*. The performance of mutants in a passive avoidance task was unimpaired *(82)*. However, this single task does not cover the wide range of cognitive abilities, and no final conclusions can be drawn about possible learning and memory impairments in *Agtr2*-null mutant mice.

2.7. L1 Mutations and the CRASH Syndrome

2.7.1. Human Genetics and Disease

CRASH syndrome (previously designated as hydrocephalus due to stenosis of the Sylvius [HSAS] syndrome) is a recessive X-linked disease characterized by corpus callosum hypoplasia, mental retardation, adducted thumbs, spastic paraplegia and hydrocephalus, leading to the acronym CRASH. The estimated incidence lies around 1 in 30,000 male births. In 1990, the candidate gene whose mutation causes CRASH was located on Xq28. A few years later, the gene was identified as the cell adhesion molecule, L1 (for review, *see* ref. *83*). Subsequently, *L1* was found to be mutated in several other syndromes, including a syndrome characterized by MR, aphasia, shuffling gate, and adducted thumbs (MASA), X-linked spastic paraplegia type 1, X-linked agenesis of the corpus callosum, and MR with clasped thumbs *(83)*. All of the different clinical pictures caused by the pleiotropic effects of a mutation in *L1* are now subsumed under the term CRASH syndrome. The transmembrane neuronal cell adhesion molecule, L1, belongs to the immunoglobulin superfamily, and is abundantly expressed in the central and peripheral nervous systems. L1 plays an important role during neural development for cell migration, neurite outgrowth, and axon bundling *(84)*. L1 has also been implicated in memory processes in adult individuals *(85)*.

2.7.2. Animal Model

Staining patterns of L1 in humans and mice are qualitatively similar *(86)*, raising the possibility that at least part of the CRASH syndrome might be manifest in L1-mutated mice. Mouse models with null mutations in the *L1* gene were generated in two different laboratories *(87,88)*. *L1*-knockout mice were not only lighter in weight than their wild-type littermates, they also developed a number of abnormal morphological features, in part, resembling those observed in patients with CRASH. In such mice, ventricles were dilated and a few mice displayed severe hydrocephalus (only in the mutation generated by Dahme et al., ref. *88*), disruption of crossing over of the descending corticospinal tract, weak and uncoordinated hind limbs, and hypoplasia of the vermis and corpus callosum *(83,87–90)*. Histological analysis revealed further specific changes of forebrain structures and pathways *(89,91)*. Some of the symptoms were dependent on genetic background, e.g., corpus callosum hypoplasia visible on a 129/SvJae background *(89)* but not on a 129SvEv background *(87)*; and dilated ventricles were observed only on a C57BL/6 background, not in 129SvEvxC57BL/6 F1 mice *(88)*.

On the behavioral level, male *L1*-knockout mice showed impaired exploration patterns, characterized by stereotyped circling along walls without any tendency to either explore the center of the arena or two female mice kept in a cage inside the arena *(92)*. Further, the startle response in the knockout mice was weaker than in the wild-type mice, and prepulse inhibition at lower prepulse intensities was reduced *(93)*. Pain sensitivity was also lower in knockout mice compared with wild-type *(88,94)*.

Synaptic plasticity as revealed by hippocampal long-term potentiation was not significantly altered in *L1*-knockout mice *(95)*. This result was unexpected in light of the reduction in hippocampal volume and neuronal aberrations previously found in *L1*-knockout mice *(89)*. Memory tasks testing cognitive function revealed only a slightly impaired performance of *L1* knockouts in the water maze; in a passive avoidance task, knockout and wild-type mice performed equally well *(92)*. Small performance deficits in spatial navigation in the water maze were also manifest when ablation of the *L1* gene was restricted to brain subregions after early development *(96)*. The deficits of mice with conditional ablation of *L1* were probably caused by their use of nonspatial strategies to escape from the water maze. In contrast to mice lacking *L1*, mice overexpressing *L1* showed improved spatial learning in the water maze *(97)*, and visited new feeding places earlier when released in large outdoor pens (Vyssotski et al, unpublished data).

2.8. Mutations of ARHGEF6

2.8.1. Human Genetics and Disease

The rho guanine nucleotide exchange factor 6 *(ARHGEF6)* gene (also known as *alphaPIX* or *Cool-2*), encodes a protein with homology to guanine nucleotide exchange factors (GEF) for Rho GTPases, and was identified as the non-specific X-linked *MR* gene in a man with severe MR, hearing loss, and mild dysmorphic features, carrying a reciprocal X/21 translocation. The X-chromosomal breakpoint in Xq26 was found to disrupt the *ARHGEF6* gene. Subsequent mutation analysis of 119 patients with X-linked nonspecific MR revealed a mutation in the first intron of *ARHGEF6* (IVS1-11T>C) in all affected men in a large Dutch family. The mutation resulted in preferential skipping of exon 2, predicting a protein lacking 28 amino acids in the N-terminal calponin homology domain of ARHGEF6 *(98)*.

2.8.2. Animal Model

The orthologous mouse gene, *Arhgef6*, encodes a polypeptide of 771 amino acids with high homology to human ARHGEF6. It is present in various mouse tissues, indicating ubiquitous expression, similar to that found for the human gene *(99)*. A spatiotemporal expression analysis during embryonic development showed that *Arhgef6* is highly expressed in the ventricular zones, where neuronal progenitor cells are located. Nonetheless, in postnatal brain, expression of *Arhgef6* is enhanced in the peripheral CA3 region of the hippocampus *(100)*.

Behaviorally, *Arhgef6*-null mice were evaluated in a water maze place navigation task, in a spatial working memory procedure on the eight-arm radial maze, as well as in a set of tests assessing locomotor activity, anxiety, exploratory behavior, and reaction to novel stimuli *(101)*. *Arhgef6*-null mice displayed specific behavioral alterations suggesting deficient behavioral control and altered processing of spatial information. In the water maze, they produced navigation errors and showed spatial perseverance when required to adapt to a changed goal position. Their reaction to a novel stimulus within a familiar environment was clearly disinhibited. Spatial reference and working memory, as such, and day-to-day habituation to a novel area were intact. We found no indication of altered basal activity, anxiety, or abnormal adaptation to stressful situations. In addition, detailed analysis of locomotion and swim patterns did not reveal neurological deficits. The behavioral profile of *Arhgef6*-null mice is reminiscent of phenotypic changes observed in other mouse models with impaired function of the hippocampus and connected cortical regions.

2.9. Phenylketonuria

2.9.1. Human Genetics and Disease

Phenylketonuria is characterized by a variety of symptoms, including MR, light pigmentation, peculiarities of gait and posture, and epilepsy. It is inherited in an autosomal recessive manner and equally frequent in men and women. The incidence of phenylketonuria varies among populations, but is usually given as 1 in 10,000. If diagnosed early, phenylketonuria is treatable by a low-phenylalanine diet *(102,103)*.

Numerous different mutations of the phenylalanine hydroxylase (*PAH*) gene (12q24.1) can cause absence or marked deficiency of the liver enzyme, PAH. As a result, phenylalanine cannot be converted to tyrosine. The resulting accumulation of excess phenylalanine and its breakdown chemicals from other enzymatic routes in the body during infancy and early childhood affects the myelination of subcortical white matter and spinal cord, and the proliferation of neurons in the cerebral cortex (for review, *see* ref. *102*). How excess phenylalanine causes neuropathological changes is not yet fully understood.

2.9.2. Animal Model

McDonald and coworkers generated a mouse line with a mutation in the *Pah* gene using the germline mutagen, ethylnitrosourea *(104,105)*. Homozygous mutant mice (*Pah^{enu2}*) displayed hypopigmentation and had plasma levels of phenylalanine that were 10–20 times higher than normal. Performance of homozygous *Pah^{enu2}* mice in an olfactory discrimination task was inferior to that of heterozygous *Pah^{enu2}* mice. Homozygotes were also impaired in latent learning *(106)*. Using a similar olfactory task, Mihalick and coworkers *(107)* did not find a difference in performance between homozygous and heterozygous *Pah^{enu2}* mice. The failure to repeat the discrimination impairment was probably a result of the much smaller increase in blood phenylalanine levels of homozygous *Pah^{enu2}* mice in the latter experiment compared with the increase in the former experiment. Homozygous *Pah^{enu2}* mice were also impaired in a spatial and in a nonspatial recognition task *(108)*.

Several research groups tried to correct the defect of *Pah^{enu2}* mice with gene therapy (for review, *see* ref. *109*). An alternative to the currently used dietary restriction would be desirable because dietary restriction is complicated, unpalatable, and expensive. Earlier experiments to replace the missing function of phenylalanine hydroxylase were only partially successful *(109)*. More recent attempts succeeded in introducing a murine or human *Pah* gene using adeno-associated virus vectors *(110,111)*. Treated *Pah^{enu2}* mice had normal levels of phenylalanine and no longer showed hypopigmentation. Gene therapy also reversed the hypoactivity of Pah^{enu2} mice *(110)*. Surprisingly, however, female Pah^{enu2} mice did not respond to the therapy as well as male mice.

2.10. Neurofibromatosis Type 1

2.10.1. Human Genetics and Disease

Neurofibromatosis type 1 is an autosomal dominant disorder characterized by a predisposition to benign and malignant tumor formation, pigmentary abnormalities, and learning disabilities. Abnormalities of nervous tissue include neurofibroma, optic pathway glioma, and astrogliosis. The incidence of neurofibromatosis type 1 is approx 1 in 4000 individuals, with approx 50% of the cases being the result of new mutations (for review, *see* refs. *112–114*).

Symptoms for neurofibromatosis type 1 develop because of mutations in the neurofibromatosis type 1 (*NF1*) gene (17q22) that codes for neurofibromin. Human and mouse forms of neurofibromin are highly homologous, with a 98% sequence similarity *(115)*. The protein can accelerate the inactivation of Ras. Neurofibromin also regulates cyclic adenosine monophosphate (cAMP) generation *(112,113)*.

2.10.2. Animal Model

Several animal models for neurofibromatosis type 1 were generated to investigate the role of neurofibromin in tumorigenesis, and behavioral tests were carried out in two of them *(114)*. Although homozygous *Nf1*-null mutant mice died embryonically, heterozygous *Nf1* mutant ($Nf1^{+/-}$) mice survived *(116)*. Similar to patients, who frequently have visuo-spatial problems *(117)*, $Nf1^{+/-}$ mice showed selective learning deficits. $Nf1^{+/-}$ mice acquired a simple associate learning task (cued fear conditioning) equally well as wild-type littermates. Mutants were also not different from wild-types in an open-field test. However, mutants showed less spatial learning during the probe trial in a Morris water maze task, but extended training of mutants could overcome this deficit.

Genetic variation can influence the severity of learning deficits in patients. Therefore, $Nf1^{+/-}$ mice were crossbred with mice heterozygous for a targeted disruption of the *N*-methyl-D-aspartate receptor 1 ($Nmdr1^{+/-}$; ref. *116*). Whereas $Nmdr1^{+/-}$ mice performed normally in the water maze, the mutation exacerbated the spatial learning deficit of $Nf1^{+/-}$ mice. Cued fear conditioning remained unimpaired in the double mutant $Nf1^{+/-}/Nmdr1^{+/-}$ mice.

Learning deficits caused by lack of neurofibromin could be secondary to tumors or neurodevelopmental problems *(114)*. Therefore, mice carrying a homozygous deletion restricted to the exon 23a of *Nf1* ($Nf1^{23a-/-}$ mice) were investigated *(118)*. They were physically normal and did not have increased predisposition for tumor formation. Learning to find a hidden or visual platform in the water maze was the same in wild-type and mutant mice. However, mutant mice showed impaired spatial learning in a probe trial. Again, this deficit could be overcome with extended training. $Nf1^{23a-/-}$ mice were also impaired in a contextual discrimination task (contextual fear conditioning). Because NF1 patients show delayed acquisition of motor skills, $Nf1^{23a-/-}$ mice were tested on the accelerated rotarod apparatus. Compared with wild-types, mutants had delayed acquisition of this motor skill task. Mutants performed normally in a number of simple tasks testing muscular strength, motor activity, and exploratory behavior.

Spatial learning deficits in mutants could be caused by increased Ras activity. Therefore, Ras activity in $Nf1^{+/-}$ mice was reduced by genetic manipulations or pharmacological treatment *(119)*. $Nf1^{+/-}$ mice were again impaired in spatial performance during a probe trial in the water maze. However, when $Nf1^{+/-}$ mice were crossed with mice characterized by a reduced Ras activity ($K\text{-}ras^{+/-}$ or $N\text{-}ras^{+/-}$ mice), the double mutants performed as well as wild-type mice. Similarly, learning deficits of $Nf1^{+/-}$ mice could be rescued by pharmacologically reducing Ras signaling.

3. CONCLUSIONS

The study of mouse models of human MR has, not unsurprisingly, provided both insights and confusion. The approach has some strong points.

1. It is close to reality because it is based on a human phenotype with documented impaired cognition. Other approaches aimed at unraveling memory and cognition try to modify suspected mechanisms in the mouse (e.g., intracellular signaling pathways), and the results are speculatively extrapolated to the human brain.
2. It permits a better understanding of the gene-to-phenotype pathway, even if the mouse phenotypes do not always fully match the human symptomatology (*see* next page). This pathway is practically always pleiotropic, because there is no one-to-one gene effect on cognition caused by the omnipresent compensatory processes during brain development and even in the adult brain. Genes active during long periods of development (e.g., *L1*) produce a palette of somatic changes but also of major and minor cerebral changes, including cognitive impairment. Thus, mouse models allow insight into the nature of syndromic MR.

Mouse models also permit singling-out of potential candidates for nonsyndromic human MR. Such candidate mechanisms are typically late developing, thus not associated with major pathologies of the central nervous system, and include, for example, ultrastructure and physiology of synaptic vesicles, such as observed in the *Gdi1* mutants. One may note that the latter mutations belong to a class of mutations modulating GTPases. Out of four mouse models for human *MRX* genes affecting Rho- and Rab GTPases (oligophrenin, *OPHN1*; *ARHGEF6*; *PAK3*, and *GDI1*), three mouse models have shown behavioral deficits (*Arhgef6*, ref. *101*; *Pak3*, unpublished personal communication with C. Janus). In addition, *Ophn1* knock-down in rat hippocampal slices showed an altered spine morphology *(119)*.

Mouse models permit an experimental assessment of the role of genetic background, developmental factors, and therapeutic attempts by means of gene therapy. The ability to generate not only null but also knock-in or point mutations helps to elucidate the primary genetic deficits and to judge the role of central nervous system developmental mechanisms. The main points of confusion arise from a relative lack of knowledge regarding causal mechanisms that produce both human and mouse phenotypes, comparative neuroanatomical misinterpretations, inappropriate tests for mouse cognition, and difficulties in judging the nature of developmental compensatory processes.

Until now, it has been difficult to elucidate causative mechanisms for cognitively impaired phenotypes by delineating clear pathways from a mutation to the mental deficit. Part of this problem is that the brain mechanisms underlying normal and pathological variation of mental abilities are poorly understood, both in humans and mice. Even the relatively specific class of mutations affecting Rho- and Rab GTPases mentioned in Section 2.8. poses considerable analytical problems because these small proteins interact with several intracellular signaling pathways and developmental processes.

Mice are not shrunken humans, and are not even miniaturized rats. This is most evident in their brain development and the architecture of associative ("cognitive") brain areas, which renders the interpretation of mouse cognition and MR fairly difficult. Mice (and many other rodents) lack most of the associative neocortical areas subserving human cognition. In fact, the largest multimodal cortex in mice is the hippocampal formation, and the amygdala plays an important role in multimodal integration as well (depending on differential inputs, however). This implies that mice must use the hippocampus/amygdala for orchestrating a number of cognitive processes that are probably relegated to specialized cortical association areas in primates and humans.

This also infers that most genetic defects with widespread distribution in the neocortex (but not severely impairing neuronal function in the forebrain) will not result in overt sensorimotor disturbances, but are likely to become evident as hippocampal and amygdalar deficits. However, these are more easily discovered by behavioral tests than sensorimotor deficits—a typical "search under the street light effect." In fact, removing the entire neocortex (and even the forebrain) in rodents has much less deleterious consequences on overt behavior than in primates *(121,122)*. Thus, the challenge is to determine to which degree such changes are also affecting subtle noncognitive functional deficits in other forebrain systems that are typically not evaluated, e.g., sensory thresholds and discriminative abilities.

Because the hippocampus and the amygdala in mice are much more directly connected to secondary sensory and motor neocortical areas as well as to subcortical structures, it follows that malfunction in the mouse limbic forebrain is likely to manifest itself in impaired behavioral processes not typically subsumed under cognition but rather under species-specific behavior. For example, mice with hippocampal lesions loose the ability to dig out morsels or pebbles from a tube, an innate behavior characterizing normal mice *(123)*. The mechanisms underlying such unwillingness are not understood, but this burrowing test clearly demonstrates that mice use their hippocampus for behaviors not inferable from the typical human memory deficit. This nourishes the suspicion that many laboratory tests for cognition are not appropriately assessing hippocampal and amygdalar malfunctions in mice.

Likewise, many of the standard laboratory tests have been designed to demonstrate qualitatively the presence of a memory trace, as for example the probe trial in the water maze tests in which searching over the former platform position is a clear indicator of spatial memory. However, a deficit in probe trial scores as typically observed in many mutant mice (not only models of MR) does by no means prove a quantifiable loss in spatial memory but may be based on other impairments not necessarily related to memory. For example, some mice deficient for cAMP response element binding protein (CREB) in the forebrain and hippocampus stubbornly visit the former platform position during platform reversal learning *(124)*, indicating that CREB deficiency acts primarily on behavioral flexibility. In fact, a meta-analysis of more than 90 mouse mutant lines and strains has shown that the limiting factor in mouse water maze learning is behavioral flexibility rather than impairments of spatial memory *(125)*. Subjecting mutant mice to large test batteries is, therefore, a necessity when studying models of human MR, as evidenced by the *Gdi1* mutants, which showed deficits mainly in working memory tasks and conditioned taste aversion but seemed fairly normal in the standard water maze test. On the other hand, it will probably become necessary to design behavioral tasks that are not based on psychological criteria but, rather, designed to discover deficits predicted by the nature of the deficit, e.g., short-term memory processes in *Gdi1*-deficient mice.

Finally, potential confounds arise from the mechanisms of adult and developmental compensation. Mice can, as humans can, compensate remarkably well for loss of neurons. For example, in mice with seizure-induced loss of hippocampal neurons, deficits in water maze learning remained subtle up to a loss of 30% of pyramidal cells in hippocampal areas CA3 and CA1, and severe impairments were only observed in a group of animals with more than 60% apoptotic cells *(126)*. Likewise, the massive anomalies in hippocampal synaptic vesicles, as observed in the Gdi1-deficient mice (*see* Section 2.4.2.) seemed to entail a pattern of functional deficits not well-fitting standard concepts of hippocampal function.

A final (and often neglected) problem arises from the observation that many knockout mutations generate, pleiotropically, different behavioral phenotypes. Except in cases of massive impairments, samples of knockout mice often show mutant mice that seem normal in some tests. Statistical analysis of the within-mutant variation then shows that poor or normal performance in one test is not predictive for performance variation in other tests. In other words, a given knockout mutation generates subclasses of mice that are normal in one test but impaired in another test, whereas the comparison of groups only shows statistically significant deficits for a majority of tests. This differs from the situation in wild-type mice, in which performance variations in one task are often, at least moderately, predictive for performance in other tasks *(127)*. It would seem likely that not only the penetrance of a mutation but also the phenotype of a mutation are subject to stochastic developmental compensation. This is reminiscent of human border-like symptoms in mental capacity, e.g., after perinatal brain injuries in premature babies, for which it is possible to anticipate statistically some kind of deficit but impossible to predict its precise manifestation.

To summarize, mouse models of human MR present a powerful tool to unravel etiology and pathophysiology of cognitive processes, but one is well-advised to take into account the interpretational caveats inherent in the use of mouse models.

ACKNOWLEDGMENTS

This article was made possible by the support of the Swiss National Science Foundation, the National Competence Center of Research "Neural Plasticity and Repair," and the Telethon Foundation. We appreciate the expert help of Rosmarie Lang and Inger Drescher-Lindh in testing countless mice, the extensive analysis of *Rsk2* mutants by Sylvie Jacquot, the help of Laetitia Prut, Nada BenAbdallah, and Frieder Neuhäusser-Wespy in testing *Arhgef6*-mutant mice, and we thank Mike Galsworthy for critical reading.

REFERENCES

1. American Psychiatric Association. Diagnostic and Statistical Manual of Mental Disorders. Washington, DC: American Psychiatric Association, 1994.
2. World Health Organization. The ICD-10 Classification of Mental and Behavioural Disorders: Diagnostic Criteria for Research. Geneva: World Health Organization, 2003.
3. Luckasson R, Borthwick-Duffy S, eds. AAMR—Mental Retardation Definition, Classification, and Systems of Support. Washington, DC: American Association on Mental Retardation, 2002.
4. McLaren J, Bryson SE. Review of recent epidemiological studies of mental retardation: prevalence, associated disorders, and etiology. Am J Ment Retard 1987;92:243–254.
5. Menkes JH, Migeon BR. Biochemical and genetic aspects of mental retardation. Annu Rev Med 1966;17:407–430.
6. Neri G, Opitz JM. Sixty years of X-linked mental retardation: a historical footnote. Am J Med Genet 2000;97:228–233.
7. Lejeune J, Gautier M, Turpin R. Les chromosomes somatique des enfants mongoliens. C R Hebd Seances Acad Sci 1959;248:1721–1722.
8. Kremer EJ, Pritchard M, Lynch M, et al. Mapping of DNA instability at the fragile X to a trinucleotide repeat sequence p(CCG)n. Science 1991;252:1711–1714.
9. Verkerk AJ, Pieretti M, Sutcliffe JS, et al. Identification of a gene (*FMR-1*) containing a CGG repeat coincident with a breakpoint cluster region exhibiting length variation in fragile X syndrome. Cell 1991;65:905–914.
10. Scriver CR, Eisensmith RC, Woo SLC, Kaufman S. The hyperphenylalaninemias of man and mouse. Annu Rev Genet 1994;28:141–166.
11. Pruitt KD, Maglott DR. RefSeq and LocusLink: NCBI gene-centered resources. Nucleic Acids Res 2001;29:137–140.
12. Plomin R. The genetics of g in human and mouse. Nat Rev Neurosci 2001;2:136–141.
13. Toniolo D, D'Adamo P. X-linked non-specific mental retardation. Curr Opin Genet Dev 2000;10:280–285.
14. O'Donnell WT, Warren ST. A decade of molecular studies of fragile X syndrome. Annu Rev Neurosci 2002;25:315–338.
15. Turner G, Webb T, Wake S, Robinson H. Prevalence of fragile X syndrome. Am J Med Genet 1996;64:196–197.
16. Bakker CE, de Diego Otero Y, Bontekoe C, et al. Immunocytochemical and biochemical characterization of FMRP, FXR1P, and FXR2P in the mouse. Exp Cell Res 2000;258:162–170.

17. Ashley CT, Sutcliffe JS, Kunst CB, et al. Human and murine *FMR-1*: alternative splicing and translational initiation downstream of the CGG-repeat. Nat Genet 1993;4:244–251.
18. Hergersberg M, Matsuo K, Gassmann M, et al. Tissue-specific expression of a FMR1/beta-galactosidase fusion gene in transgenic mice. Hum Mol Genet 1995;4:359–366.
19. Bontekoe CJ, Bakker CE, Nieuwenhuizen IM, et al. Instability of a (CGG)98 repeat in the *Fmr1* promoter. Hum Mol Genet 2001;10:1693–1699.
20. Bakker CE, Verheij C, Willemsen R, et al. *Fmr1* knockout mice: a model to study fragile X mental retardation. The Dutch-Belgian Fragile X Consortium. Cell 1994;78:23–33.
21. Kooy RF, D'Hooge R, Reyniers E, et al. Transgenic mouse model for the fragile X syndrome. Am J Med Genet 1996;64:241–245.
22. Slegtenhorst-Eegdeman KE, de Rooij DG, Verhoef-Post M, et al. Macroorchidism in *FMR1* knockout mice is caused by increased Sertoli cell proliferation during testicular development. Endocrinology 1998;139:156–162.
23. Beckel-Mitchener A, Greenough WT. Correlates across the structural, functional, and molecular phenotypes of fragile X syndrome. Ment Retard Dev Disabil Res Rev 2004;10:53–59.
24. Weiler IJ, Spangler CC, Klintsova AY, et al. Fragile X mental retardation protein is necessary for neurotransmitter-activated protein translation at synapses. Proc Natl Acad Sci USA 2004;101:17,504–17,509.
25. Steward O, Bakker CE, Willems PJ, Oostra BA. No evidence for disruption of normal patterns of mRNA localization in dendrites or dendritic transport of recently synthesized mRNA in FMR1 knockout mice, a model for human fragile-X mental retardation syndrome. Neuroreport 1998;9:477–481.
26. Ivanco TL, Greenough WT. Altered mossy fiber distributions in adult *Fmr1* (FVB) knockout mice. Hippocampus 2002;12:47–54.
27. Mineur YS, Sluyter F, de Wit S, Oostra BA, Crusio WE. Behavioral and neuroanatomical characterization of the *Fmr1* knockout mouse. Hippocampus 2002;12:39–46.
28. Godfraind JM, Reyniers E, De Boulle K, et al. Long-term potentiation in the hippocampus of fragile X knockout mice. Am J Med Genet 1996;64:246–251.
29. Paradee W, Melikian HE, Rasmussen DL, Kenneson A, Conn PJ, Warren ST. Fragile X mouse: strain effects of knockout phenotype and evidence suggesting deficient amygdala function. Neuroscience 1999;94:185–192.
30. Li J, Pelletier MR, Perez Velazquez JL, Carlen PL. Reduced cortical synaptic plasticity and GluR1 expression associated with fragile X mental retardation protein deficiency. Mol Cell Neurosci 2002;19:138–151.
31. Huber KM, Gallagher SM, Warren ST, Bear MF. Altered synaptic plasticity in a mouse model of fragile X mental retardation. Proc Natl Acad Sci USA 2002;99:7746–7750.
32. Chen RZ, Akbarian S, Tudor M, Jaenisch R. Deficiency of methyl-CpG binding protein-2 in CNS neurons results in a Rett-like phenotype in mice. Nat Genet 2001;27:327–331.
33. Incorpora G, Sorge G, Sorge A, Pavone L. Epilepsy in fragile X syndrome. Brain Dev 2002;24:766–769.
34. Nielsen DM, Derber WJ, McClellan DA, Crnic LS. Alterations in the auditory startle response in *Fmr1* targeted mutant mouse models of fragile X syndrome. Brain Res 2002;927:8–17.
35. Chen L, Toth M. Fragile X mice develop sensory hyperreactivity to auditory stimuli. Neuroscience 2001;103:1043–1050.
36. Frankland PW, Wang Y, Rosner B, et al. Sensorimotor gating abnormalities in young males with fragile X syndrome and *Fmr1*-knockout mice. Mol Psychiatry 2004;9:417–425.
37. Musumeci SA, Bosco P, Calabrese G, et al. Audiogenic seizures susceptibility in transgenic mice with fragile X syndrome. Epilepsia 2000;41:19–23.
38. Todd PK, Mack KJ. Sensory stimulation increases cortical expression of the fragile X mental retardation protein in vivo. Brain Res Mol Brain Res 2000;80:17–25.
39. Peier AM, McIlwain KL, Kenneson A, Warren ST, Paylor R, Nelson DL. (Over)correction of FMR1 deficiency with YAC transgenics: behavioral and physical features. Hum Mol Genet 2000;9:1145–1159.

40. Qin M, Kang J, Smith CB. Increased rates of cerebral glucose metabolism in a mouse model of fragile X mental retardation. Proc Natl Acad Sci USA 2002;99:15,758–15,763.
41. Dobkin C, Rabe A, Dumas R, El Idrissi A, Haubenstock H, Brown WT. *Fmr1* knockout mouse has a distinctive strain-specific learning impairment. Neuroscience 2000;100:423–429.
42. Van Dam D, D'Hooge R, Hauben E, et al. Spatial learning, contextual fear conditioning and conditioned emotional response in *Fmr1* knockout mice. Behav Brain Res 2000;117:127–136.
43. D'Hooge R, Nagels G, Franck F, et al. Mildly impaired water maze performance in male *Fmr1* knockout mice. Neuroscience 1997;76:367–376.
44. Fisch GS, Hao HK, Bakker C, Oostra BA. Learning and memory in the *FMR1* knockout mouse. Am J Med Genet 1999;84:277–282.
45. Mientjes EJ, Willemsen R, Kirkpatrick LL, et al. Fxr1 knockout mice show a striated muscle phenotype: implications for Fxr1p function in vivo. Hum Mol Genet 2004;13:1291–1302.
46. Bontekoe CJ, McIlwain KL, Nieuwenhuizen IM, et al. Knockout mouse model for *Fxr2*: a model for mental retardation. Hum Mol Genet 2002;11:487–498.
47. Knight SJ, Flannery AV, Hirst MC, et al. Trinucleotide repeat amplification and hypermethylation of a CpG island in *FRAXE* mental retardation. Cell 1993;74:127–134.
48. Gu Y, Nelson DL. FMR2 function: insight from a mouse knockout model. Cytogenet Genome Res 2003;100:129–139.
49. Gecz J. The *FMR2* gene, FRAXE and non-specific X-linked mental retardation: clinical and molecular aspects. Ann Hum Genet 2000;64:95–106.
50. Hillman MA, Gecz J. Fragile XE-associated familial mental retardation protein 2 (FMR2) acts as a potent transcription activator. J Hum Genet 2001;46:251–259.
51. Chakrabarti L, Bristulf J, Foss GS, Davies KE. Expression of the murine homologue of FMR2 in mouse brain and during development. Hum Mol Genet 1998;7:441–448.
52. Gu Y, McIlwain KL, Weeber EJ, et al. Impaired conditioned fear and enhanced long-term potentiation in *Fmr2* knock-out mice. J Neurosci 2002;22:2753–2763.
53. Hanauer A, Young ID. Coffin-Lowry syndrome: clinical and molecular features. J Med Genet 2002;39:705–713.
54. Jacquot S, Zeniou M, Touraine R, Hanauer A. X-linked Coffin-Lowry syndrome (CLS, MIM 303600, *RPS6KA3* gene, protein product known under various names: pp90(rsk2), RSK2, ISPK, MAPKAP1). Eur J Hum Genet 2002;10:2–5.
55. Trivier E, De Cesare D, Jacquot S, et al. Mutations in the kinase Rsk-2 associated with Coffin-Lowry syndrome. Nature 1996;384:567–570.
56. Delaunoy J, Abidi F, Zeniou M, et al. Mutations in the X-linked RSK2 gene (*RPS6KA3*) in patients with Coffin-Lowry syndrome. Hum Mutat 2001;17:103–116.
57. Zeniou M, Ding T, Trivier E, Hanauer A. Expression analysis of *RSK* gene family members: the *RSK2* gene, mutated in Coffin-Lowry syndrome, is prominently expressed in brain structures essential for cognitive function and learning. Hum Mol Genet 2002;11:2929–2940.
58. Dufresne SD, Bjorbaek C, El-Haschimi K, et al. Altered extracellular signal-regulated kinase signaling and glycogen metabolism in skeletal muscle from p90 ribosomal S6 kinase 2 knockout mice. Mol Cell Biol 2001;21:81–87.
59. El-Haschimi K, Dufresne SD, Hirshman MF, Flier JS, Goodyear LJ, Bjorbaek C. Insulin resistance and lipodystrophy in mice lacking ribosomal S6 kinase 2. Diabetes 2003;52:1340–1346.
60. Jacquot S, Zeniou M, Usiello A, et al. Behavior analysis of RSK2 deficient mice: an animal model for the cognitive impairment in the Coffin-Lowry syndrome. Forum European Neuroscience. Paris, 2002 (Abstract).
61. Putz G, Bertolucci F, Raabe T, Zars T, Heisenberg M. The S6KII (rsk) Gene of *Drosophila melanogaster* differentially affects an operant and a classical learning task. J Neurosci 2004;24:9745–9751.
62. D'Adamo P, Menegon A, Lo Nigro C, et al. Mutations in *GDI1* are responsible for X-linked non-specific mental retardation. Nat Genet 1998;19:134–139.
63. Novick P, Zerial M. The diversity of Rab proteins in vesicle transport. Curr Opin Cell Biol 1997;9:496–504.

64. Ishizaki H, Miyoshi J, Kamiya H, et al. Role of Rab GDP dissociation inhibitor alpha in regulating plasticity of hippocampal neurotransmission. Proc Natl Acad Sci USA 2000;97:11,587–11,592.
65. D'Adamo P, Welzl H, Papadimitriou S, et al. Deletion of the mental retardation gene *Gdi1* impairs associative memory and alters social behavior in mice. Hum Mol Genet 2002;11:2567–2580.
66. Umbricht D, Vyssotky D, Latanov A, et al. Midlatency auditory event-related potentials in mice: comparison to midlatency auditory ERPs in humans. Brain Res 2004;1019:189–200.
67. D'Adamo P, Meskenaite V, Ziegler U, Wolfer DP, Toniolo D, Lipp H-P. Tracking the roots of human mental retardation: cognitive impairments in *Gdi1* knockout mice are associated with anomalous synaptic vesicles. In: Society for Neuroscience, 33rd Annual Meeting, 2003; New Orleans, LA (Abstract).
68. D'Adamo P, Wolfer DP, Kopp C, Tobler I, Toniolo D, Lipp HP. Mice deficient for the synaptic vesicle protein Rab3a show impaired spatial reversal learning and increased explorative activity but none of the behavioral changes shown by mice deficient for the Rab3a regulator Gdi1. Eur J Neurosci 2004;19:1895–1905.
69. Shahbazian MD, Zoghbi HY. Rett syndrome and MeCP2: linking epigenetics and neuronal function. Am J Hum Genet 2002;71:1259–1272.
70. Guy J, Hendrich B, Holmes M, Martin JE, Bird A. A mouse *Mecp2*-null mutation causes neurological symptoms that mimic Rett syndrome. Nat Genet 2001;27:322–326.
71. Shahbazian M, Young J, Yuva-Paylor L, et al. Mice with truncated MeCP2 recapitulate many Rett syndrome features and display hyperacetylation of histone H3. Neuron 2002;35:243–254.
72. Luikenhuis S, Giacometti E, Beard C, Jaenisch R. Expression of MeCP2 in postmitotic neurons rescues Rett syndrome in mice. Proc Natl Acad Sci USA 2004;101:6033–6038.
73. Collins AL, Levenson JM, Vilaythong AP, et al. Mild overexpression of MeCP2 causes a progressive neurological disorder in mice. Hum Mol Genet 2004;13:2679–2689.
74. Young J, Zoghbi H. X-chromosome inactivation patterns are unbalanced and affect the phenotypic outcome in a mouse model of rett syndrome. Am J Hum Genet 2004;74:511–520.
75. Bienvenu T, Poirier K, Van Esch H, et al. Rare polymorphic variants of the *AGTR2* gene in boys with non-specific mental retardation. J Med Genet 2003;40:357–359.
76. Vervoort VS, Beachem MA, Edwards PS, et al. *AGTR2* mutations in X-linked mental retardation. Science 2002;296:2401–2403.
77. Ylisaukko-oja T, Rehnstrom K, Vanhala R, Tengstrom C, Lahdetie J, Jarvela I. Identification of two AGTR2 mutations in male patients with non-syndromic mental retardation. Hum Genet 2004;114:211–213.
78. Erdmann J, Dahmlow S, Guse M, Hetzer R, Regitz-Zagrosek V. The assertion that a G21V mutation in *AGTR2* causes mental retardation is not supported by other studies. Hum Genet 2004;114:396; author reply, 397.
79. Hein L, Dzau VJ, Barsh GS. Linkage mapping of the angiotensin AT2 receptor gene (*Agtr2*) to the mouse X chromosome. Genomics 1995;30:369–371.
80. Ichiki T, Labosky PA, Shiota C, et al. Effects on blood pressure and exploratory behaviour of mice lacking angiotensin II type-2 receptor. Nature 1995;377:748–750.
81. Okuyama S, Sakagawa T, Chaki S, Imagawa Y, Ichiki T, Inagami T. Anxiety-like behavior in mice lacking the angiotensin II type-2 receptor. Brain Res 1999;821:150–159.
82. Sakagawa T, Okuyama S, Kawashima N, et al. Pain threshold, learning and formation of brain edema in mice lacking the angiotensin II type 2 receptor. Life Sci 2000;67:2577–2585.
83. Fransen E, Van Camp G, Vits L, Willems PJ. L1-associated diseases: clinical geneticists divide, molecular geneticists unite. Hum Mol Genet 1997;6:1625–1632.
84. Brümmendorf T, Kenwrick S, Rathjen FG. Neural cell recognition molecule L1: from cell biology to human hereditary brain malformations. Curr Opin Neurobiol 1998;8:87–97.
85. Welzl H, Stork O. Cell adhesion molecules: key players in memory consolidation? News Physiol Sci 2003;18:147–150.
86. Miller PD, Chung WW, Lagenaur CF, DeKosky ST. Regional distribution of neural cell adhesion molecule (N-CAM) and L1 in human and rodent hippocampus. J Comp Neurol 1993; 327:341–349.

87. Cohen NR, Taylor JS, Scott LB, Guillery RW, Soriano P, Furley AJ. Errors in corticospinal axon guidance in mice lacking the neural cell adhesion molecule L1. Curr Biol 1998;8:26–33.
88. Dahme M, Bartsch U, Martini R, Anliker B, Schachner M, Mantei N. Disruption of the mouse L1 gene leads to malformations of the nervous system. Nat Genet 1997;17:346–349.
89. Demyanenko GP, Tsai AY, Maness PF. Abnormalities in neuronal process extension, hippocampal development, and the ventricular system of L1 knockout mice. J Neurosci 1999;19: 4907–4920.
90. Rolf B, Kutsche M, Bartsch U. Severe hydrocephalus in L1-deficient mice. Brain Res 2001;891:247–252.
91. Wiencken-Barger AE, Mavity-Hudson J, Bartsch U, Schachner M, Casagrande VA. The role of L1 in axon pathfinding and fasciculation. Cereb Cortex 2004;14:121–131.
92. Fransen E, D'Hooge R, Van Camp G, et al. L1 knockout mice show dilated ventricles, vermis hypoplasia and impaired exploration patterns. Hum Mol Genet 1998;7:999–1009.
93. Irintchev A, Koch M, Needham LK, Maness P, Schachner M. Impairment of sensorimotor gating in mice deficient in the cell adhesion molecule L1 or its close homologue, CHL1. Brain Res 2004;1029:131–134.
94. Thelin J, Waldenstrom A, Bartsch U, Schachner M, Schouenborg J. Heat nociception is severely reduced in a mutant mouse deficient for the L1 adhesion molecule. Brain Res 2003;965:75–82.
95. Bliss T, Errington M, Fransen E, et al. Long-term potentiation in mice lacking the neural cell adhesion molecule L1. Curr Biol 2000;10:1607–1610.
96. Law JW, Lee AY, Sun M, et al. Decreased anxiety, altered place learning, and increased CA1 basal excitatory synaptic transmission in mice with conditional ablation of the neural cell adhesion molecule L1. J Neurosci 2003;23:10,419–10,432.
97. Wolfer DP, Mohajeri HM, Lipp HP, Schachner M. Increased flexibility and selectivity in spatial learning of transgenic mice ectopically expressing the neural cell adhesion molecule L1 in astrocytes. Eur J Neurosci 1998;10:708–717.
98. Kutsche K, Yntema H, Brandt A, et al. Mutations in ARHGEF6, encoding a guanine nucleotide exchange factor for Rho GTPases, in patients with X-linked mental retardation. Nat Genet 2000;26:247–250.
99. Kutsche K, Gal A. The mouse Arhgef6 gene: cDNA sequence, expression analysis, and chromosome assignment. Cytogenet Cell Genet 2001;95:196–201.
100. Kohn M, Steinbach P, Hameister H, Kehrer-Sawatzki H. A comparative expression analysis of four MRX genes regulating intracellular signalling via small GTPases. Eur J Hum Genet 2004;12:29–37.
101. Wolfer DP, Kuchenbecker K, Prut L, Neuhaeusser-Wespy F, Kutsche K, Lipp HP. Impaired behavioral control and altered processing of spatial information in mice deficient for the x-chromosomal mental retardation gene Arhgef6. 7th IBANGS meeting, 2005, Sitges, Spain (Abstract).
102. Huttenlocher PR. The neuropathology of phenylketonuria: human and animal studies. Eur J Pediatr 2000;159(Suppl 2):S102–106.
103. Kahler SG, Fahey MC. Metabolic disorders and mental retardation. Am J Med Genet 2003;117C:31–41.
104. McDonald JD, Bode VC, Dove WF, Shedlovsky A. Pahhph-5: a mouse mutant deficient in phenylalanine hydroxylase. Proc Natl Acad Sci USA 1990;87:1965–1967.
105. Shedlovsky A, McDonald JD, Symula D, Dove WF. Mouse models of human phenylketonuria. Genetics 1993;134:1205–1210.
106. Zagreda L, Goodman J, Druin DP, McDonald D, Diamond A. Cognitive deficits in a genetic mouse model of the most common biochemical cause of human mental retardation. J Neurosci 1999;19:6175–6182.
107. Mihalick SM, Langlois JC, Krienke JD, Dube WV. An olfactory discrimination procedure for mice. J Exp Anal Behav 2000;73:305–318.
108. Cabib S, Pascucci T, Ventura R, Romano V, Puglisi-Allegra S. The behavioral profile of severe mental retardation in a genetic mouse model of phenylketonuria. Behav Genet 2003;33:301–310.

109. Ding Z, Harding CO, Thony B. State-of-the-art 2003 on PKU gene therapy. Mol Genet Metab 2004;81:3–8.
110. Mochizuki S, Mizukami H, Ogura T, et al. Long-term correction of hyperphenylalaninemia by AAV-mediated gene transfer leads to behavioral recovery in phenylketonuria mice. Gene Ther 2004;11:1081–1086.
111. Oh HJ, Park ES, Kang S, Jo I, Jung SC. Long-term enzymatic and phenotypic correction in the phenylketonuria mouse model by adeno-associated virus vector-mediated gene transfer. Pediatr Res 2004;56:278–284.
112. Arun D, Gutmann DH. Recent advances in neurofibromatosis type 1. Curr Opin Neurol 2004;17:101–105.
113. Dasgupta B, Gutmann DH. Neurofibromatosis 1: closing the GAP between mice and men. Curr Opin Genet Dev 2003;13:20–27.
114. Costa RM, Silva AJ. Mouse models of neurofibromatosis type I: bridging the GAP. Trends Mol Med 2003;9:19–23.
115. Bernards A, Snijders AJ, Hannigan GE, Murthy AE, Gusella JF. Mouse neurofibromatosis type 1 cDNA sequence reveals high degree of conservation of both coding and non-coding mRNA segments. Hum Mol Genet 1993;2:645–650.
116. Silva AJ, Frankland PW, Marowitz Z, et al. A mouse model for the learning and memory deficits associated with neurofibromatosis type I. Nat Genet 1997;15:281–284.
117. North K. Neurofibromatosis type 1. Am J Med Genet 2000;97:119–127.
118. Costa RM, Yang T, Huynh DP, et al. Learning deficits, but normal development and tumor predisposition, in mice lacking exon 23a of *Nf1*. Nat Genet 2001;27:399–405.
119. Costa R, Federov N, Kogan J, et al. Mechanism for the learning deficits in a mouse model of neurofibromatosis type 1. Nature 2002;415:526–530.
120. Govek EE, Newey SE, Akerman CJ, Cross JR, Van der Veken L, Van Aelst L. The X-linked mental retardation protein oligophrenin-1 is required for dendritic spine morphogenesis. Nat Neurosci 2004;7:364–372.
121. Huston JP, Borbely AA. The thalamic rat: general behavior, operant learning with rewarding hypothalamic stimulation, and effects of amphetamine. Physiol Behav 1974;12:433–448.
122. Huston JP, Tomaz C, Fix I. Avoidance learning in rats devoid of the telencephalon plus thalamus. Behav Brain Res 1985;17:87–95.
123. Deacon RM, Croucher A, Rawlins JN. Hippocampal cytotoxic lesion effects on species-typical behaviours in mice. Behav Brain Res 2002;132:203–213.
124. Balschun D, Wolfer DP, Gass P, et al. Does cAMP response element-binding protein (CREB) have a pivotal role in hippocampal synaptic plasticity and hippocampus-dependent memory? J Neurosci 2003;23:6304–6314.
125. Wolfer DP, Lipp H-P. Meta-analysis of strategy choice by 85 mutant mouse lines in a standardized place navigation task identifies behavioral flexibility as performance limiting factor. In: Society for Neuroscience, 33rd Annual Meeting, 2003; New Orleans LA (Abstract).
126. Mohajeri MH, Saini K, Li H, et al. Intact spatial memory in mice with seizure-induced partial loss of hippocampal pyramidal neurons. Neurobiol Dis 2003;12:174–181.
127. Galsworthy MJ, Paya-Cano JL, Monleon S, Plomin R. Evidence for general cognitive ability (g) in heterogeneous stock mice and an analysis of potential confounds. Genes Brain Behav 2002; 1:88–95.

7
How Can Studies of Animals Help to Uncover the Roles of Genes Implicated in Human Speech and Language Disorders?

Simon E. Fisher

Summary

The mysterious human propensity for acquiring speech and language has fascinated scientists for decades. A substantial body of evidence suggests that this capacity is rooted in aspects of neurodevelopment that are specified at the genomic level. Researchers have begun to identify genetic factors that increase susceptibility to developmental disorders of speech and language, thereby offering the first molecular entry points into neuronal mechanisms underlying human vocal communication. The identification of genetic variants influencing language acquisition facilitates the analysis of animal models in which the corresponding orthologs are disrupted. At face value, the situation raises a perplexing question: if speech and language are uniquely human, can any relevant insights be gained from investigations of gene function in other species? This chapter addresses the question using the example of *FOXP2*, a gene implicated in a severe monogenic speech and language disorder. *FOXP2* encodes a transcription factor that is highly conserved in vertebrate species, both in terms of protein sequence and expression patterns. Current data suggest that an earlier version of this gene, present in the common ancestor of humans, rodents, and birds, was already involved in establishing neuronal circuits underlying sensory–motor integration and learning of complex motor sequences. This may have represented one of the factors providing a permissive neural environment for subsequent evolution of vocal learning. Thus, dissection of neuromolecular pathways regulated by *Foxp2* in nonlinguistic species is a necessary prerequisite for understanding the role of the human version of the gene in speech and language.

Key Words: Speech; language; *FOXP2*; evolution; vocal learning; verbal dyspraxia; Broca's area; caudate nucleus; cerebellum; song system.

1. INTRODUCTION

For many years, scientists have been intrigued by the aptitude of human children for acquiring highly intricate communication skills with little effort and without any need for formal instruction. A prevailing view is that this extraordinary capacity for imbibing language depends on particular features of brain structure and/or

From: *Contemporary Clinical Neuroscience: Transgenic and Knockout Models of Neuropsychiatric Disorders*
Edited by: G. S. Fisch and J. Flint © Humana Press Inc., Totowa, NJ

neurological processing that are innately encoded *(1)*. This position is supported by a wealth of indirect evidence from a diverse range of sources, including identification of so-called universal aspects of linguistic structure *(2,3)*, data from neuropsychological studies of language acquisition *(4–6)*, adult cases of disrupted speech and/or language caused by brain lesions in particular regions *(7–9)*, mapping of functionally active neural sites during language-related tasks *(10–12)*, and comparisons between our species and other primates, who, despite being closely related, do not share the same capacities for vocal learning *(13–15)*. In the search for more direct evidence that might support a role for genetic factors in this most elusive of human skills, some researchers have recently begun to focus on studies of the small (but significant) proportion of children who are unable to acquire normal speech and language abilities, although they possess adequate intelligence and are exposed to the usual level of linguistic input in their environment *(16)*. These cases of developmental speech and language disorder may potentially have a variety of causes, but it has clearly been established that a large part of susceptibility is accounted for by genetic influences *(17)*. There is considerable effort being invested worldwide with the intention of precisely identifying allelic variants that predispose to speech and language impairment *(17)*, and other language-related disorders, such as developmental dyslexia *(18)* and autism *(19)*. Progress has been impeded by complexity at phenotypic and genotypic levels *(17–19)*. As such, despite considerable success in localizing genetic effects to roughly defined chromosomal intervals, it has proved difficult to pinpoint specific allelic variants that are unambiguously involved in disordered language development *(17)*. However, the pace of research is accelerating and it is likely that this situation will be remedied in the course of the next decade.

At present, we know of just one gene whose mutation or disruption is clearly implicated in impaired development of speech and language, while leaving other faculties relatively spared *(20)*. The discovery of this gene, known as *FOXP2*, has opened up a number of exciting new avenues for exploration of neurogenetic influences on vocal communication *(21)*. One of the most powerful and flexible systems for studying the involvement of a gene in the development and function of the brain is to exploit animal models in which the gene of interest is disrupted *(22)*. However, this leads us to something of a paradox. Speech and language are widely considered to be unique to our species (which is one of the reasons why humans find this research topic so intriguing) *(1)*. How can one effectively study the potential role of a gene in speech and language by investigating its function in a nonlinguistic species? Mice, currently the geneticist's mammal of choice for gene knockouts, are notoriously poor at making conversation, and even worse at conjugating verbs. The problem seems to be pervasive, because we cannot even solve it by moving to our closest evolutionary relatives, other primates (an effective recourse when studying neurodegenerative disease, for example; *see* ref. *23*) and the prospects for investigating gene function during human brain development are severely limited at present, for a host of ethical and technical reasons.

The aim of this chapter is to persuade the reader that, however counterintuitive this may seem at the outset, studies of gene function in other animals (and even birds) may be critical for yielding insights into innate aspects of vocal communication in humans. The evidence to support this will, by necessity, come predominantly from studies of *FOXP2*, because this is currently the only known molecular inroad into the relevant pathways. Nevertheless, the *FOXP2* findings serve to illustrate some general points that are likely to apply more widely in studies of the genetic bases of human neurodevelopment. Crucially, any informed discussion regarding the potential role of genetic factors in human speech and language depends on a clear definition of what is meant by this term, accompanied by insight into exactly how our communication skills do differ from those of other species.

2. DISSECTING SPEECH AND LANGUAGE: SHARED VS UNIQUE CHARACTERISTICS

In the broad sense, the human capacity for speech and language is best viewed as a multicomponential system enabling rapid verbal communication of an infinite array of complex meaningful ideas from one human being to another *(24)*. I am concerned here primarily with vocal communication; considerations of features of written language that developed in the past 2000 yr or aspects of modern forms of sign language are not directly relevant to the present discussion. The average human is able to make very rapid and precisely synchronized movements of articulators (tongue, lips, jaws, and so on), coordinated with time of onset/offset for vibrations of the larynx, to produce the particular sound sequences that are peculiar to speech *(25)*. The combinatorial nature of speech—a finite number of subunits can be assembled to form an unlimited range of unique utterances—is a crucial feature of human vocal communication *(1)*. This is constrained by rules of language (syntax) that govern the ways in which units can be combined, at word, phrase, and sentence levels, to encode the intended meaning (semantics) *(2,3)*. The stream of speech sounds that are produced by the speaker has to be perceived by the auditory system of the listener, and somehow decoded (in a way that is independent of irrelevant factors such as variation in vocal tract anatomy from one speaker to the next), thus, yielding the particular sequence of phonetic units that were originally intended by the speaker *(26)*. The phonetic sequence must then be parsed by the listener, using the same multilevel syntactic rules that would have been adopted by the speaker, and assuming an equivalent vocabulary of lexical items.

It is obvious from this already grossly oversimplified view of the workings of speech and language, that human vocal communication depends on multiple processes and involves a variety of cognitive, motor, and sensory substrates. It should also be clear that, if the system is to be effective in correctly conveying meaning from speaker to listener, each of these processes must be placing substantial constraints on the workings of the others. In other words, the relationships between different elements of the speech and language system must be finely tuned to provide parity, and it is difficult to explain the evolution of human vocal communication without assuming a considerable degree of bootstrapping *(27)*. However, there are conflicting opinions regarding the specificity of each of the sensory, motor, and cognitive substrates for the capacity for language (e.g., refs. *24,27*). At the heart of this debate, lies the following question: are there aspects of the human vocal communication system that are unique to our species (i.e., one or more qualitatively distinct mechanisms) or do the capabilities of our species just reflect honing of features that are also present in nonspeaking animals (i.e., quantitative variation of pre-existing mechanisms)? To answer this question, scientists have tried to assess whether particular sensory, motor, or cognitive features of the human vocal communication system are present in a comparable form in other animals and/or birds. Comparative studies represent a burgeoning field, therefore, I will confine my discussion to three of the more prominent areas of research: speech production (vocal tract anatomy/motor control), speech perception, and syntactical processing. The emerging findings illustrate that animals and birds, despite their lack of speech and language, can provide important clues to the neural basis of human vocal communication. This is an idea that will be extended to the domains of developmental genetics and molecular neuroscience later in this chapter.

2.1. Speech Production

In the majority of mammalian species, and in human infants, the high position of the larynx in the throat allows it to be engaged into the nasal passages, such that the animal can breathe and swallow (or suckle) at the same time *(25)*. During the first 3–4 yr of human life, the larynx undergoes a gradual descent, and its adult location is notably lower than that found in adults of the other primates, including our closest relative, the chimpanzee. This arrangement enables the human tongue to make a greater variety of movements, both horizontally and vertically, within the vocal tract. The concomitant widening of the repertoire of possible vocal-tract shapes results in a dramatically expanded range of distinguishable speech sounds (phonemes) that can be produced *(28)*. There is no doubt that the wider phonetic repertoire conferred by this laryngeal descent is an important feature of human communication, and that its evolution was a key step for emergence of speech and language. What is less clear is why this unusual vocal tract anatomy evolved in the first place *(25)*. Evidence has recently emerged for independent evolution of a descended larynx in some other mammalian species (including red and fallow deer), indicating that this kind of configuration can evolve for reasons other than increased phonetic diversity *(29)*. One suggested explanation is that the vocalizations of an animal with an elongated vocal tract may convey the impression to the listener that the animal is larger than it really is, and this ability to exaggerate size could provide a selective advantage *(29)*.

Regardless of the original reasons for evolution of a descended larynx in humans (among other anatomical differences that are not discussed here), a salient point is that the distinctive vocal-tract anatomy of humans, as compared with other primates, must be under genetic control; hence, at least a subset of the genes that are

Animals, Speech, and Language

critical for human speech and language will have little or nothing to do with brain function, but will instead be related to anatomical constraints *(21)*. The anatomical data also imply that even a chimpanzee with a fully humanized brain would still be severely limited in its abilities to converse using human speech and language. Similarly, it is extremely unlikely that geneticists will ever be able to engineer a mouse that could learn to talk, however sophisticated our techniques of manipulation may become, simply because the anatomical constraints of the mouse vocal tract do not allow it. Curiously, anatomical constraints do not seem to be a problem for certain birds, such as parrots, which show remarkable abilities to mimic human speech.

Even with the appropriate vocal tract anatomy, the complexity of human speech places elaborate demands on the neural systems that control movement. The articulatory gestures that underlie speech involve finely coordinated sequences of movements by the tongue, lips, jaws, and so on, closely synchronized with vibrations of the larynx; such feats of motor control are rarely (if ever) required by any other system in the body. As noted by Fitch *(25)*, minutely subtle changes in relative timing of different articulators or onset of laryngeal vibrations are enough to yield dramatic phonetic distinctions; with a simple change in timing of just tens of milliseconds the word 'pat' may be turned into 'bat'. Although it has yet to be established whether nonhuman species are capable of comparable levels of fine motor control, based on current data it seems likely that the evolution of human speech depended to some extent on the honing of such abilities *(30)*. This might have involved peripheral mechanisms (for example higher numbers of motor neurons innervating the muscle fibers of the articulators; ref. *31*) or higher-order reorganization in the brain (such as increased connectivity of neural circuitry underlying motor coordination; ref. *30*), or a combination of the two. I will revisit the key issue of motor control in Section 4. of this chapter because it has recently become particularly relevant for gene-driven studies of precursors of speech in nonhuman species.

2.2. Speech Perception

An effective communication system must possess the property of parity, processes of production must somehow match up with processes for perception if meaning is to be accurately conveyed from one individual to another *(27)*. Indeed, humans are born with capacities that are already well attuned to the properties of speech *(26)*. From 4 d of age, a human newborn can discriminate phonemes in a categorical manner *(32)* and can detect differences between languages that have distinctive rhythmic structures, but not if the speech sequences are played backwards *(33)*. However, it has been robustly demonstrated that capacities such as phoneme discrimination and sensitivity to speech rhythm are not actually unique to humans. A wide variety of animal and bird species are able to distinguish different human speech sounds; for example, Japanese quail can be taught to discriminate between consonants /b/, /d/, and /g/ even if they are placed before novel vowel sounds that they have not encountered *(34)*. Moreover, adult tamarin monkeys, similar to newborn humans, can discriminate the distinctive rhythmic properties of

Dutch and Japanese; an ability that, as for newborn humans, is lost when the speech stream is reversed *(35)*. Of course, this does not mean that monkeys can understand Dutch or Japanese. It is important to appreciate the limitations of this kind of comparative study, which may be open to a number of alternative interpretations *(36)*. For example, the adult tamarins that have been studied received a lifetime of exposure to human speech patterns, whereas the only exposure for the human neonates would have been in the womb (albeit at a period of brain development with a dramatic level of plasticity), and the underlying mechanisms for rhythm discrimination might be radically different in the two species *(36)*. Nevertheless, this work does indicate that we should not simply assume that perceptual capabilities attuned to properties of speech are necessarily unique to humans, even if they do appear very early in development.

2.3. Syntactical Processing

Currently, the best candidate for a qualitatively distinct process that makes a unique contribution to human vocal processing is the capacity for generating an unlimited array of meaningful utterances using a finite set of lexical units, and a system of rules (syntax) regarding how they can be combined *(1–3)*. At the cognitive level, syntactical processing is able to act independently, as illustrated by Noam Chomsky's much quoted sentence "colourless green ideas sleep furiously," which can be easily judged as grammatically correct even though it lacks any real meaning *(1–3)*. Compare this with another example with the same words in a different order, such as "furiously sleep ideas green colourless," which is obviously ungrammatical. During the first few years of life, a human child is exposed to only a tiny subset of possible sentences, but is nevertheless able to extract correctly the syntactic rules that are inherent to his or her native language *(1–5)*. The use of hierarchical rules (in which one phrase can be embedded within another) is a core feature of syntax that gives it enormous generative power *(2,3,15)*. It has been proposed that monkeys are able to learn simple rule systems, but that, among the primates, the ability to perform abstract hierarchical processing is only found in humans *(15)*. It remains to be seen whether this ability is seen in primitive form anywhere else in the animal kingdom. Finally, it is worth noting that syntactical processing is, in large part, a question of assembling sequences of language elements, and may, thus, be closely related to motor sequencing; interestingly, classic language regions of the brain, such as Broca's area, have been implicated both in syntactical processing and motor sequencing *(30)*.

3. THE HUMAN *FOXP2* GENE: A FOOT IN THE DOOR

The bulk of the evidence from comparative studies supports a view in which many of the processes underlying human vocal communication exploit neural substrates that were already present (at least in primitive form) in our nonspeaking ancestors. Such a perspective predicts that elucidation of neuromolecular mechanisms in animals will provide an important basis for understanding the potential roles of genetic factors in human speech and language. We are now in a position to evaluate this prediction directly. As outlined in the introduction, studies of human language-related disorders are promising to deliver genes that are implicated in the process of speech and language acquisition *(16–21)*. *FOXP2*, the first of these genes to emerge, provides compelling empirical confirmation of the relevance of ancestral neurogenetic pathways for modern human vocal communication.

3.1. What is FOXP2?

The *FOXP2* gene encodes a protein belonging to the forkhead family of transcription factors *(20,37)*, so-called because mutations of the founding member in fruit flies cause homeotic transformation of terminal portions of the gut into ectopic head structures *(38)*. The human genome encodes more than 40 different forkhead proteins, each of which includes a highly conserved DNA-binding domain (the forkhead-box or FOX domain) of 84–110 residues *(37)*. All forkhead proteins seem to be activators or repressors of gene transcription, but they can play many diverse roles at cellular and developmental levels, influencing cell cycling, signaling, differentiation, apoptosis, and so on, with some proteins having multiple functions in different tissues or at different times *(37)*. Many forkhead proteins influence developmental pathways during embryogenesis, and mutations in the genes encoding them are known to cause a variety of inherited human and mouse disorders, including glaucoma, immune deficiency, ovarian failure, diabetes, and hearing impairment *(39)*. Alterations in dosage (the number of functional gene copies) of individual forkhead genes can perturb development in a striking manner *(37,39)*. For example, heterozygous missense and nonsense mutations of *FOXC1* have been found in subjects with autosomal dominant eye disorders; functional and structural analyses indicate that the mutant *FOXC1* alleles lead to loss-of-function of the encoded protein *(40)*, supporting a haploinsufficiency mechanism, i.e., the quantity of functional protein made from a single gene copy is insufficient to allow normal development.

FOXP2 is a member of a specific sub-branch of forkhead proteins (FOXP1-4), which are defined by an unusual variant of the DNA-binding domain spanning only 84 residues, as compared with the more than 100-residue domains usually found in other forkhead proteins (Fig. 1) *(20,41–43)*. Outside the distinctive DNA-binding domain, the FOXP proteins share a glutamine-rich C-terminus, and a highly conserved region containing a zinc finger and a leucine zipper *(20,41–43)*. The latter mediates homodimerization and heterodimerization, which seem to be necessary for DNA-binding and transcription factor function *(42,43)*. The requirement for dimerization is another feature that makes these proteins distinct from other forkheads, most of which are thought to act as monomers *(37)*.

3.2. Why is FOXP2 Relevant for Human Speech and Language?

Point mutations in the human form of *FOXP2* (Fig. 1), which maps to cytogenetic band 7q31, cause a rare monogenic form of speech and language disorder that is inherited in an autosomal dominant manner *(20,44–46)*. In one well-studied

```
HUMAN  MMQESATETISNSSMNQNGMSTLSSQLDAGSRDGRSSGDTSSEVSTVELLHLQQQALAARQLLLQQQTSGLKSPKSSDKQRPLQVPVSVAMMTPQVIT  100
MOUSE  MMQESATETISNSSMNQNGMSTLSSQLDAGSRDGRSSGDTSSEVSTVELLHLQQQALQQALQQQTSGLKSPKSSEKQRPLQVPVSVAMMTPQVIT  100
                                                                            ====PolyQ====       =

HUMAN  PQQMQQILQQQVLSPQQLQALLQQQQAVMLQQQQLQEFYKKQQEQLHLQLLQQQQQQQQQQQQQQQQQQQQQHPGKQAKEQ  200
MOUSE  PQQMQQILQQQVLSPQQLQALLQQQQAVMLQQQQLQEFYKKQQEQLHLQLLQQQQQQQQQQQQQQQQQQQHPGKQAKEQ   200
       =PolyQ==

HUMAN  QQQQQQQQLAAQQLVFQQQLLQMQQLQQQHLLSLQRQGLISIPPGAALPVQSLPQAGLSPAEIQQLMKEVTGVHSMEDNGIKHGGLDLTTNNSSSTT  300
MOUSE  QQQQQQQ-LAAQQLVFQQQLLQMQQLQQQHLLSLQRQGLISIPPGAALPVQSLPQAGLSPAEIQQLMKEVTGVHSMEDNGIKHGGLDLTTNNSSSTT  299
                                                                                          =====LeuZ=====

HUMAN  SSNTSKASPPITHHSIVNGQSSVLSARRDSSSHEETGASHTLYGHGVCKWPGCESICEDFGQFLKHLNNEHALDDRSTAQCRVQMQVVQQLEIQLSKERE  400
MOUSE  SSTTSKASPPITHHSIVNGQSSVLNARRDSSSHEETGASHTLYGHGVCKWPGCESICEDFGQFLKHLNNEHALDDRSTAQCRVQMQVVQQLEIQLSKERE  399
       *                                =========ZnF==========                                    
                                                                  H

HUMAN  RLQAMMTHLHMRPSEPKPSPKPLNLVSSVTMSKNMLETSPQSLPQTPTTPAPVTPITQGPSVITPASVPNVGAIRRRHSDKYNIPMSSEIAPNYEFYKN  500
MOUSE  RLQAMMTHLHMRPSEPKPSPKPLNLVSSVTMSKNMLETSPQSLPQTPTTPAPVTPITQGPSVITPASVPNVGAIRRRHSDKYNIPMSSEIAPNYEFYKN  499
                                                          ========FOX========

HUMAN  ADVRPPFTYATLIRQAIMESSDRQLTLNEIYSWFTRTFAYFRRNAATWKNAVRHNLSLHKCFVRVENVKGAVWTVDEVEYQKRRSQKITGSPTLVKNIPT  600
MOUSE  ADVRPPFTYATLIRQAIMESSDRQLTLNEIYSWFTRTFAYFRRNAATWKNAVRHNLSLHKCFVRVENVKGAVWTVDEVEYQKRRSQKITGSPTLVKNIPT  599
       ===========================                                              ======Acidic======

HUMAN  SLGYGAALNASLQAALAESSLPLLSNPGLINNASSGLLQAVHEDLNGSLDHIDSNGNSSPGCSPQPHIHSIHVKEEPVIAEDEDCPMSLVTTANHSPELE  700
MOUSE  SLGYGAALNASLQAALAESSLPLLSNPGLINNASSGLLQAVHEDLNGSLDHIDSNGNSSPGCSPQPHIHSIHVKEEPVIAEDEDCPMSLVTTANHSPELE  699
       ====================

HUMAN  DDREIEEPLSEDLE  715
MOUSE  DDREIEEPLSEDLE  714
```

family, known as KE *(44,45)*, a heterozygous missense mutation altering a highly conserved residue in the DNA-binding domain (arginine-to-histidine; R553H) was found to co-segregate with disorder in 15 affected individuals across three generations *(20)*. The mutation was not present in any unaffected members of the family, or in a large panel of normal controls *(20)*. An equivalent R-to-H substitution at the corresponding position in the DNA-binding domain of another human forkhead protein, FOXC1, causes a developmental eye disorder, and in vitro assays indicate that it dramatically disrupts function *(40)*. Thus, there is strong evidence that the R553H change is of etiological significance for the KE family. In an unrelated small nuclear family with similar impairment in speech and language abilities, a heterozygous nonsense mutation (R328X) was found in two affected siblings, as well as their mother, who had a history of speech problems *(46)*. The mutation, which was not detected in screening of more than 250 control chromosomes, leads to severe truncation of the product encoded by this allele (Fig. 1). The resulting FOXP2 protein is predicted to lack essential functional motifs, including the entire zinc finger/leucine zipper region and forkhead-box domain, and is, thus, unlikely to be able to dimerize or bind to DNA *(46)*. Moreover, the resulting FOXP2 protein has lost critical nuclear localization signals, and, thus, may be inappropriately targeted to the cytoplasm, further hindering its function.

In addition to these mutations, gross chromosomal rearrangements disrupting *FOXP2* have been identified in isolated cases of speech and language delay. For example, the *FOXP2* locus is directly interrupted by a chromosomal breakpoint in an affected child with a *de novo* balanced translocation involving chromosomes 7 and 5 *(20,47)*, and is hemizygous in cases of disorder associated with interstitial deletion encompassing 7q31 *(48, 49)*. Thus, the evidence from point mutations and chromosomal abnormalities indicate that loss-of-function of one copy of the *FOXP2* gene leads to developmental deficits in speech and language acquisition. Based on these findings, it is likely that *FOXP2*-associated disorder results from a mechanism of reduced functional gene dosage during early development.

Fig. 1. (*opposite page*) Alignment of amino acid sequences of human and mouse FOXP2/Foxp2 proteins (main isoform), as inferred from complementary DNA sequence. Key protein domains include polyglutamine tracts (PolyQ), a zinc-finger motif (ZnF), a leucine zipper (LeuZ), a forkhead domain (FOX), and a C-terminal acidic region. Disease mutations: two independent point mutations causing speech and language disorder have been identified in the human *FOXP2* gene; one leads to an R553H substitution in the forkhead domain, the other truncates the protein at R328, yielding a severely truncated product lacking ZnF, LeuZ, FOX, and acidic domains. The positions of the changes caused by these two mutations are indicated by shaded text beneath the alignment. Comparative genomics: the second polyglutamine tract of the mouse protein is one residue shorter than that in human. Elsewhere, only three amino acids differ between human and mouse (indicated by * above the alignment); none of these occur in known functional domains. Two of these changes (in the region upstream of the ZnF domain) are specific to the human lineage. *See* Section 4.3. for further details.

3.3. Insights From Humans: Defining the Phenotype

Human speech and language depend on a complex system of interconnected sensory, motor, and cognitive neural substrates. Impaired acquisition of speech and language skills might potentially result from a range of different abnormalities at the neural level. As such, it is important to probe carefully the phenotypic profile of a disorder to shed light on this issue, and to establish whether impairment is confined to language systems or extends into more general domains. Regarding the disorder caused by disruption of *FOXP2*, the vast majority of evidence has so far come from the KE family *(44,50–57)*, which may limit the conclusions that can be drawn, but data are now beginning to surface from other cases *(49)*. An early, much-publicized report proposed that affected subjects in the KE family suffered from a specific deficit in syntactical processing *(50)*, such that this disorder might, for the first time, give a genetic handle to a uniquely human cognitive subprocess *(1)*. However, detailed phenotypic evaluations carried out over the past decade cast doubt on this characterization *(51–57)*. Rather than involving just one aspect of syntax *(50)*, the *FOXP2*-associated disorder broadly affects a range of language-related skills, impacting both the production and comprehension of language *(51,52,55)*. Moreover, in a subset of affected subjects within the KE family, language impairment is accompanied by a less significant reduction in nonverbal abilities *(51)*, although the evidence suggests that these nonverbal deficits are not direct effects of *FOXP2* disruption *(55)*. It has been generally established that presence of language impairment during early development puts children at higher risk of wider cognitive deficits and behavioral problems later in life *(58,59)*.

In fact, the most overt feature of disorder transmitted through the KE family, and one that unambiguously distinguishes between affected and unaffected subjects, is a persistent problem in coordinating sequences of mouth movements underlying speech *(44,52,54)*. This kind of articulatory impairment is often referred to as developmental verbal dyspraxia, and it also affects production of nonspeech sounds *(52,54)*. Notably, etiological variants of *FOXP2* have not yet been found in children with typical forms of autism, specific language impairment, or dyslexia *(60–63)*, which are all language-related disorders that do not usually involve verbal dyspraxia as a primary feature. In contrast, FOXP2 coding changes, although still a rare cause of impairment, may be present in up to 6% of children in whom verbal dyspraxia is the most prominent aspect of disorder *(46)*.

It remains open to question whether all of the wider deficits in language ability in affected subjects stem from a primary deficit in articulation *(21)*. An attractive alternative hypothesis is that a single core deficit in learning/production of sequences is directly responsible both for problems with speech and impairment in grammatical processing *(53)*. A third possibility is that *FOXP2* disruption simultaneously impacts multiple different neural substrates to yield the wide spectrum of language problems in affected subjects. It is worth noting here that the linguistic dysfunctions observed in these subjects are not confined to oral output (which is obviously strongly influenced by articulatory difficulties), but are also apparent in written language *(55)*. More detailed discussion of issues regarding the phenotypic profile of the KE family can be found elsewhere *(17,21)*.

3.4. Insights From Humans: Brain Imaging Studies

The brains of affected members of the KE family seem to be radiologically normal *(52)*. However, detailed structural neuroimaging of multiple individuals from this family has revealed several sites in which the density of gray matter is significantly different between affected and unaffected members *(52,56)*. Because the disorder is developmental in nature, it was hypothesized that structural pathology must be present in both hemispheres; otherwise the early plasticity of the brain would allow the subjects to compensate by recruiting circuitry in the unaffected hemisphere *(52,56)*. Damage to just one hemisphere in young children rarely leads to language impairment, because of the ability of the brain to reorganize during early development. Notably, a bilateral reduction in gray matter density has been found in the inferior frontal gyrus (which contains Broca's area, a classic language region of the brain), the caudate nucleus, the precentral gyrus, the temporal pole, and the cerebellum *(52,56)*. Other regions, including the posterior superior temporal gyrus, angular gyrus, and putamen, have been shown to have bilateral increases in density *(52,56)*.

These kinds of neuroanatomical studies can help to point to potential sites mediating dysfunction in subjects with *FOXP2* disruption, but they are not direct indicators of abnormalities in neurological mechanisms. In recent years, functional magnetic resonance imaging (fMRI) approaches have been used to ask whether the brains of affected members of the KE family are able to process language in a comparable way to their unaffected relatives *(57)*. In normal right-handed individuals, a task involving generation of verbs in response to aurally presented nouns leads to activation of Broca's area in the left hemisphere, which can be clearly detected using fMRI. For members of the KE family who have the R553H FOXP2 change, there is significant underactivation of Broca's area, and other language-related regions, when carrying out this verb generation task. Instead, these subjects show diffuse bilateral activation of regions of the cortex that are not usually associated with speech- or language-based tasks *(57)*. The affected subjects are able to perform the task adequately, but the pattern of neural activation seems to be highly atypical, suggesting that they may be compensating for the underlying genetic deficit to a certain extent via reorganization of neural circuitry during development (a common feature for neurodevelopmental disorders such as dyslexia and specific language impairment) *(18,19)*. For the overt scenario, in which test subjects are asked to give spoken responses during the verb generation task, the underactivation of Broca's area in affected KE family members may be simply related to abnormalities in execution of speech output, because this region of the brain is also implicitly involved in motor sequencing *(30)*. Therefore, perhaps this finding might be explained purely in terms of peripheral speech mechanisms. However, the same underactivation was observed for covert verb generation, in which the subjects were instructed to think their responses but not produce any speech output *(57)*. Thus, these fMRI experiments provide the strongest support so far for the hypothesis that *FOXP2* disruption leads to abnormalities in the neural circuitry underlying language processing (specifically, semantic retrieval and articulatory planning), rather than impacting only on domain-general mechanisms involved in controlling fine muscle movements. Subcortical regions of the brain, particularly areas of the striatum, have also been implicated by functional imaging studies. Thus, both structural and functional approaches indicate that there is distributed pathology associated with *FOXP2* disruption, involving cortical and subcortical structures, as opposed to damage of one specific focal region of the central nervous system (CNS) *(49,52,56,57)*.

4. *FOXP2* IN ORGANISMS THAT LACK LANGUAGE

Studies of human subjects with *FOXP2* disruption have been of great importance for increasing understanding of the nature of the resulting disorder, but they hit limitations on two separate fronts. First, mutation of *FOXP2* is extremely rare in the human population, therefore, conclusions must be drawn from extensive testing of very small numbers of individuals; as stated above in Section 3.3., almost all phenotypic and neuroimaging investigations have thus far focused on only one allelic variant, which is found in just a single pedigree (R553H in the KE family) *(50–57)*. Second, there are (at least currently) severe practical and ethical restrictions to the scope of human-based research, precluding the possibility of proper in vivo investigations of human CNS function at the neuromolecular level. Unfortunately, investigations of gene function in human neuronal-like cell lines, although certainly of value, are not able to model the complexities and subtleties of brain architecture and function. A tried and tested solution to this problem is to gain insights into function of the human gene by studying orthologs found in other species, with techniques that enable investigation and correlation of data at multiple levels (anatomical, developmental, molecular, and so on). However, if human *FOXP2* is intimately involved in speech and language acquisition, aspects of our make-up that are supposedly unique to our species, then what could the orthologs of this gene be doing in nonlinguistic species? Although the full answer to this question is not yet known, a number of provocative clues have already emerged.

4.1. Foxp2 *in Rodents*

A naïve perspective of "speech genes" might predict that *FOXP2* would not be present in the genomes of nonlinguistic species, or that it may differ greatly in sequence, expression, or function. Studies of mammalian orthologs of *FOXP2* have contradicted this simplistic view in a spectacular manner. In terms of coding sequence, the *FOXP2* gene represents one of the most highly conserved loci in the evolutionary history of humans and rodents (Fig. 1) *(64,65)*. Of the 715 amino acid residues in the main isoform of *FOXP2*, the mouse version differs at only four positions (three amino acid substitutions, and a single residue reduction in length of one of the polyglutamine tracts). In other words, the mouse protein is identical in sequence to the human ortholog for approx 99.5% of its length. Moreover, the amino acid substitutions lie outside the currently known functional domains of the protein, thus it is not clear what impact they might have on behavior of this transcription factor.

Could the presence of a highly conserved version of *Foxp2* in nonspeaking rodents be explained in terms of a role or roles outside the CNS, in the development or functioning of other tissues? In part, the answer to this question is yes. During embryogenesis, murine *Foxp2* is expressed in defined regions of multiple tissues, including the distal alveolar lung epithelium, the outer mesodermal intestinal layer, and the outflow tract and atrium of the cardiovascular system *(41)*. Studies have shown that the Foxp2 protein is able to repress lung-specific target genes in vitro *(41–43)*. Similarly, human *FOXP2* is expressed in multiple tissues during fetal development and in adulthood *(20)*. The recruitment of the same transcription factor to multiple pathways in different tissues and at distinctive times in the life of an organism is a characteristic feature of this class of proteins *(37)*, thus, FOXP2 is not at all unusual in this respect. Therefore, FOXP2 is indeed likely to have evolutionarily conserved roles that have no relation whatsoever to brain function or speech. However, when we look to expression patterns that are found in the CNS, the story becomes even more complex, and more intriguing.

In mice, rats, and humans, *Foxp2*/*FOXP2* messenger RNA (mRNA) is expressed in neurons (but is not detected in glial cells) in a variety of different brain structures during embryogenesis *(66–68)*, supporting a possible role in aspects of early neurodevelopment. In the same way that disruption of *FOXP2* in humans does not lead to damage to just one specific region of the brain, expression of the gene is not confined to any one CNS structure. In fact, significant levels of mRNA and protein have been detected in multiple regions of the forebrain, midbrain, and hindbrain *(66–68)*. However, the peculiar spatiotemporal expression patterns found in each region indicate that this gene is subject to very precise regulation and give hints regarding potential mechanisms that *FOXP2* may be controlling during neurodevelopment. Three cases of sublocalization merit discussion:

1. In the developing and mature cortex, *Foxp2* expression is restricted to the deepest layers (mainly layer 6), consisting of early-born pyramidal neurons that project back to subcortical regions, such as the dorsal thalamus, also a site of Foxp2 expression *(66–68)*. Other transcription factors, such as *Tbr1* and *Otx1*, show similar deep layer-specific (or enriched) expression patterns in the mammalian cortex; mutation of genes encoding these transcription factors results in disruption to corticothalamic connectivity *(69,70)*.
2. A well-documented subcortical site of high *FOXP2*/*Foxp2* expression is the striatum (consisting of the caudate nucleus and the putamen), with levels that are highest during embryogenesis but persist throughout development *(66–68)*. In rats, *Foxp2* shows differential postnatal expression in the two different compartments that make up the striatum, which are known as the striosomes and the matrix *(68)*. These compartments are distinguished by differing profiles of expression for receptors, neurotransmitters, signaling molecules and so on, and by distinct neurodevelopmental origins and pat-

terns of connectivity with other brain regions *(71)*. Rat *Foxp2* expression seems to be restricted to the striosomal compartment throughout the life of the animal *(68)*, although it has not been determined whether this observation applies to other mammalian species. Studies have shown that perturbations in the balance between striosomes and matrix can impact motor behavior *(72)*, which may be relevant to the motor aspects of the phenotype associated with human *FOXP2* disruption.

3. The cerebellum is a complex multilayered structure in the hindbrain with many different cell types, including granule, Golgi, basket, stellate, and Purkinje cells. In this brain region, *FOXP2/Foxp2* is detected specifically in Purkinje cells and deep cerebellar nuclei. In addition, strong persistent expression is found in the inferior olive *(67)*, a precerebellar nucleus providing direct input to these neurons. During late embryogenesis, climbing fibers from the inferior olives innervate Purkinje cells *(73)* and organize their topography in an ordered fashion, perhaps in response to polarity cues *(74)*.

Overall, the expression studies that have been conducted thus far in mammalian species support two key conclusions. First, the spatiotemporal patterns observed in humans and rodents are strikingly similar (Fig. 2). Although the human studies are inevitably more limited in scope because of restricted availability of CNS tissue at later stages of embryonic development, such that some potential expression differences may be missed, the regulation of *FOXP2/Foxp2* at comparable stages is remarkably concordant in different species *(67)*. Second, the expression data are most compatible with this gene having a role (or roles) in establishing and/or maintaining neural circuitry involved in motor control and sensorimotor integration. In particular, corticostriatal and olivocerebellar circuits (key sites of *FOXP2/Foxp2* expression identified by multiple studies) are fundamental for motor control; the basal ganglia modulate activity of premotor and prefrontal cortex via complex connections projecting through the globus pallidus, substantia nigra, and thalamus *(75)*, and the Purkinje cells of the cerebellum play a key role in regulating motor coordination, receiving strong synaptic excitation from inferior olivary climbing fibers *(76)*. Of note, there is growing appreciation that circuits involving striatal and cerebellar structures are not limited to motor function, but are also essential for aspects of cognition, underpinning aspects of many complex learned behaviors *(30,75)*.

4.2. Foxp2 *in Birds*

Although humans are the only organisms known to speak, we are not alone in our capacity to acquire new vocalizations via imitation of our peers. This ability (known as vocal learning) is nevertheless an extremely rare trait among animals and birds, found only in a handful of species, in contrast to the more pervasive use of innately specified calls (such as alarm calls made on encountering a predator) *(77)*. The only known vocal learning species are found in three groups of birds—parrots, hummingbirds, and songbirds—and three groups of mammals—humans, cetaceans (whales and dolphins), and bats. In both birds and mammals, different vocal-learning species are phylogenetically separated from each other by nonlearners, suggesting that there may have been independent evolution of this trait on at least six different occasions *(77)*. The human communication system can be viewed as a highly specialized form of vocal learning; acknowledging that the

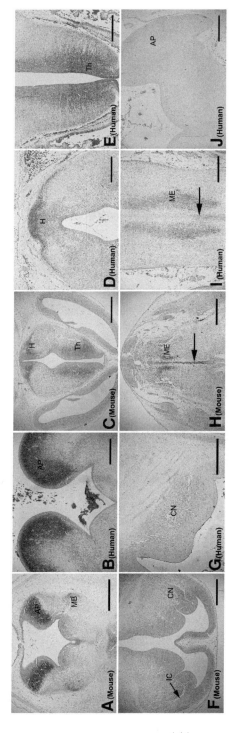

Fig. 2. Conserved expression of *Foxp2/FOXP2* in the developing central nervous system of mouse and human embryos. *In situ* hybridizations to detect *Foxp2/FOXP2* messenger RNA expression in the developing mouse brain at E13.5 and in a human brain at a comparable embryonic stage (Carnegie stage 23). (A) In the mouse, intense signal is detected in the alar plate (AP) of the developing cerebellum. Weaker signal is also observed in the mantle layer of the midbrain (MB). (B) A very similar expression pattern is seen in a corresponding section from a human brain at CS23. (C–E) In the mouse and human diencephalon, *Foxp2/FOXP2* is expressed in the mammillary area of the hypothalamus (H) and the dorsal thalamus (Th). (F,G) Diffuse signal is found in the caudate nucleus (CN), adjacent to the internal capsule (IC) in both species. (H,I) In the mouse and human hindbrain, *Foxp2/FOXP2* transcripts are detected in the medullary raphe (arrows) and the medulla oblongata (ME). (J) No signal is seen in control hybridizations. Scale bars: mouse panels, 1 mm; human panels, 0.5 mm. (Adapted from ref. *67*).

communication skills of parrots, hummingbirds, and songbirds are not as sophisticated as human language (for example it is not known whether avian vocal learners have the capacity to use hierarchical systems of syntax).

Neuroanatomical and functional genetic investigations of different bird species have indicated the presence of seven discrete forebrain structures involved in vocal control (learning and production), which seem to be common to all vocal-learners. These so-called song nuclei are absent from the brains of nonlearning birds, or perhaps present in such a rudimentary form that they have escaped detection *(77)*. Elegant neurobiological studies demonstrate that gene expression patterns in the song system change during song learning and production *(78,79)*. If the song nuclei are indeed unique to vocal learners, then one explanation would be the independent evolution of a set of seven similar forebrain structures on three separate occasions in the ancestry of modern birds, thus indicating severe epigenetic constraints on evolution of vocal control structures. Conversely, the song system may have been present in the common ancestor of all avian vocal learners, with loss of the entire set of forebrain control structures on four separate occasions during bird evolution. The maintenance of a vocal learning system in the brain places considerable demands on an organism's resources, and thus may only represent a selective advantage in certain types of environment *(77)*. Regardless of these evolutionary conundrums, which are yet to be resolved, studies of birds offer a unique window on neurogenetic mechanisms involved in vocal learning, and the resulting insights may be generally relevant for our understanding of this trait in the form that is found in our own species *(79)*. Notably, the anterior nuclei of the song system participate in neural pathways comparable to mammalian corticostriatal circuitry, active in the learning and maintenance of motor sequences dependent on sensorimotor integration *(77–79)*. Thus, given the emerging evidence that mammalian *FOXP2* may be involved in motor-related circuitry, it becomes of immediate interest to evaluate the potential role of avian *FoxP2* in the development and/or behavior of the song system.

Two recent studies have made the first attempts to address this issue *(80,81)*. *FoxP2* in songbirds is very similar to the human and rodent orthologs; the zebra finch protein sequence is approx 98% identical to the mammalian proteins. Moreover, for this gene there is a remarkable concordance in the spatiotemporal CNS expression patterns found in birds and mammals. High levels of *FoxP2* mRNA and its protein product have been robustly detected in the striatum, dorsal thalamus, Purkinje cells, and inferior olives, with weaker expression in a number of related brain nuclei *(80,81)*. This pattern has been found to be consistent in eleven different species of bird, regardless of sex or song-learning ability *(81)*. As in mammals, expression starts early in embryogenesis, and persists through development and into adulthood. The levels of expression in these regions are comparable across multiple avian species, but when it comes to vocal control structures (only present in vocal learners), *FoxP2* expression seems to show species-specific differences, which might be related to variability in vocal plasticity *(81)*. For example, in zebra finches, the striatal nucleus known as Area X shows higher *FoxP2* levels than that found in surrounding tissue, but only at the point in development when the birds are learning to imitate song. In adult canaries, Area X expression seems to vary with the season; *FoxP2* levels are highest in the months when song shows most plasticity.

Considered together, the data from humans, rodents, and birds clearly support an ancient role for this gene in establishing and maintaining sensorimotor circuitry of the CNS in all vertebrates, regardless of vocal-learning abilities. Thus, it has been suggested that the original CNS function of *FoxP2* was not specific for vocal learning, but that it created a so-called permissive environment in the brain, on which vocal learning might evolve, conditional on other factors *(82)*. Modification and elaboration of sensorimotor-related circuitry is likely to have been a key feature in the independent evolution of complex learned behaviors in different species, including the development of song learning in subsets of avian species and the acquisition of a capacity for speech and language in humans. It should be stressed that this argument does not assume a direct equivalence for birdsong and human language, but is, nevertheless, based on striking parallels in these differing vocal communication systems.

4.3. Foxp2 *in Primates*

In contrast to the wealth of expression data already accumulated for the CNS of birds and rodents (and even human embryos) little is known about the spatiotemporal patterning of *Foxp2* during brain development of nonhuman primates. Investigations of the gene in nonspeaking primates have approached the question of species difference from another direction, that of comparative DNA sequencing *(64,65)*. These studies revealed that, despite the notable conservation in protein sequence of FOXP2 and its orthologs in multiple vertebrate species, something peculiar seems to have happened during primate evolution. Specifically, out of the three amino acid substitutions that differentiate between the rodent and human orthologous proteins, two seemed to have occurred on the lineage that led to modern Homo sapiens, after it split from the lineage leading to chimpanzees (Fig. 1) *(64,65)*. In other words, regarding the sequence of the FOXP2 protein, a chimpanzee is more similar to a mouse than to a human, even though the evolutionary distance that separates chimpanzees and mice is much larger than that between chimpanzees and humans. Examination of noncoding intronic sequences in the vicinity of the exon carrying the two human-specific amino acid substitutions suggested that these changes had been subject to positive selection on the human lineage, probably within the last 200,000 yr *(64,65)*. Although these data might point to some functional impact for one, or both, of the amino acid changes in question, with respect to recently evolved human traits (perhaps relating to speech), there is, as yet, no corroborating data from protein studies. In fact, the changes occurred in a region of FOXP2 that is little understood, outside the known domains, and, although one of them creates a potential phosphorylation site that could affect posttranslational regulation, the same change occurred independently during the evolution of carnivores *(65)*.

Because data from evolutionary anthropology suggest that human speech evolved in the past 200,000 yr *(83)*, it might be tempting to speculate that positive selection of amino acid substitutions of FOXP2 was the sole driving force for the appearance of a functioning human vocal communication system. In fact, such a scenario is highly unlikely, especially given what we now know about the conserved ancestral functions of the gene in the vertebrate CNS. Instead, it is more appropriate to view the changes in this gene as one element in a complex network of events that led to our modern speech and language capacity. The picture that is emerging is one in which *Foxp2* was already playing a role in the common ancestor of humans and rodents, probably in patterning of sensorimotor neural circuitry underpinning behaviors such as sensory-mediated learning of motor sequences. At a time when vocal learning was becoming important in human society, modifications to the gene that improved such behaviors could have rapidly spread through the population by positive selection and become fixed. This hypothesis is similar to the idea that *FoxP2* helped to provide a permissive neural environment for vocal learning in avian species *(82)*, although there are no reports demonstrating evidence for positive selection of the gene in avian evolution.

5. TALKING TO THE ANIMALS

The *FOXP2* gene has provided us with the first glimpse into neurogenetic mechanisms that contribute to our species' capacity for speech and language. It is impossible to say at this stage whether other genes related to human vocal communication will be akin to this example, but the *FOXP2* story suggests an optimistic outlook for animal studies in this area. Studies of expression patterns in nonspeaking species have already yielded critical new insights into the potential role of this gene in the brain *(66–68,80,81)*, and indicate that the available technologies for working with model organisms will be powerful for future research into the gene's function. A particularly valuable approach will be to carry out targeted knockouts of the *Foxp2* gene in mice. These studies are already underway by a number of different research groups. It would be absurd to suggest that this will create a murine model of human speech and language disorder, although there is certainly no question that the resulting mice would be unable to speak. Instead, these mice will provide molecular, developmental, anatomical, and behavioral insights into gene function that might never be revealed by studies of humans, and they will facilitate approaches for establishing in vivo targets and interaction partners of the gene in question.

What exactly might scientists be able to discover from mice with a *Foxp2* disruption? From a molecular neuroscience perspective, it will be possible to use methods such as birth dating of neurons, tracing of axonal tracts, and labeling with region-specific markers (e.g., refs. *69,70*) to discover precisely how *Foxp2* influences the development of corticostriatal and olivocerebellar circuitry in the mammalian CNS. These techniques might answer questions such as: is the gene involved in proliferation, migration, differentiation, axon growth, or cell death? The expression data suggests a postmigratory function in a number of CNS regions, perhaps in connectivity *(66–68)*. Does it have distinct roles in different regions and at different times? Does reduced *Foxp2* dosage in mice disrupt development in all of the known regions of expression, or just a subset? What happens to striosome/matrix compartmentalization *(71,72)* or ordering of olivocerebellar connections *(73,74)* when *Foxp2* dosage is altered? None of these questions can be reliably addressed by studying humans. At the anatomical level, it will be interesting to determine whether structural anomalies observed in humans with *FOXP2* disruption (such as the reduction of gray matter in the caudate nucleus; ref. *56*) might be similarly present in mutant mice, and, if they are, for the first time, it will be possible to establish the molecular and cellular bases of this finding. For behavioral studies, it will be feasible to test whether abnormalities in *Foxp2* have consequences for complex learned behaviors in nonspeaking species. We cannot hope to ever study linguistics and grammar in mice with *Foxp2* disruption, but this is something that can be very effectively studied in humans.

Of course, knockout studies have their own well-documented limitations; for example, phenotypic consequences of gene disruption may be highly dependent on modifier genes in the background, or on environmental factors. One common problem can be that genes such as *Foxp2*, which are normally expressed in multiple tissues, may lead to embryonic lethality when disrupted *(84)*. Conditional techniques in which the gene is knocked out in only a subset of tissues, or at a specific point in development *(84)*, will ameliorate this in many cases, although not always. Conditional targeting might also allow a dissection of region-specific roles of *Foxp2* in the brain, by disrupting the gene in some CNS structures, while allowing it to be normally expressed in others. The bottom line is that we cannot expect to obtain all the answers from just one approach. An understanding of the way that *FOXP2* impacts vocal communication must rely on integration of data from multiple distinct types of endeavor, but it is crucial that animal work is one of these.

In conclusion, studies of the first known example of a gene that influences speech and language development indicate that key aspects of human vocal communication involve modifications of neurogenetic mechanisms that were already present in our nonspeaking ancestors. It remains possible that other genes exist with functions that are confined to our species, but the discovery of *FOXP2* illustrates that this need not be the case for all language-related genes. Consequently, investigations of such genes in animals promise to be highly fruitful for our future understanding of the neural basis of speech and language. *FOXP2* itself is likely to influence development of cortical and subcortical circuits involved in sensorimotor functions, not just in humans, but also in all vertebrates. A comprehensive account of the neuromolecular pathways regulated by the ancestral form of this gene is a prerequisite for gaining insight into the role of modern human *FOXP2* in speech and language development.

ACKNOWLEDGMENTS

Dr Fisher is a Royal Society Research Fellow. His research is also funded by project grants from the Wellcome Trust and the Medical Research Council.

REFERENCES

1. Pinker S. The Language Instinct. London, UK: Allen Lane, 1994.
2. Chomsky N. Aspects of the Theory of Syntax. Cambridge, MA: MIT Press, 1965.
3. Chomsky N. Rules and Representations. Columbia University Press, New York, NY, 1980.
4. Saffran JR, Aslin RN, Newport EL. Statistical learning by 8-month-old infants. Science 1996;274:1926–1928.
5. Marcus GF, Vijayan S, Bandi Rao S, Vishton PM. Rule learning by seven-month-old infants. Science 1999;283:77–80.
6. Peña M, Bonatti LL, Nespor M, Mehler J. Signal-driven computations in speech processing. Science 2002;298:604–607.
7. Damasio AR. Aphasia. N Engl J Med 1992;326:531–539.
8. Dronkers NF. A new brain region for coordinating speech articulation. Nature 1996;384:159–161.
9. Poeppel D, Hickok G. Towards a new functional anatomy of language. Cognition 2004;92:1–12.
10. Wise RJ, Greene J, Buchel C, Scott SK. Brain regions involved in articulation. Lancet 1999;353:1057–1061.
11. Musso M, Moro A, Glauche V, et al. Broca's area and the language instinct. Nat Neurosci 2003;6:774–781.
12. Peña M, Maki A, Kovačić D, et al. Sounds and silence: an optical topography study of language recognition at birth. Proc Natl Acad Sci USA 2003;100:11,702–11,705.
13. Terrace HS, Petitto LA, Sanders RJ, Bever TG. Can an ape create a sentence? Science 1979;206:891–902.
14. Brockelman WY, Schilling D. Inheritance of stereotyped gibbon calls. Nature 1984;312:634–636.
15. Fitch WT, Hauser MD. Computational constraints on syntactic processing in a nonhuman primate. Science 2004;303:377–380.
16. Bishop DVM. Genetic and environmental risks for specific language impairment in children. Phil Trans Biol Sci 2001;356:369–380.
17. Fisher SE, Lai CSL, Monaco AP. Deciphering the genetic basis of speech and language disorders. Ann Rev Neurosci 2003;26:57–80.
18. Fisher SE, DeFries JC. Developmental dyslexia: genetic dissection of a complex cognitive trait. Nat Rev Neurosci 2002;3:767–780.
19. Folstein SE, Rosen-Sheidley B. Genetics of autism: complex aetiology for a heterogeneous disorder. Nat Rev Genet 2001;2:943–955.
20. Lai CSL, Fisher SE, Hurst JA, Vargha-Khadem F, Monaco AP. A novel forkhead-domain gene is mutated in a severe speech and language disorder. Nature 2001;413:519–523
21. Marcus GF, Fisher SE. FOXP2 in focus: what can genes tell us about speech and language? Trends Cog Sci 2003;7:257–262.
22. Zaki PA, Quinn JC, Price DJ. Mouse models of telencephalic development. Curr Opin Genet Dev 2003;13:423–437.
23. Kirik D, Annett LE, Burger C, Muzyczka N, Mandel RJ, Bjorklund A. Nigrostriatal alpha-synucleinopathy induced by viral vector-mediated overexpression of human alpha-synuclein: a new primate model of Parkinson's disease. Proc Natl Acad Sci USA 2003;100:2884–2889.
24. Hauser MD, Chomsky N, Fitch WT. The faculty of language: what is it, who has it, and how did it evolve? Science 2002;298:1569–1579.
25. Fitch WT. The evolution of speech: a comparative review. Trends Cogn Sci 2000;4:258–267.
26. Diehl RL, Lotto AJ, Holt LL. Speech perception. Annu Rev Psychol 2004;55:149–179.
27. Liberman AM, Whalen DH. On the relation of speech to language. Trends Cogn Sci 2000;4:187–196.
28. Lieberman PH, Klatt DH, Wilson WH. Vocal tract limitations on the vowel repertoires of rhesus monkey and other nonhuman primates. Science 1969;164:1185–1187.
29. Fitch WT, Reby D. The descended larynx is not uniquely human. Proc R Soc Lond B Biol Sci 2001;268:1669–1675.
30. Lieberman P. On the nature and evolution of the neural bases of human language. Am J Phys Anthropol 2002(Suppl);35:36–62.

31. Kay RF, Cartmill M, Balow M. The hypoglossal canal and the origin of human vocal behavior. Proc Natl Acad Sci USA 1998;95:5417–5419.
32. Eimas PD, Siqueland ER, Jusczyk P, Vigorito J. Speech perception in infants. Science 1971;171:303–306.
33. Mehler J, Jusczyk P, Lambertz G, Halsted N, Bertoncini J, Amiel-Tison C. A precursor of language acquisition iNYoung infants. Cognition 1988;29:143–178.
34. Kluender KR, Diehl RL, Killeen PR. Japanese quail can learn phonetic categories. Science 1987;237:1195–1197.
35. Ramus F, Hauser MD, Miller C, Morris D, Mehler J. Language discrimination by human newborns and by cotton-top tamarin monkeys. Science 2000;288:349–351.
36. Werker JF, Vouloumanos A. Language. Who's got rhythm? Science 2000;288, 280–281.
37. Carlsson P, Mahlapuu M. Forkhead transcription factors: key players in development and metabolism. Dev Biol 2002;250:1–23.
38. Weigel D, Jurgens G, Kuttner F, Seifert E, Jackle H. The homeotic gene fork head encodes a nuclear protein and is expressed in the terminal regions of the Drosophila embryo. Cell 1989;57:645–658.
39. Lehmann OJ, Sowden JC, Carlsson P, Jordan T, Bhattacharya SS. Fox's in development and disease. Trends Genet 2003;19:339–344.
40. Saleem RA, Banerjee-Basu S, Berry FB, Baxevanis AD, Walter MA. Structural and functional analyses of disease-causing missense mutations in the forkhead domain of FOXC1. Hum Mol Genet 2003;12:2993–3005.
41. Shu W, Yang H, Zhang L, Lu MM, Morrisey EE. Characterization of a new subfamily of winged-helix/forkhead (Fox) genes that are expressed in the lung and act as transcriptional repressors. J Biol Chem 2001;276:27,488–27,497.
42. Wang B, Lin D, Li C, Tucker P. Multiple domains define the expression and regulatory properties of Foxp1 forkhead transcriptional repressors. J Biol Chem 2003;278:24,259–24,268.
43. Li S, Weidenfeld J, Morrisey EE. Transcriptional and DNA binding activity of the Foxp1/2/4 family is modulated by heterotypic and homotypic protein interactions. Mol Cell Biol 2004;24:809–822.
44. Hurst JA, Baraitser M, Auger E, Graham F, Norell S. An extended family with a dominantly inherited speech disorder. Dev Med Child Neurol 1990;32:347–355.
45. Fisher SE, Vargha-Khadem F, Watkins KE, Monaco AP, Pembrey ME. Localisation of a gene implicated in a severe speech and language disorder. Nature Genet 1998;18:168–170.
46. MacDermot KD, Bonora E, Sykes N, et al. Identification of FOXP2 truncation as a novel cause of developmental speech deficits. Am J Hum Genet 2005;76:1074–1080.
47. Lai CSL, Fisher SE, Hurst JA, et al. The SPCH1 region on human 7q31: genomic characterization of the critical interval and localization of translocations associated with speech and language disorder. Am J Hum Genet 2000;67:357–368.
48. Sarda P, Turleau C, Cabanis MO, Jalaguier J, de Grouchy J, Bonnet H. Interstitial deletion in the long arm of chromosome 7. Ann de Genet 1988;31:258–261.
49. Liègeois F, Belton E, Lai CSL, et al. Comparable behavioural and brain abnormalities in cases with deletion or mutation of the FOXP2 gene, submitted.
50. Gopnik M. Feature-blind grammar and dysphasia. Nature 1990;344:715 [correspondence].
51. Vargha-Khadem F, Watkins K, Alcock K, Fletcher P, Passingham R. Praxic and nonverbal cognitive deficits in a large family with a genetically transmitted speech and language disorder. Proc Natl Acad Sci USA 1995;92:930–933.
52. Vargha-Khadem F, Watkins KE, Price CJ, et al. Neural basis of an inherited speech and language disorder. Proc Natl Acad Sci USA 1998;95:12,695–12,700.
53. Ullman MT, Gopnik M. Inflectional morphology in a family with inherited specific language impairment. Appl Psycholing 1999;20:51–117.
54. Alcock KJ, Passingham RE, Watkins KE, Vargha-Khadem F. Oral dyspraxia in inherited speech and language impairment and acquired dysphasia. Brain Lang 2000;75:17–33.

55. Watkins KE, Dronkers NF, Vargha-Khadem F. Behavioural analysis of an inherited speech and language disorder: comparison with acquired aphasia. Brain 2002;125:452–464.
56. Belton E, Salmond CH, Watkins KE, Vargha-Khadem F, Gadian DG. Bilateral brain abnormalities associated with dominantly inherited verbal and orofacial dyspraxia. Hum Brain Map 2003;18:194–200.
57. Liègeois F, Baldeweg T, Connelly A, Gadian DG, Mishkin M, Vargha-Khadem F. Language fMRI abnormalities associated with FOXP2 gene mutation. Nature Neurosci 2003;6:1230–1237.
58. Rutter M, Mawhood L. The Long-Term Psychosocial Sequelae of Specific Developmental Disorders of Speech and Language. In: Rutter M, Casaer P, eds. Biological Risk Factors for Psychosocial Disorders. Cambridge, UK: Cambridge University Press, 1991, pp. 233–259.
59. Tallal P, Townsend J, Curtiss S, Wulfeck B. Phenotypic profiles of language-impaired children based on genetic/family history. Brain Lang 1991;41:81–95.
60. Newbury DF, Bonora E, Lamb JA, et al., and the International Molecular Genetic Study of Autism Consortium. FOXP2 is not a major susceptibility gene for autism or Specific Language Impairment (SLI). Am J Hum Genet 2002;70:1318–1327.
61. Wassink TH, Piven J, Vieland VJ, et al. Evaluation of FOXP2 as an autism susceptibility gene. Am J Med Genet 2002;114:566–569.
62. Gauthier J, Joober R, Mottron L, et al. Mutation screening of FOXP2 in individuals diagnosed with autistic disorder. Am J Med Genet 2003;118A:172–175.
63. Kaminen N, Hannula-Jouppi K, Kestila M, et al. A genome scan for developmental dyslexia confirms linkage to chromosome 2p11 and suggests a new locus on 7q32. J Med Genet 2003;40:340–345.
64. Enard W, Przeworski M, Fisher SE, et al. Molecular evolution of FOXP2, a gene involved in speech and language. Nature 2002;418:869–872.
65. Zhang J, Webb DM, Podlaha O. Accelerated protein evolution and origins of human-specific features. Foxp2 as an example. Genetics 2002;162:1825–1835.
66. Ferland RJ, Cherry TJ, Preware PO, Morrisey EE, Walsh CA. Characterization of Foxp2 and Foxp1 mRNA and protein in the developing and mature brain. J Comp Neurol 2003;460:266–279.
67. Lai CSL, Gerrelli D, Monaco AP, Fisher SE, Copp AJ. FOXP2 expression during brain development coincides with adult sites of pathology in a severe speech and language disorder. Brain 2003;126:2455–2462.
68. Takahashi K, Liu FC, Hirokawa K, Takahashi H. Expression of Foxp2, a gene involved in speech and language, in the developing and adult striatum. J Neurosci Res 2003;73:61–72.
69. Hevner RF, Shi L, Justice N, et al. Tbr1 regulates differentiation of the preplate and layer 6. Neuron 2001;29:353–366.
70. Weimann JM, Zhang YA, Levin ME, Devine WP, Brulet P, McConnell SK Cortical neurons require Otx1 for the refinement of exuberant axonal projections to subcortical targets. Neuron 1999;24:819–831.
71. Jain M, Armstrong RJ, Barker RA, Rosser AE. Cellular and molecular aspects of striatal development. Brain Res Bull 2001;55:533–540.
72. Canales JJ, Graybiel AM. A measure of striatal function predicts motor stereotypy. Nature Neurosci 2000;3:377–383.
73. Wang VY, Zoghbi HY. Genetic regulation of cerebellar development. Nature Rev Neurosci 2001;2:484–491.
74. Chedotal A, Bloch-Gallego E, Sotelo C. The embryonic cerebellum contains topographic cues that guide developing inferior olivary axons. Development 1997;124:861–870.
75. Middleton FA, Strick PL. Basal ganglia and cerebellar loops: motor and cognitive circuits. Brain Res Brain Res Rev 2000;31:236–250.
76. Welsh JP, Lang EJ, Suglhara I, Llinas R. Dynamic organization of motor control within the olivocerebellar system. Nature 1995;374:453–457.

77. Jarvis ED, Ribeiro S, da Silva ML, Ventura D, Vielliard J, Mello CV. Behaviourally driven gene expression reveals song nuclei in hummingbird brain. Nature 2000;406:628–632.
78. Jarvis ED, Scharff C, Grossman MR, Ramos JA, Nottebohm F. For whom the bird sings: context-dependent gene expression. Neuron 1998;21:775–788.
79. Brainard MS, Doupe AJ. What songbirds teach us about learning. Nature 2002;417:351–358.
80. Teramitsu I, Kudo LC, London SE, Geschwind DH, White SA. Parallel FoxP1 and FoxP2 expression in songbird and human brain predicts functional interaction. J Neurosci 2004;24:3152–3163.
81. Haesler S, Wada K, Nshdejan A, et al. FoxP2 expression in avian vocal learners and non-learners. J Neurosci 2004;24:3164–3175.
82. Scharff C, White SA. Genetic components of vocal learning. Ann NY Acad Sci 2004;1016:325–347.
83. Boyd R, Silk JB. How humans evolved. New York: W.W. Norton, 2000.
84. Kuhn R, Schwenk F, Aguet M, Rajewsky K. Inducible gene targeting in mice. Science 1995;269:1427–1429.

8
Animal Models of Autism
Proposed Behavioral Paradigms and Biological Studies

Thomas Bourgeron, Stéphane Jamain, and Sylvie Granon

Summary

Autism is a developmental disorder characterized by impaired social interaction and communication, in addition to a restricted range of interests and activities. Although this syndrome may seem specific to humans, the very few genes known to be associated with autism are also present in other species. Therefore, studies of animals bearing mutations in these genes may give crucial information for the understanding of the biological pathways involved in the development of autism in humans. These animal models of autism may also shed light on the complex development of higher cognitive functions during evolution (i.e., language or theory of mind). Because of the multifaceted behavior and cognitive profiles present in individuals with autism, there is a need for adequate behavioral paradigms and biological studies to investigate the role of specific genes in different aspects of the autism spectrum. This chapter, focusing mainly on findings reported in humans and in mice, is divided into three parts. The first part summarizes the current findings obtained from studies of animals harboring mutations associated with autism spectrum disorders. The second part proposes behavioral tests as paradigms for diagnostic items for autism. Finally, the third part discusses how associated clinical or biological findings observed in individuals with autism could be manifested in animal models. This chapter should not be considered as a strict guideline, but rather as a framework to study the present and future animal models of autism and related disorders.

Key Words: Autism; mice; behavioral paradigms; animal models; fragile X; Rett syndrome; Tuberous sclerosis; neurofibromatosis; epilepsy; synaptogenesis.

1. INTRODUCTION

Autism is a neurodevelopmental disorder that appears in children before 3 yr of age. It is characterized by qualitative impairments in reciprocal social interaction and communication, associated with restrictive interests and repetitive behaviors [1]. It affects approx 1 in 1000 persons, and has a broad range of severity [2]. Because of the lack of biological markers, autism is diagnosed uniquely on a clinical examination of the individual behavior. A genetic etiology is strongly supported

From: *Contemporary Clinical Neuroscience: Transgenic and Knockout Models of Neuropsychiatric Disorders*
Edited by: G. S. Fisch and J. Flint © Humana Press Inc., Totowa, NJ

by family studies showing that the recurrence risk among siblings of autistic individuals is approx 45 times higher than in the general population. Moreover, twin studies have shown a markedly increased concordance rate in monozygotic twins compared with dizygotic twins *(3)*. Finally, approx 15–25% of autism cases co-occur with known genetic disorders, such as fragile X syndrome (FXS), Rett syndrome (RTT), neurofibromatosis, or tuberous sclerosis, or with certain viral diseases, such as congenital rubella and cytomegalovirus *(2)*. In the remaining cases, the etiology remains largely unknown.

The strategies to identify autism susceptibility genes include linkage studies and chromosomal rearrangements analyses followed by mutation screening in candidate genes *(4)* (Fig. 1). The next step consists of obtaining animal models to dissect the role of specific biological pathways and to investigate the interactions between the pathways and the environment in the development of autism *(5,6)*. To date, no convincing mouse model exits for idiopathic autism. In addition to the complexity of the syndrome, the absence of a model is mainly caused by two additional problems. First, until recently, no gene was reported to be clearly involved in autism *(7,8)*. Several positive associations have been identified between an allelic form of a gene and the syndrome, but without formal proof and replication in independent studies. Another difficulty comes from the possible involvement of several susceptibility genes for each individual. Indeed, the difference between concordance rates in monozygotic and dizygotic twins suggests a multi-locus model involving two or more interacting genes *(9)*. Although this polygenic model of autism is currently accepted, it may not be true for all affected individuals. In some cases, a single gene could be responsible for the cognitive deficits, and differences in the severity of the symptoms may be explained by the action of modifier genes and/or the environment *(7)*.

This chapter reviews recent behavioral and biological findings, reported using animal models for genetic diseases associated with autism. In the second part, for each item of the autism diagnosis, we propose behavioral paradigms and biological tests to perform on animal models of autism (AMA). Finally, the third part discusses the clinical findings observed in individuals with autism that should be regarded in AMA.

2. BEHAVIOR AND BIOLOGICAL FINDINGS OBSERVED IN AMA

2.1. Animal Models With Genetic Diseases Associated With Autism

In 15–25% of the cases, autism is associated with known genetic diseases, such as FXS, RTT, neurofibromatosis type 1 (NF1), or tuberous sclerosis complex (TSC) *(2)*. The reason why some individuals suffering from these syndromes also have autism, whereas others do not, remains largely unknown. Analyzing these differences may be highly relevant for the understanding of the susceptibility to autism and for identifying modifier genes involved in this syndrome. Therefore, studies of animals with targeted mutations in genes such as the fragile X gene (*Fmr1*), the methyl-CpG-binding protein 2 gene (*Mecp2*), *Nf1*, *Tsc1*, and *Tsc2* are important to understand the etiology of autism in humans.

Animal Models of Autism 153

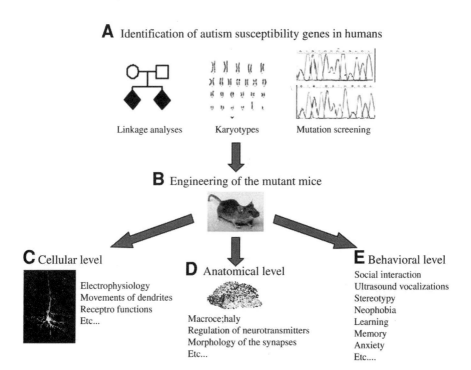

Fig. 1. Strategies used to identify autism susceptibility genes in humans and an example of functional studies in animal models. The strategies to identify susceptibility genes for autism include linkage and chromosomal rearrangement analyses followed by mutation screening of the candidate genes in affected individuals. When susceptibility genes are identified in humans, inactivation, replacement, or overexpression of the gene can be performed in mice. Such animal models will allow us to study the normal function of the gene as well as the effect of mutations in these genes at the cellular, anatomical, and behavioral levels.

2.1.1. Fragile X Syndrome

To date, the mouse model for FXS, together with *MeCP2*-mutant mice, may be considered as the best AMAs. FXS results from the absence of functional fragile X mental retardation protein (FMRP), encoded by the fragile X gene (*FMR1*). People affected with FXS are males with moderate to severe mental retardation, macroorchidism in adults, large ears, and prominent jaw *(10)*. In 25–40% of cases, affected individuals also have autistic behaviors, making this syndrome one of the most frequently associated with autism. Indeed, it has been suggested that more than 2% of autistic individuals have mutations in the *FMR1* gene *(11)*.

Mice lacking normal Fmrp show macroorchidism, learning deficits, and hyperactivity *(12,13)*. This model has the main characteristics of the human syndrome and has been a critical tool in elucidating the physiological role of FMRP and mechanisms involved in some behavioral symptoms *(14)*. Patients with FXS or autism are relatively inflexible. Mice lacking the *Fmr1* gene have mild learning deficits in tests that are dependent on spatial learning, especially in phases involving a sudden change *(14)*. Indeed, in the Morris water maze test, they have difficulties in finding the hidden platform when its position is suddenly changed during the test *(12,15,16)*. Other symptoms, similar to those of FXS and autism, have also been reported in mice deficient for *Fmr1*. The magnitude of the startle reflex of the *Fmr1* knockout mice is lower than that of control littermates and prepulse inhibition (PPI) is more efficient *(17,18)*, reminiscent of the reaction deficiency to external stimuli observed in autism. Finally, epilepsy is present in approx 30% of autistic individuals and patients with FXS *(19)*. Consistent with this, mice mutant for *Fmr1* have an increased susceptibility to epileptic seizures elicited by auditory stimuli *(18)*.

At the molecular level, FMRP interacts with at least two other proteins, fragile X 1 and 2 (FXR1 and FXR2), as well as non-translated RNAs, to regulate the translation of specific messenger RNAs (mRNAs). Using microarrays on mutant mice, 432 associated mRNAs have been identified in mouse brain *(20)*, some of which encode proteins located at synapses *(21)*. Lack of FMRP leads to abnormal synaptic connections with an excess of long, thin, and immature dendritic spines in humans and mice *(22,23)*. These results converge with our recent data concerning the involvement of neuroligins in autism *(7)*, which are likely to be involved in synapse formation *(24,25)*. Therefore, we hypothesize that FMRP may be a translational regulator of neuroligins (or other components of the same pathway), and its absence leads to autistic disorders in patients with FXS. Finally, animal models may also be used to determine whether FXS or autism could be potentially treatable disorders. Several attempts have been made to rescue the silenced murine *Fmr1* gene using a transgenic human *FMR1* gene *(26)*. These experiments restored partially a normal phenotype to mutant mice, which is encouraging for future treatment of these diseases.

2.1.2. Rett Syndrome

RTT is a neurodegenerative disorder affecting girls. After a period of normal function after birth, affected girls manifest mental retardation, autistic disorders, microcephaly, and loss of purposeful hand skills. In approx 80% of cases, RTT is caused by a heterozygous mutation in the *MECP2* gene *(27)*. At least two autistic girls without RTT symptoms have been reported with mutations in *MECP2*, suggesting that this gene may be involved in autism in girls *(28)*.

Several mouse models exist for RTT, corresponding either to absence of Mecp2 *(29,30)* or to a truncating mutation identified in patients *(31)*. These mice exhibit phenotypes similar to those observed in RTT patients. *Mecp2*-null mice are normal up to 5 wk of age, when they begin to develop disease, leading to death between 6 and 12 wk. Female *Mecp2*-null-mice are small, hypoactive, clasp their hind limbs, have breathing abnormalities, and die by 3 mo of age. As with human patients, they also have smaller brain size and a general reduction of neuronal cell size. Heterozygous mutant females for the null mutation seem normal for the first 4 mo, but begin to show symptoms such as weight gain, reduced activity, and ataxic gait at a later age. However, the best models remain those with alteration in the *Mecp2* gene similar to human mutations *(31)*. These mice have normal motor function for approx 6 wk, but then develop a progressive neurological phenotype that includes the main features of RTT, such as tremors, motor impairment, hypoactivity, increased anxiety-related behavior, seizures, and stereotypic forelimb motions *(31)*. At the molecular level, MECP2 binds to methylated sites in genomic DNA and facilitates gene silencing *(32,33)*. In humans, there is a continuous increase of MECP2 in cortical neurons throughout childhood *(34)*. This points to a dynamic regulation of this protein, suggesting a role during synaptic connection *(35)*. Indeed, the loss of learning skills during development suggests that MECP2 is not essential for promoting synapses but may be critical for maintaining or modulating them. In the near future, mouse models for RTT should allow the identification of targets of MECP2, revealing whether it regulates neuroligins or other genes associated with autism.

2.1.3. Tuberous Sclerosis Complex

TSC is a dominantly inherited disease clinically characterized by epilepsy, learning difficulties, behavioral disorders, and lesions in the brain, skin, and kidney tissues. *TSC1* (encoding hamartin) or *TSC2* (encoding tuberin), the genes frequently mutated in TSC, are tumor suppressor genes involved in the formation of benign tumors. Between 43 and 86% of TSC patients have developmental disorders similar to autism *(36)*. Moreover, 0.4–3% of autistic individuals have TSC; this increases to 14% when only epileptic patients are considered. This association suggests that mutations in *TSC1* and *TSC2*, or in a signal transduction pathway involving these genes, may trigger autistic disorders in affected individuals *(37)*.

There are several mouse models for TSC, produced by gene targeting *(38–41)*. Although used mainly in studies for carcinogenesis, mice in which the *Tsc1* gene was specifically inactivated in astrocytes demonstrate an age-dependent progression of increased astrocyte proliferation, abnormal neuronal organization in the hippocampus, seizures, and death, suggesting that these genes are important growth regulators for astrocytes *(41)*. These results may help in the description of the neuropathology of TSC as well as autism, which are hampered in humans by difficulties in obtaining cerebral tissues from affected individuals.

2.1.4. Neurofibromatosis Type 1

NF1 patients display learning disabilities (30–45% of patients), macrocephaly, and brain tumors, as well as abnormalities that can be detected by magnetic resonance imaging and that are postulated to result from abnormal myelination *(42,43)*. A 104- to 190-fold increased risk of neurofibromatosis was reported in individuals with autism *(44)*. The *NF1* gene encodes neurofibromin, a multi-domain molecule with the capacity to regulate several intracellular processes, including the extracellular signal-regulated kinase–mitogen-activated protein kinase cascade, adenylate cyclase, and cytoskeletal assembly *(45,46)*. *Nf1* mutations have also been found to affect learning and memory in drosophila and mice *(47–49)*. Interestingly, similar to the *Tsc1* and *Tsc2* mutant mice, $Nf1^{+/-}$ mice show increased numbers of astrocytes in cortex, hippocampus, and brain stem. As in humans, the learning and memory deficits of $Nf1^{+/-}$ mice are restricted to specific types of learning, such as visuospatial learning, attention, and motor coordination, whereas other forms of learning, such as classical conditioning, seem to be intact *(50,51)*. Moreover, only 40–60% of the mutant mice are affected, and remedial training can alleviate the learning deficits *(52)*. $Nf1^{+/-}$ mice also have increased γ-amino butyric acid-mediated inhibition and specific deficits in long-term potentiation. Interestingly, learning deficits of $Nf1^{+/-}$ mice can be rescued by pharmacological manipulations that decrease Ras function *(52)*. Therefore, the possibility that this biological pathway could be involved in autism should be studied in AMA and this may have important implications for the development of treatment.

2.2. AMA Based on Animal Behavior

Independent of the genetic findings, several animals have been proposed as potential AMA based only on their behavior *(53–62)*. These animal models should be considered with great care because of difficulties in comparing human and mouse pathology, especially for psychiatric disorders of unknown cause. A recent review has listed the different AMA present in the literature *(63)*. Most of these AMA derive from lesions of the brain leading to a behavior resembling autism. This includes amygdala lesions in monkeys or rats, early medial cerebellar lesion of rats, and borna virus-infected rats. Amygdala lesions have diverse consequences on socioemotional behavior, including social interactions, changes in facial and body expression, and development of stereotypic behaviors *(64,65)*. However, recent data from studies on the effects of amygdala lesions in macaque monkeys do not support a fundamental role for the amygdala in social behavior *(53)*. Damage to the amygdala does have an effect on a monkey's response to normally fear-inducing stimuli, such as snakes, and removes a natural reluctance to engage novel conspecifics in social interactions. These findings lead to the conclusion that an important role for the amygdala is in the detection of threats and mobilizing an appropriate behavioral response, part of which is fear. Interestingly, an important comorbid feature of autism is anxiety *(66)*. Therefore, if the amygdala is pathological in subjects with autism, it may contribute to their abnormal fears and increased anxiety rather than to their abnormal social behavior *(53)*. Rats infected with the borna virus present with a loss of cerebellar neurons and decreased play behavior *(67)*.

An interesting additional model concerns the oxytocin–vasopressin system *(68)*. The nonapeptides oxytocin and vasopressin are hormones/neurotransmitters involved in communication, ritual, and social behavior. Central administration of oxytocin and vasopressin reduces the isolation calls of infant rats, and induces stereotypic behaviors in mice. Consistently, mice lacking the oxytocin gene fail to develop social memory, whereas other forms of memory and learning seem intact *(68)*. To our knowledge, no mutation has been reported in the oxytocin gene, but, interestingly, levels of oxytocin in children with autism was found to be significantly lower than in controls *(69)*.

3. PROPOSED BEHAVIORAL PARADIGMS FOR AMA

3.1. General Remarks

What can be compared between mice and humans? Comparison of cellular pathways may be suitable, but when the level of anatomy of the brain or the behavior is considered, the differences between humans and other animals make accurate comparisons difficult. Consistent with this, an AMA may surely not satisfy the whole spectrum of autistic features, but may display intermediate traits or endophenotypes present in autism *(5,6)*. In rodents, the ability to orient in space is the brain function that has been evaluated most successfully. It is studied mainly because it implies the management of a wide range of cognitive processes, such as the general ability to learn a rule, the use of visual information to infer a position, and working or reference memory. These processes can be thus modeled in animals for which spatial orientation is naturally a significant behavior *(70)*. Knowledge about the organization and processing of spatial learning has been obtained mainly in rat models and, afterward, extended to other rodents. However, a strict comparison between spatial abilities in rat and mouse has scarcely been performed, and even comparison between mice strains shows major differences concerning emotional, sensory, and cognitive modalities *(71,72)*. In addition, investigation of AMA should be regarded not only at the adult stage, but also during development. This is a highly difficult task because there are not many reports on normal behaviors of young mice *(73,74)*.

Overall, these remarks underline the necessity of establishing a standard framework for behavioral experimentation in mice models of pathologies involving cognitive impairments (*see* ref. 75 for comparison between laboratories). In this context, we propose paradigms (developed for adult mice) to test AMA for almost all phenotypic aspects that have been identified in individuals with autism (social interaction, social communication, and stereotypy).

3.2. Social Symptoms

The first item of the autism diagnose in the Diagnostic and Statistical Manual of Mental Disorders, 4th Edition (DSM-IV) concerns a qualitative impairment in social interaction, as manifested by at least two of the following:

1. Marked impairment in the use of multiple nonverbal behaviors, such as eye-to-eye gaze, facial expression, body postures, and gestures to regulate social interaction.
2. Failure to develop peer relationships appropriate to developmental level.
3. A lack of spontaneous seeking to share enjoyment, interests, or achievements with other people.
4. Lack of social or emotional reciprocity.

Two paradigms, the social interaction with a new conspecific and social transmission of food cues, have been designed to test for social behavior in rodents. Together, they include the almost complete behavioral repertoire observed in adult mice.

3.2.1. Social Interaction With a New Conspecific

The social-interaction test measures the way two mice of the same sex, which have never met before, interact in a new environment. The test resident mouse is isolated for at least 4 wk, whereas the second animal, the social intruder, is reared in a social cage. The test resident mouse is allowed to explore the environment for 30 min before the introduction of the intruder. The protocol is described in Fig. 2A (for details, *see* ref. *76*). Interactions can be divided into four types, depending on which animal initiated the social contact and the escape behaviors. The frequency of each interaction type is scored, allowing an almost complete investigation of the natural repertoire and sequences of social interactions.

3.2.2. Social Transmission of Food Cues

The test of social transmission of food cues is based on ethological observations that mice, similar to other rodents, display normal neophobia toward novel food, therefore, avoiding it in a first instance (*see* Subheading 3.4.2. for specific tests of neophobic behaviors). Within a natural mice population, the recognition that a novel food is safe depends on the experience of any member of the group *(77)*. This paradigm (Fig. 2B) includes social interaction and neophobia, two important features of autism. It should be noted that performance in this test is influenced by the ability to show normal odor learning/discrimination. Therefore, this should be tested as a control measure, for example, in a maze paradigm.

3.2.3. Theory of Mind?

The term *theory of mind* (ToM) was coined by Premack and Woodruff in 1978 and is often used to refer to the ability to attribute mental states to others *(78)*. This ability seems to be deficient or delayed in the majority of individuals with autism *(79)*. The specificity of ToM to humans is still a question of debate and, to our knowledge, no study has addressed the question of ToM in rodents. Although, no behavioral paradigms are available to test for this ability in mice, there is some, although scant, data suggesting that other animals, such as dogs, parrots, or dolphins may have evolved similar abilities, such as secondary representational skill *(80)*. Therefore, these studies should deserve special attention in the future for the analysis of AMA.

3.3. Communication Symptoms

According to the DSM-IV, individuals with autism must have qualitative impairments in communication, as manifested by at least one of the following:

1. Delay in, or total lack of, the development of spoken language.
2. In individuals with adequate speech, marked impairment in the ability to initiate or sustain a conversation with others.
3. Stereotyped and repetitive use of language or idiosyncratic language.
4. Lack of varied, spontaneous make-believe play or social imitative play appropriate to developmental level.

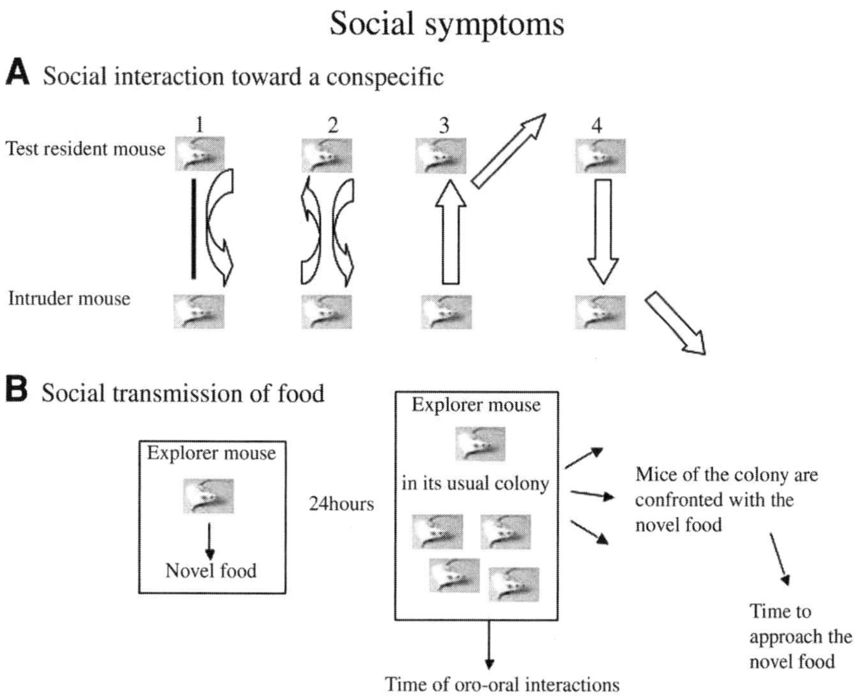

Social symptoms

A Social interaction toward a conspecific

Test resident mouse

1 2 3 4

Intruder mouse

B Social transmission of food

Explorer mouse → Novel food

24 hours

Explorer mouse in its usual colony → Mice of the colony are confronted with the novel food → Time to approach the novel food

Time of oro-oral interactions

Fig. 2. (**A**) Social symptoms. Social interaction toward a conspecific. The environment consists of a large transparent Plexiglas® box containing clean sawdust. The test lasts 4 min. The sequences of social interactions are divided into four types: (1) the resident animal initiates approach and the escape behaviors, whereas the intruder shows a neutral reaction, (2) the resident animal and the intruder both initiate approach and escape behaviors, (3) the intruder initiates the approach and the resident shows escape behavior, and (4) the resident initiates the approach and the intruder shows escape behavior. Other variables, such as sniffing time (sniffings oriented toward the intruder by the resident), nonsocial exploration (rearing against the walls), or aggressiveness (number of bites and domination attempts) can also be recorded. (**B**) Social transmission of food. A food-restricted mouse—the explorer mouse—is put in contact with a particular novel food (e.g., chocolate cereals) for 24 h. The explorer mouse is then put back in its home cage with its usual conspecifics. It will be extensively investigated—via oral sniffing—by the other members of the colony. After this investigation, the test consists of placing each mouse of the colony in contact with novel foods of different scents, among which, the food sampled by the explorer mouse. A control group should be used, with a mouse put in contact with its usual food. Novel foods should be counterbalanced between groups so that natural preference for some scents do not bias the test (*102*). As control measures, it is necessary to score the sniffing time—oral interaction—between conspecifics, because this measure will greatly influence the information transmitted by the explorer mouse to the rest of the colony.

3.3.1. Language?

What is language and how to test for the basic components of language in mice? Recently, Hauser et al. proposed a distinction between the faculty of language in the broad sense (FLB) and in the narrow sense (FLN) *(81)*. FLB includes a sensorimotor system, a conceptual–intentional system, and the computational mechanisms for recursion, providing the capacity to generate an infinite range of expressions from a finite set of elements. The authors hypothesize that FLN includes exclusively recursion and is the only uniquely human component of the faculty of language *(81)*.

Therefore, one component of the FLB present in mice could be the ability to discriminate between different sounds. Phylogenetically, hearing in vertebrates is a late development *(82)*. In all species, there is an optimal frequency range, usually the one that is most vital for communication. In humans, the optimal frequency range is the one that is most important for transmitting speech sounds. In mice, ultrasound vocalizations emitted by isolated mouse pups and pairs of adult male and female mice have communicative significance *(83)*. An analysis of the natural variability in their spectral content, median frequency, duration, and repetition period reveals acoustic structure that could be used for recognizing the calls. For example, it has been shown that wriggling-call perception in mice is comparable to unconditioned vowel discrimination and perception in prelinguistic human infants *(83)*. Interestingly, pup isolation calls develop systematically between postnatal d-5 and -12 toward a more stereotyped vocalization, contracting from a wide range of values into narrower clusters of frequency and duration, and shifting from longer to shorter repetition periods. Longer lead or lag times significantly reduce the maternal responsiveness. Similar to humans, central inhibition seems to be responsible for setting this time window.

These data suggest that ability to process, learn, and recognize sounds is present in mice and points to common rules for time-critical spectral integration, perception of acoustical objects, and auditory streaming in mice and humans *(84,85)*. More strikingly, in mice, similar to humans, the calls are also preferentially recognized by the left hemisphere. In female mice with no experience of pups that have been trained to respond to the same ultrasonic calls by conditioning, no advantage for one hemisphere is detected. The results suggest that lateralization of this function evolved early in mammals and emphasize that an innate predisposition for perceiving communication sounds is connected with a left-hemisphere advantage in processing them *(86)*.

These lateralized auditory brain functions, and the ability to discriminate different sounds and to use sounds to communicate cannot be directly considered as a language. However, it would be a significant breakthrough if specific alteration in these processes were reported in AMA, suggesting a link between these proto-language abilities of mice and autism.

3.4. Behavior Symptoms

Individuals with autism show restricted, repetitive, and stereotyped patterns of behavior, interests, and activities, as manifested by at least one of the following:

1. Encompassing preoccupation with one or more stereotyped and restricted patterns of interest that is abnormal in either intensity or focus.
2. Apparently inflexible adherence to specific, nonfunctional routines or rituals.
3. Stereotyped and repetitive motor mannerisms.
4. Persistent preoccupation with parts of objects.

To address the presence of behaviors resembling these symptoms in AMA, we propose three paradigms to test for stereotypy and for the lack of new interest associated with neophobia (Fig. 3).

3.4.1. Stereotypic Behaviors

Stereotypies are defined as repetitive actions that display fixed patterns and that serve no obvious purpose *(87)*. It has been hypothesized that stereotypies serve either as a coping function to reduce stress, or as a compensatory function to get enough sensory stimulation, suggesting disruption of gating functions *(88,89)*. In both cases, stereotypies represent an alternative, although inadequate and incomplete, way of processing information. Stereotypic behaviors (Fig. 3) can be evaluated in exploratory paradigms *(70,89,90)*. However, these paradigms require other cognitive and emotional processes that should be controlled in subsequent tests (*see* Section 3.6.). For example, a circular environment (Fig. 3A1) will promote anxiety, because rodents favor exploration of angles and naturally avoid the center of circular environments. Therefore, circular open fields also give an index of anxiety by scoring the number of entries and time spent in the center of the arena.

Alternatively, stereotypic behaviors may be measured in operant conditioning tasks (Fig. 3A2). Conditional tasks are behavioral tasks in which animals learn to associate a stimulus, generally either visual or auditory, with an operant response to get a reward, e.g., food or water. Although originally designed to target attentional processes, the five-choice serial reaction time task gives the number of perseverative responses, among several other types of responses *(91)*. For example, using this procedure, it has recently been shown that the disconnection of the prefrontal cortex and of the core region of the nucleus accumbens results in a specific increase in perseverative responding, leaving attentional processes unaffected *(92)*. Using the same type of approach, but in a different task, the same research group showed the critical role of the orbitofrontal cortex in the control of perseverative responding *(93)*.

3.4.2. Lack of New Interest: Neophobia

In rodents, the normal behavior toward unknown objects consists of a cautious rapid exploration—by nose or whiskers contact—followed by a rapid escape of the object area. With time, the nose contacts stay longer and, after a few minutes, the animal starts sniffing, rearing, circling around the object, and sometimes climbing

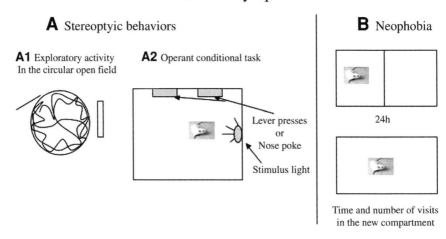

Fig. 3. (A) Behavioral symptoms. Stereotypic behaviors. (A1) Exploratory activity in the circular open field. The apparatus consists of an empty open field, either circular or square, above which is placed a system designed to track the animal's trajectory, thereby allowing the experimenter to record and observe the behavior out of the sight of the animals. Mice trajectories should be recorded for enough time to observe habituation processes, i.e., a spontaneous shift from fast navigatory moves to slow exploration, which occur between 10 and 30 min. The tracking system should allow the discrimination between both types of movements. (A2) Operant conditional task. Automated boxes allow the scoring of repetitive responses, such as lever presses or nose pokes, either correlated or not, with the reward. The number of repeated lever presses (or nose pokes) gives an index of perseverative responding. (B) Neophobia. The mouse is familiarized with one side of a two-compartment box during 24 h. After 24 h, the temporary partition is removed. Several behavioral measures, such as the numbers of visits to the novel part of the box and the time spent in the novel area, reflect an index of neophobia. With this paradigm, a decrease in the time of contact the animal makes with an item can be measured. This slow decrease in exploratory behavior, called habituation *(70)*, relies on the memory of the object's features, such as size, shape, spatial position, and so on. In this context, neophobia relies on the spontaneous avoidance of a new environment or a novel item of the environment. This item can consist of an object, novel food, or an unknown conspecific animal. Therefore, latency to the initiation of exploratory activity of a new open field, to consume a novel food, or to approach an unknown object is scored as an index of neophobia.

on it *(94)* (an example of the protocol [Fig. 3B] is described in detail in ref. *95*). This measure constitutes of an index of the interest for novelty and represents an essential control measure for most of the learning paradigms that rely on food motivation or apparatus visit. It should be noted, however, that the literature reports different levels of neophobic response in the presence of a novel environment between mice strains; differences that are hypothesized to be related to melatonin levels *(96)*. This important variation between strains reinforces the need for comparing any mutant animal with its wild-type littermates *(97)*.

3.5. Learning and Memory

Impairment, or at the opposite extreme, exceptional skills, in specific learning and memory do not characterize autism *per se*. However, mental retardation is present in 70% of individuals with autism, and higher memory or skills are sometime associated with Asperger's syndrome. As mentioned above, most of the behavioral paradigms to test learning and memory are based on spatial learning. The two major paradigms (Fig. 4A,B) that are used are the radial-arm maze and the Morris water maze *(98,99)*. Both tasks involve spatial orientation and spatial memory. The identification of impaired or improved learning performance, therefore, requires a rigorous control of sensory abilities and of strategies. This is particularly important for memory testing in one of the most commonly used paradigms, the radial-arm maze, because, in mice, this maze paradigm is known to favor the development of motor biases and strategies (for description of the protocol, *see* ref. *100*).

3.5.1. Spatial Learning

We recently designed a test (Fig. 4C) to investigate spatial learning ability independently of aversive motivation *(99)*. The test is, in several points, comparable to the Morris water maze, except that the Morris water maze is based on the inference that emotional processes, such as anxiety, do not prevent animals from learning the task. Indeed, some mice strains may be very sensitive to anxiogenic environments *(101)*, and it seems important to offer an alternative to the Morris water maze in spatial learning. Details of the paradigm are presented in Fig. 4C.

3.5.2. Learning of Odor Discrimination

Odor discrimination is an easy way of evaluating the memory span *(102)*. Regarding object discrimination, animals can memorize a list of olfactory items (Fig. 4D) and this test can be used as a control measure for social interaction and neophobia.

3.6. Complementary and Control Measures

Additional paradigms must be used to delimit, as much as possible, the specific behavior of the AMA. Deficits in areas such as visual discrimination and attention, sensory gating, exploratory behavior, and anxiety may be observed in some individuals with autism, and the question of whether these behaviors are present in AMA is, therefore, important to also address.

3.6.1. Anxiety

Anxiety may be a particular comorbid symptom of autism *(66)*. It may affect the execution, and, therefore, the interpretation, of any of the behavioral tests described above. Therefore, it is essential to obtain a measure of the anxiety levels in the

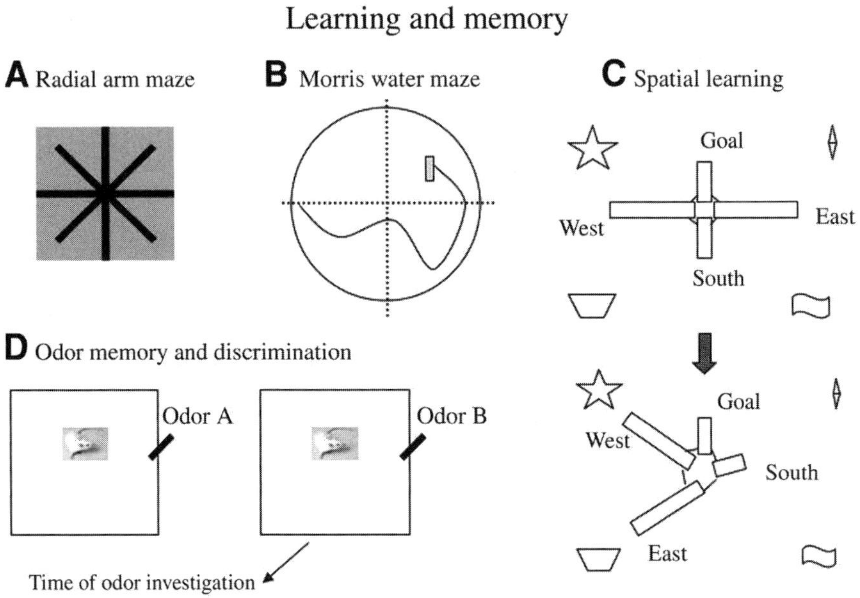

Learning and memory

A Radial arm maze   **B** Morris water maze   **C** Spatial learning

D Odor memory and discrimination

Fig. 4. Learning and memory. (**A**) The radial-arm maze. The radial-arm maze consists of a Plexiglas maze made of eight arms connecting a central platform. The aim of the task is to collect the food at the end of each arm without visiting the same arm twice. Therefore, within a single trial, this test requires working memory. A version of this task consists of bating only four arms of the eight, therefore, also involving long-term memory when tested over days. (**B**) The Morris water maze. In this task, the animal has to reach, from various starting points, a submerged (therefore invisible) platform in painted water. (**C**) Spatial location learning. Mice have to learn a food location. The apparatus consists of a transparent Plexiglas maze equipped with four dissociable arms creating a cross. A cup containing the animal's usual food is presented at the end of the north arm of the maze, whereas a similar cup with unreachable food is placed at the end of the south arm to prevent the animals from being guided by the odor of the food. Each animal receives three trials a day. The starting arm is either the west or east arm and is randomly chosen. The variables collected per day are: number of entries into the arms, time to reach the food, and rearings. After stable learning, animals are submitted to a procedure that challenges the way they manage the task. During that phase, the task consists of reaching the goal, in its usual position, from the usual arms located in modified places. (**D**) Odor memory and discrimination. This can be evaluated in the animal's own cage. Mice are presented with different odors dissolved in mineral oil through a Pasteur pipette introduced on the side of the animal's cage. During the familiarization phase (3 d), mice are exposed to odor A. For the test, mice are presented twice with odor B during 5 min, with a variable interval between both presentations (from 30 to 480 min). For presentation one and two, the odor can be the same or different. If the odor is different for both presentations, a normal animal will spontaneously explore the novel odor more than the familiar one, therefore giving an index of odor recognition. If the odor is identical for both presentations, a significant decrease in investigation time during the second presentation (habituation) gives a delay-dependent index of memory for odor.

animals. The plus-maze test (described in Fig. 5A) has long been used to test anxiety processes and pharmacotherapy in rodents (for review, see refs. *101* and *103*). Furthermore, as in many anxiety tests, there is a clear interaction between the habituation of the animal to the experimenter and the score on the behavioral test. Therefore, anxiety tests should be conducted first when planning a series of behavioral experiments with the same animals *(103)*.

3.6.2. Visual Discrimination and Visual Attention

A protocol to evaluate the ability of mice to discriminate between visual patterns and to pay attention to visual stimuli is particularly important, because most behavioral protocols require the use of visual information. Visual discrimination can be tested simply by requiring the mice to identify visual patterns (Fig. 5B). In addition, visual attention can be tested by the paradigm created by Durkin and colleagues in 2000, which requires the mice to choose, among five arms of a maze, the single lit and baited arm *(104)*. This simple procedure requires the animal to be able to visually discriminate places with rewards from those without rewards.

3.6.3. Sensory Gating

PPI, the reduction in startle produced by a prepulse stimulus, is diminished in patients with schizophrenia *(105)*. This abnormal PPI in schizophrenia has been related to the loss of sensorimotor gating that may lead to sensory flooding and cognitive fragmentation. In autism, to our knowledge, no difference has been found between affected children and the control children regarding absolute positive-polarity response at 50 msec (P50) amplitudes and P50 suppression *(106)*. Nevertheless, even if the relevance to the syndrome is still unclear, PPI is the classic paradigm designed for measuring the ability to integrate external stimuli, and, therefore, it should be tested in AMA (Fig. 5C).

3.6.4. Exploratory Behavior

Although exploratory behavior in the open field provides measures of locomotor activity, it has been criticized for confounding exploration, locomotion, and anxiety *(107)*. Therefore, it may be desirable to evaluate exploratory and locomotor activity in separate paradigms. The first control could consist of using the same open field, but changing the lighting conditions, therefore, reducing the anxiety level of the mice. These experimental conditions require the use of a digital camera specific to infrared light. Alternatively, another paradigm, the hole-board test (Fig. 5D), was introduced and thereafter modified to investigate novelty exploration and recognition *(101)*. This paradigm tests a spontaneous behavior, because rodents naturally tend to approach holes. Therefore, it is regarded as a less anxiogenic test than the open-field test.

4. PROPOSED BIOLOGICAL STUDIES FOR AMA

In addition to the behavioral phenotypes, AMA will also be highly important in the identification of the molecular factors and biological pathways involved in the development of autism. To target this research, we emphasize several clinical and

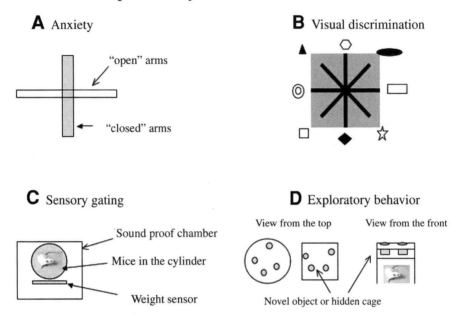

Fig. 5. Complementary and control measures. (**A**) Anxiety. The elevated plus maze is in the form of a cross with two 70-cm arms with no walls opposite to each other and two closed arms, of the same dimensions, with walls facing each other. The whole apparatus is elevated 70 cm above the floor. Therefore, the apparatus allows the evaluation of novelty behavior, exploratory of open vs closed arms, i.e., the anxiety for openness, and the phobia for elevation. The ratio of the time spent in the open arms over the time spent in the closed arms, as well as the number of entries into the open arms, give an index of anxiety level within 10 min. The number of entries into closed arms gives an index of locomotor/exploratory activity in a stressful environment. Some authors suggest conducting this test in two separate 5-min trials, the first trial testing mainly anxiety and the second trial testing phobia for elevation. (**B**) Visual discrimination. Mice have to discriminate four visual patterns of eight, randomly associated with rewards in a radial-arm maze (*see* ref. *130* for detailed protocol). As in the classic radial-maze protocol, mice learn to avoid visiting the arms that are identified as not containing rewards, therefore, reducing the time to consume the total amount of food in the maze. (**C**) Sensory gating. The startle reflex and prepulse inhibition (PPI) are measured in a soundproof chamber. Mice are first left undisturbed for 5 min in a small cylinder placed on a platform sensitive to fine weight differences. Then, the startle stimulus (from 110 to 120 dB) is presented several times in a pseudo-random manner. The prepulse sound (74 or 78 dB) is presented 100 ms before the startle stimulus. Several combinations of startle and prepulse stimulus are presented (no pulse/startle 110 dB; no pulse/120 dB startle; 74 dB PPI /110 dB startle; 78 dB PPI/110 dB startle; 74 dB PPI/120 dB startle; 78 dB PPI/120 dB startle). (**D**) Exploratory behavior. The apparatus consists of an elevated circular or square floor containing four holes. Under each hole, the experimenter can introduce an object, either novel or familiar. The time of head-dipping, normally longer for initial exposure,

biological features observed in individuals with autism that should be regarded in AMA.

4.1. Epilepsy

As reviewed recently, there is an increased, but variable, risk of epilepsy in autism *(19)*. Three main factors (age, cognitive level, and type of language disorder) account for the variability in the reported prevalence of epilepsy. The prevalence is highest in individuals with moderate to severe mental retardation and in individuals with severe receptive language deficits. The association of autism with clinical or subclinical epilepsy might denote common genetic factors in some cases. Therefore, the presence of, or greater sensitivity to, epileptic seizures should be regarded in AMA. The epilepsies are characterized by spontaneous recurrent epileptic seizures, which are caused by generalized paroxysmal or partial (focal) discharges in the brain. Generalized seizures may be induced by intermittent light stimulation, noise, movement, and stress *(108)*. Models for simple partial seizures include focal microapplication of topical convulsants, such as kainate on the cerebral cortex. Chronic seizure models develop after the application of metals such as alumina hydroxide, cobalt, tungsten, or zinc. Spontaneous recurrent seizures may appear 2 mo after the injection, and persist for several years in some cases. Finally, animal models of *status epilepticus* may be produced via techniques such as administration of chemical convulsant or electrical stimulation.

4.2. Abnormal Brain Anatomy and Imaging

Macrocephaly of postnatal onset was observed in 24% of 100 patients with autism *(109)* and is a common feature in autism *(110–113)*. Furthermore, family history of macrocephaly occurred in 62% of macrocephaly–autism cases *(114)*. Structural brain-imaging studies performed in autistic patients have reported abnormalities such as increased total brain volume and cerebellar abnormalities *(115–117)*. However, because of the heterogeneity of the syndrome, these differences were not observed in all individuals with autism. Functional brain imaging studies conducted with high-functioning adults with autism or Asperger's syndrome highlight the patterns of decreased activation in ventromedial prefrontal cortex, temporoparietal junction, amygdala, and periamygdaloid cortex, along with aberrantly increased activation in primary sensory cortices *(118)*. Recently, with improved technology, bilateral hypoperfusion of the temporal lobes has been reported in children with autism *(119,120)*. In addition, using perceptive and cognitive paradigms, activation studies have shown an abnormal pattern of cortical acti-

---

Fig. 5. (*continued from opposite page*) reflects the interest for novelty, because a normal animal would exhibit a longer exploration of a novel object than of a familiar object. This measure also reflects the memory for objects if the test is repeated. In the spatial learning version, a small cage with clean sawdust is hidden in one of holes, so that the animals learn to locate this nonfood goal.

vation in autistic patients *(121)*. These results suggest that different connections between particular cortical regions could exist in autism. In addition to classic brain imaging methods that have been successfully used in mice *(122)*, recent techniques, such as diffusion tensor imaging have been developed to measure the efficiency of axonal connectivity *(123)*. Although yet not used in mice, we can expect these methods to be available in rodents soon.

4.3. Synaptogenesis and Neuronal Networks

Reduced dendritic arborizations were observed in hippocampal pyramidal cells of autistic individuals *(124)*. Indeed, one emerging picture coming from the genetic studies of autism is the key role of synaptogenesis and synaptic plasticity in the development of this syndrome *(7,125,126)*. We have recently identified mutations in genes encoding the neuroligins genes, *NLGN3* and *NLGN4*, in two independent families with autism *(7)*. This has been confirmed in an independent study that reports a similar *NLGN4* mutation in a large, five-generation family with mental retardation and/or autism *(127)*. These mutations lead to abnormal processing of the neuroligin proteins *(128,129)*, which seems to be sufficient to trigger cognitive deficits in affected children. These results, in addition to the structural findings in RTT and FXS, should encourage researchers to characterize the morphology of the synapses and the electrophysiology of the neurons in AMA.

5. CONCLUSION

To date, very few genes associated with autism have been identified. Nevertheless, chromosomal regions (2q, 3q, 6q, 7q, 13q, 15q, 17q, and X) and genes (*Reelin*, *WNT2*, *GRIK2*, *LAMB1*, *Rho-GEF*, and so on) are investigated by molecular biologists, and, obviously, more susceptibility genes for autism will be identified soon. After the discovery of these genes, studies to reveal their specific roles in the development of the syndrome may be more difficult to achieve; especially because the consequences of specific human variations on behavior may be impossible to evaluate in mice. At the cellular level, mice mutant for *Fmr1*, *Mecp2*, *Tsc1*, *Tsc2*, and *Nf1* already indicate that chromatin structure and synaptogenesis play major roles in genetic diseases associated with autism. Crucial information on the molecular pathways involved in autism will also come from the genetics and functional studies in other animal models, such as zebrafish, *Drosophila melanogaster*, and *Caenorhabditis elegans (54)*. At the behavioral level, this task will evidently be more difficult. However, in the near future, the possibility of studying the behaviors of AMA will certainly lead to a better characterization of the relation between the genome and the environment in the course of normal or pathological development of the brain. Finally, results in this field may also shed light on the unexplained evolution of cognitive functions in different species and will offer a unique opportunity to discover and to test new knowledge-based therapies for autism.

ACKNOWLEDGMENTS

We thank Patricia Baldacci, Bernard Lakowski, and Jonas Melke for helpful comments on the manuscript. This work was funded by The Pasteur Institute, The University Paris 7, France Telecom foundation, Fondation de France, Cure Autism Now, NRJ foundation, and by the European Molecular Biology Organization (EMBO) with a postdoctoral fellowship (S. Jamain).

REFERENCES

1. American Psychiatric Association. Diagnostic and Statistical Manual of Mental Disorders, 4th Ed. Washington DC: American Psychiatric Press, 1994.
2. Folstein SE, Rosen-Sheidley B. Genetics of autism: complex aetiology for a heterogeneous disorder. Nat Rev Genet 2001;2:943–955.
3. Bailey A, Le Couleur A, Grottesman I, et al. Autism as a strongly genetic disorder: evidence from a British twin study. Psychological Medicine 1995;25:63–77.
4. Bourgeron T, Giros B. Genetic markers in psychiatric genetics. Methods Mol Med 2003;77:63–98.
5. Belmonte MK, Cook EH, Anderson GM, et al. Autism as a disorder of neural information processing: directions for research and targets for therapy. Mol Psychiatry, 2004.
6. Seong E, Seasholtz AF, Burmeister M. Mouse models for psychiatric disorders. Trends Genet 2002;18:643–650.
7. Jamain S, Quach H, Betancur C, et al. Mutations of the X-linked genes encoding neuroligins NLGN3 and NLGN4 are associated with autism. Nat Genet 2003;34:27–29.
8. Veenstra-VanderWeele J, Cook EH. Molecular genetics of autism spectrum disorder. Mol Psychiatry, 2004.
9. Pickles A, Bolton P, Macdonald H, et al. Latent-class analysis of recurrence risks for complex phenotypes with selection and measurement error: a twin and family history study of autism. Am J Hum Genet 1995;57:717–726.
10. Bardoni B, Mandel JL. Advances in understanding of fragile X pathogenesis and FMRP function, and in identification of X linked mental retardation genes. Curr Opin Genet Dev 2002;12:284–293.
11. Wassink TH, Piven J, Vieland VJ, et al. Evidence supporting WNT2 as an autism susceptibility gene. Am J Med Genet 2001;105:406–413.
12. The Dutch-Belgian Fragile X Consortium. Fmr1 knockout mice: a model to study fragile X mental retardation. Cell 1994;78:23–33.
13. Bakker CE, Oostra BA. Understanding fragile X syndrome: insights from animal models. Cytogenet Genome Res 2003;100:111–123.
14. Kooy RF. Of mice and the fragile X syndrome. Trends Genet 2003;19:148–154.
15. D'Hooge R, Nagels G, Franck F, et al. Mildly impaired water maze performance in male Fmr1 knockout mice. Neuroscience 1997;76:367–376.
16. Dobkin C, Rabe A, Dumas R, et al. Fmr1 knockout mouse has a distinctive strain-specific learning impairment. Neuroscience 2000;100:423–429.
17. Nielsen DM, Derber WJ, McClellan DA, Crnic LS. Alterations in the auditory startle response in Fmr1 targeted mutant mouse models of fragile X syndrome. Brain Res 2002;927:8–17.
18. Chen L, Toth M. Fragile X mice develop sensory hyperreactivity to auditory stimuli. Neuroscience 2001;103:1043–1050.
19. Tuchman R, Rapin I. Epilepsy in autism. Lancet Neurol 2002;1:352–358.
20. Brown V, Jin P, Ceman S, et al. Microarray identification of FMRP-associated brain mRNAs and altered mRNA translational profiles in fragile X syndrome. Cell 2001;107:477–487.
21. Zalfa F, Giorgi M, Primerano B, et al. The fragile X syndrome protein FMRP associates with BC1 RNA and regulates the translation of specific mRNAs at synapses. Cell 2003;112:317–327.
22. Irwin SA, Patel B, Idupulapati M, et al. Abnormal dendritic spine characteristics in the temporal and visual cortices of patients with fragile-X syndrome: a quantitative examination. Am J Med Genet 2001;98:161–167.

23. Nimchinsky EA, Oberlander AM, Svoboda K. Abnormal development of dendritic spines in FMR1 knock-out mice. J Neurosci 2001;21:5139–5146.
24. Scheiffele P, Fan JH, Choih J, Fetter R, Serafini T. Neuroligin expressed in nonneuronal cells triggers presynaptic development in contacting axons. Cell 2000;101:657–669.
25. Dean C, Scholl FG, Choih J, et al. Neurexin mediates the assembly of presynaptic terminals. Nat Neurosci 2003;6:708–716.
26. Peier AM, McIlwain KL, Kenneson A, et al. (Over)correction of FMR1 deficiency with YAC transgenics: behavioral and physical features. Hum Mol Genet 2000;9:1145–1159.
27. Amir RE, Van den Veyver IB, Wan M, et al. Rett syndrome is caused by mutations in X-linked MECP2, encoding methyl-CpG-binding protein 2. Nat Genet 1999;23:185–188.
28. Ashley-Koch AE, Carrey RJ, Wolpert CM, et al. Screening for MECP2 mutations in females with autistic disorder. In: IXth World Congress of Psychiatric Genetics. St. Louis, 2001.
29. Chen RZ, Akbarian S, Tudor M, Jaenisch R. Deficiency of methyl-CpG binding protein-2 in CNS neurons results in a Rett-like phenotype in mice. Nat Genet 2001;27:327–331.
30. Guy J, Hendrich B, Holmes M, Martin JE, Bird A. A mouse Mecp2-null mutation causes neurological symptoms that mimic Rett syndrome. Nat Genet 2001;27:322–326.
31. Shahbazian M, Young J, Yuva-Paylor L, et al. Mice with truncated MeCP2 recapitulate many Rett syndrome features and display hyperacetylation of histone H3. Neuron 2002;35:243–254.
32. Nan X, Campoy FJ, Bird A. MeCP2 is a transcriptional repressor with abundant binding sites in genomic chromatin. Cell 1997;88:471–481.
33. Nan X, Ng HH, Johnson CA, et al. Transcriptional repression by the methyl-CpG-binding protein MeCP2 involves a histone deacetylase complex. Nature 1998;393:386–389.
34. Shahbazian MD, Antalffy B, Armstrong DL, Zoghbi HY. Insight into Rett syndrome: MeCP2 levels display tissue- and cell-specific differences and correlate with neuronal maturation. Hum Mol Genet 2002;11:115–124.
35. Zoghbi HY. Postnatal neurodevelopmental disorders: meeting at the synapse? Science 2003;302:826–830.
36. Smalley SL. Autism and tuberous sclerosis. J Autism Dev Disord 1998;28:407–414.
37. Serajee FJ, Nabi R, Zhong H, Mahbubul Huq AH. Association of INPP1, PIK3CG, and TSC2 gene variants with autistic disorder: implications for phosphatidylinositol signalling in autism. J Med Genet 2003;40:e119.
38. Onda H, Lueck A, Marks PW, Warren HB, Kwiatkowski DJ. Tsc2(+/-) mice develop tumors in multiple sites that express gelsolin and are influenced by genetic background. J Clin Invest 1999;104:687–695.
39. Kobayashi T, Minowa O, Kuno J, et al. Renal carcinogenesis, hepatic hemangiomatosis, and embryonic lethality caused by a germ-line Tsc2 mutation in mice. Cancer Res 1999;59:1 206–1211.
40. Kobayashi T, Minowa O, Sugitani Y, et al. A germ-line Tsc1 mutation causes tumor development and embryonic lethality that are similar, but not identical to, those caused by Tsc2 mutation in mice. Proc Natl Acad Sci USA 2001;98:8762–8767.
41. Uhlmann EJ, Wong M, Baldwin RL, et al. Astrocyte-specific TSC1 conditional knockout mice exhibit abnormal neuronal organization and seizures. Ann Neurol 2002;52:285–296.
42. Rosser TL, Packer RJ. Neurocognitive dysfunction in children with neurofibromatosis type 1. Curr Neurol Neurosci Rep 2003;3:129–136.
43. Arun D, Gutmann DH. Recent advances in neurofibromatosis type 1. Curr Opin Neurol 2004;17:101–105.
44. Gillberg C, Forsell C. Childhood psychosis and neurofibromatosis—more than a coincidence? J Autism Dev Disord 1984;14:1–8.
45. Dasgupta B, Gutmann DH. Neurofibromatosis 1: closing the GAP between mice and men. Curr Opin Genet Dev 2003;13:20–27.
46. Costa RM, Silva AJ. Mouse models of neurofibromatosis type I: bridging the GAP. Trends Mol Med 2003;9:19–23.
47. Jacks T, Shih TS, Schmitt EM, et al. Tumour predisposition in mice heterozygous for a targeted mutation in Nf1. Nat Genet 1994;7:353–361.

48. Guo HF, Tong J, Hannan F, Luo L, Zhong Y. A neurofibromatosis-1-regulated pathway is required for learning in Drosophila. Nature 2000;403:895–898.
49. Zhu Y, Romero MI, Ghosh P, et al. Ablation of NF1 function in neurons induces abnormal development of cerebral cortex and reactive gliosis in the brain. Genes Dev 2001;15:859–876.
50. Costa RM, Yang T, Huynh DP, et al. Learning deficits, but normal development and tumor predisposition, in mice lacking exon 23a of Nf1. Nat Genet 2001;27:399–405.
51. Silva AJ, Frankland PW, Marowitz Z, et al. A mouse model for the learning and memory deficits associated with neurofibromatosis type I. Nat Genet 1997;15:281–284.
52. Costa RM, Federov NB, Kogan JH, et al. Mechanism for the learning deficits in a mouse model of neurofibromatosis type 1. Nature 2002;415:526–530.
53. Amaral DG, Bauman MD, Schumann CM. The amygdala and autism: implications from non-human primate studies. Genes Brain Behav 2003;2:295–302.
54. Tropepe V, Sive HL. Can zebrafish be used as a model to study the neurodevelopmental causes of autism? Genes Brain Behav 2003;2:268–281.
55. Garner JP, Meehan CL, Mench JA. Stereotypies in caged parrots, schizophrenia and autism: evidence for a common mechanism. Behav Brain Res 2003;145:125–134.
56. Teitelbaum P. A proposed primate animal model of autism. Eur Child Adolesc Psychiatry 2003;12:48–49.
57. Machado CJ, Bachevalier J. Non-human primate models of childhood psychopathology: the promise and the limitations. J Child Psychol Psychiatry 2003;44:64–87.
58. Young LJ, Pitkow LJ, Ferguson JN. Neuropeptides and social behavior: animal models relevant to autism. Mol Psychiatry 2002;7(Suppl 2):S38–39.
59. Pletnikov MV, Moran TH, Carbone KM. Borna disease virus infection of the neonatal rat: developmental brain injury model of autism spectrum disorders. Front Biosci 2002;7:d593–607.
60. Pletnikov MV, Jones ML, Rubin SA, Moran TH, Carbone KM. Rat model of autism spectrum disorders. Genetic background effects on Borna disease virus-induced developmental brain damage. Ann NY Acad Sci 2001;939:318–319.
61. Young LJ. Oxytocin and vasopressin as candidate genes for psychiatric disorders: lessons from animal models. Am J Med Genet 2001;105:53–54.
62. Insel TR. Mouse models for autism: report from a meeting. Mamm Genome 2001;12:755–757.
63. Andres C, Beeri R, Friedman A, et al. Acetylcholinesterase-transgenic mice display embryonic modulations in spinal cord choline acetyltransferase and neurexin i-beta gene expression followed by late-onset neuromotor deterioration. Proc Natl Acad Sci USA 1997;94:8173–8178.
64. Bachevalier J. Brief report: medial temporal lobe and autism: a putative animal model in primates. J Autism Dev Disord 1996;26:217–220.
65. Wolterink G, Daenen LE, Dubbeldam S, et al. Early amygdala damage in the rat as a model for neurodevelopmental psychopathological disorders. Eur Neuropsychopharmacol 2001;11:51–59.
66. Muris P, Steerneman P, Merckelbach H, Holdrinet I, Meesters C. Comorbid anxiety symptoms in children with pervasive developmental disorders. J Anxiety Disord 1998;12:387–393.
67. Pletnikov MV, Rubin SA, Vasudevan K, Moran TH, Carbone KM. Developmental brain injury associated with abnormal play behavior in neonatally Borna disease virus-infected Lewis rats: a model of autism. Behav Brain Res 1999;100:43–50.
68. Insel TR, Young LJ. The neurobiology of attachment. Nat Rev Neurosci 2001;2:129–136.
69. Modahl C, Green L, Fein D, et al. Plasma oxytocin levels in autistic children. Biol Psychiatry 1998;43:270–277.
70. Thinus-Blanc C. Animal Spatial Cognition. Behavioral & Neural Approaches. Singapore, New Jersey, London, Hong Kong: World Scientific Publishing, 1996.
71. Whishaw IQ. A comparison of rats and mice in a swimming pool place task and matching to place task: some surprising differences. Physiol Behav 1995;58:687–693.
72. Wolff M, Savova M, Malleret G, Segu L, Buhot MC. Differential learning abilities of 129T2/Sv and C57BL/6J mice as assessed in three water maze protocols. Behav Brain Res 2002;136:463–474.
73. Laviola G, Terranova ML. The developmental psychobiology of behavioural plasticity in mice: the role of social experiences in the family unit. Neurosci Biobehav Rev 1998;23:197–213.

74. Laviola G, Macri S, Morley-Fletcher S, Adriani W. Risk-taking behavior in adolescent mice: psychobiological determinants and early epigenetic influence. Neurosci Biobehav Rev 2003;27:19–31.
75. Crabbe JC, Wahlsten D, Dudek BC. Genetics of mouse behavior: interactions with laboratory environment. Science 1999;284:1670–1672.
76. Granon S, Faure P, Changeux JP. Executive and social behaviors under nicotinic receptor regulation. Proc Natl Acad Sci USA 2003;100:9596–9601.
77. Valsecchi P, Bartolomucci A, Aversano M, Visalberghi E. Learning to cope with two different food distributions: the performance of house mice (Mus musculus domesticus). J Comp Psychol 2000;114:272–280.
78. Premak J, Woodruff G. Does the chimpanzee have a theory of mind? Behav Brain Sci 1978;4:515–526.
79. Baron-Cohen S, Leslie AM, Frith U. Does the autistic child have a "theory of mind"? Cognition 1985;21:37–46.
80. Suddendorf T, Whiten A. Mental evolution and development: evidence for secondary representation in children, great ages, and other animals. Psychol Bull 2001;127:629–650.
81. Hauser MD, Chomsky N, Fitch WT. The faculty of language: what is it, who has it, and how did it evolve? Science 2002;298:1569–1579.
82. Popper AN, Fay RR. Evolution of the ear and hearing: issues and questions. Brain Behav Evol 1997;50:213–221.
83. Liu RC, Miller KD, Merzenich MM, Schreiner CE. Acoustic variability and distinguishability among mouse ultrasound vocalizations. J Acoust Soc Am 2003;114:3412–3422.
84. Ehret G, Riecke S. Mice and humans perceive multiharmonic communication sounds in the same way. Proc Natl Acad Sci USA 2002;99:479–482.
85. Geissler DB, Ehret G. Time-critical integration of formants for perception of communication calls in mice. Proc Natl Acad Sci USA 2002;99:9021–9025.
86. Ehret G. Left hemisphere advantage in the mouse brain for recognizing ultrasonic communication calls. Nature 1987;325:249–251.
87. Danzer R. Stress, stereotypies and welfare. Behavioral Processes 1991;25:95–102.
88. Valenstein ES. Stereotyped Behaviour and Stress. New York: Plenum, 1976, pp. 116–124.
89. Robbins TW, Koob GF. Selective disruption of displacement behaviour by lesions of the mesolimbic dopamine system. Nature 1980;285:409–412.
90. Harkin A, Kelly JP, Frawley J, O'Donnell JM, Leonard BE. Test conditions influence the response to a drug challenge in rodents. Pharmacol Biochem Behav 2000;65:389–398.
91. Humby T, Laird FM, Davies W, Wilkinson LS. Visuospatial attentional functioning in mice: interactions between cholinergic manipulations and genotype. Eur J Neurosci 1999;11:2813–2823.
92. Christakou A, Robbins TW, Everitt BJ. Prefrontal cortical-ventral striatal interactions involved in affective modulation of attentional performance: implications for corticostriatal circuit function. J Neurosci 2004;24:773–780.
93. Chudasama Y, Robbins TW. Dissociable contributions of the orbitofrontal and infralimbic cortex to pavlovian autoshaping and discrimination reversal learning: further evidence for the functional heterogeneity of the rodent frontal cortex. J Neurosci 2003;23:8771–8780.
94. Granon S, Save E, Buhot MC, Poucet B. Effortful information processing in a spontaneous spatial situation by rats with medial prefrontal lesions. Behav Brain Res 1996;78:147–154.
95. Agmo A, Belzung C. The role of subtypes of the opioid receptor in the anxiolytic action of chlordiazepoxide. Neuropharmacology 1998;37:223–232.
96. Kopp C, Vogel E, Rettori MC, et al. Effects of melatonin on neophobic responses in different strains of mice. Pharmacol Biochem Behav 1999;63:521–526.
97. Madani R, Kozlov S, Akhmedov A, et al. Impaired explorative behavior and neophobia in genetically modified mice lacking or overexpressing the extracellular serine protease inhibitor neuroserpin. Mol Cell Neurosci 2003;23:473–494.

98. Morris RGM. Spatial localization does not require the presence of local cues. Learn Motiv 1981;12:239–260.
99. Cressant A, Granon S. Definition of a new maze paradigm for the study of spatial behavior in rats. Brain Res Brain Res Protoc 2003;12:116–124.
100. Roullet P, Lassalle JM. Radial maze learning using exclusively distant visual cues reveals learners and nonlearners among inbred mouse strains. Physiol Behav 1995;58:1189–1195.
101. File SE. Factors controlling measures of anxiety and responses to novelty in the mouse. Behav Brain Res 2001;125:151–157.
102. Rochefort C, Gheusi G, Vincent JD, Lledo PM. Enriched odor exposure increases the number of newborn neurons in the adult olfactory bulb and improves odor memory. J Neurosci 2002;22:2679–2689.
103. Hogg S. A review of the validity and variability of the elevated plus-maze as an animal model of anxiety. Pharmacol Biochem Behav 1996;54:21–30.
104. Durkin TP, Beaufort C, Leblond L, Maviel T. A 5-arm maze enables parallel measures of sustained visuo-spatial attention and spatial working memory in mice. Behav Brain Res 2000;116:39–53.
105. Geyer MA, McIlwain KL, Paylor R. Mouse genetic models for prepulse inhibition: an early review. Mol Psychiatry 2002;7:1039–1053.
106. Kemner C, Oranje B, Verbaten MN, van Engeland H. Normal P50 gating in children with autism. J Clin Psychiatry 2002;63:214–217.
107. Archer J. Tests for emotionality in rats and mice: a review. Anim Behav 1973;21:205–235.
108. Sarkisian MR. Overview of the current animal models for human seizure and epileptic disorders. Epilepsy Behav 2001;2:201–216.
109. Stevenson RE, Schroer RJ, Skinner C, Fender D, Simensen RJ. Autism and macrocephaly. Lancet 1997;349:1744–1745.
110. Bigler ED, Tate DF, Neely ES, et al. Temporal lobe, autism, and macrocephaly. Am J Neuroradiol 2003;24:2066–2076.
111. Steiner CE, Guerreiro MM, Marques-de-Faria AP. On macrocephaly, epilepsy, autism, specific facial features, and mental retardation. Am J Med Genet 2003;120A:564–565.
112. Bolton PF, Roobol M, Allsopp L, Pickles A. Association between idiopathic infantile macrocephaly and autism spectrum disorders. Lancet 2001;358:726–727.
113. Miles JH, Hadden LL, Takahashi TN, Hillman RE. Head circumference is an independent clinical finding associated with autism. Am J Med Genet 2000;95:339–350.
114. Fidler DJ, Bailey JN, Smalley SL. Macrocephaly in autism and other pervasive developmental disorders. Dev Med Child Neurol 2000;42:737–740.
115. Palmen SJ, Van Engeland H. Review on structural neuroimaging findings in autism. J Neural Transm 2004;111:903–929.
116. Sokol DK, Edwards-Brown M. Neuroimaging in autistic spectrum disorder (ASD). J Neuroimaging 2004;14:8–15.
117. Brambilla P, Hardan A, di Nemi SU, et al. Brain anatomy and development in autism: review of structural MRI studies. Brain Res Bull 2003;61:557–569.
118. Di Martino A, Castellanos FX. Functional neuroimaging of social cognition in pervasive developmental disorders: a brief review. Ann NY Acad Sci 2003;1008:256–260.
119. Boddaert N, Zilbovicius M. Functional neuroimaging and childhood autism. Pediatr Radiol 2002;32:1–7.
120. Zilbovicius M, Boddaert N, Belin P, et al. Temporal lobe dysfunction in childhood autism: a PET study. Positron emission tomography. Am J Psychiatry 2000;157:1988–1993.
121. Boddaert N, Belin P, Chabane N, et al. Perception of complex sounds: abnormal pattern of cortical activation in autism. Am J Psychiatry 2003;160:2057–2060.
122. Small SA, Wu EX, Bartsch D, et al. Imaging physiologic dysfunction of individual hippocampal subregions in humans and genetically modified mice. Neuron 2000;28:653–64.
123. Barnea-Goraly N, Kwon H, Menon V, et al. White matter structure in autism: preliminary evidence from diffusion tensor imaging. Biol Psychiatry 2004;55:323–326.

124. Raymond GV, Bauman ML, Kemper TL. Hippocampus in autism: a Golgi analysis. Acta Neuropathologica 1996;91:117–119.
125. Ahlsen G, Rosengren L, Belfrage M, et al. Glial fibrillary acidic protein in the cerebrospinal fluid of children with autism and other neuropsychiatric disorders. Biol Psychiatry 1993;33:734–743.
126. Purcell AE, Jeon OH, Zimmerman AW, Blue ME, Pevsner J. Postmortem brain abnormalities of the glutamate neurotransmitter system in autism. Neurology 2001;57:1618–1628.
127. Laumonnier F, Bonnet-Brihault F, Gomot M, et al. X-linked mental retardation and autism are associated with a mutation in the NLGN4 gene, a member of the neuroligin family. Am J Hum Genet 2004;74:552–557.
128. Comoletti D, De Jaco A, Jennings LL, et al. The Arg451Cys-neuroligin-3 mutation associated with autism reveals a defect in protein processing. J Neurosci 2004;24:4889–4893.
129. Chih B, Afridi SK, Clark L, Scheiffele P. Disorder-associated mutations lead to functional inactivation of neuroligins. Hum Mol Genet 2004;13:1471–1477.
130. Pastinen T, Kurg A, Metspalu A, Peltonen L, Syvanen AC. Minisequencing—a specific tool for DNA analysis and diagnostics on oligonucleotide arrays. Genome Research 1997;7:606–614.

III Transgenic and Knockout Models of Neuropsychiatric Dysfunction

9
Genetic Mouse Models of Psychiatric Disorders
Advantages, Limitations, and Challenges

Joseph A. Gogos and Maria Karayiorgou

Summary

The mental well-being of humans depends on the discovery of the causes of mental illnesses and the use of this knowledge to direct the generation of new treatments and the development of preventive measures. In this context, defining how we can exploit the power of animal models in investigative strategies designed to understand and manipulate candidate causal factors remains a critical challenge. The fact that mental illnesses are uniquely human disorders does not negate the feasibility of developing and using relevant animal models, but only defines the challenge and sets the limitations of an animal model. Because the field is still in its infancy, addressing the roles and targets of animal models of mental illnesses effectively and responsibly will require additional empirical data, as well as critical thinking from scientists, journal editors, and funding agencies. In this chapter, we discuss some general guidelines for the development of genetic mouse models of psychiatric disorders and offer a theoretical framework for the interpretation of their analysis. At the end, we discuss some results and practical issues emerging from our ongoing work on a genetic mouse model of schizophrenia.

Key Words: Psychosis; schizophrenia; del22q11; mouse models; susceptibility genes; endophenotypes; sensorimotor gating.

1. MOUSE MODELS OF PSYCHIATRIC DISORDERS

Three major constraints make the generation of *bona fide* mouse models for most psychiatric disorders highly unlikely.

1. Constraints imposed by the complex polygenic nature of human psychiatric disorders. According to the predominating common disease/common allele hypothesis *(1)*, individual susceptibility alleles are neither necessary nor sufficient to cause a psychiatric disorder. Therefore, a mouse gene-targeting model for an individual candidate gene is unlikely to serve as a model for the entire complexity of the disorder. In addition, the precise contribution of environmental factors to the pathophysiology of these disorders remains poorly understood, precluding the production of convincing models through environmental manipulation.

From: *Contemporary Clinical Neuroscience: Transgenic and Knockout Models of Neuropsychiatric Disorders*
Edited by: G. S. Fisch and J. Flint © Humana Press Inc., Totowa, NJ

2. Constraints imposed by the magnitude and pattern of change during hominid brain evolution. Brain morphological variation and divergence is thought to be a product of changes in the spatiotemporal deployment of regulatory genes and the evolution of genetic regulatory networks *(2)*. Therefore, a susceptibility gene may be "used" in different contexts in the human and mouse brain, and a mutation in such a gene may have very different impact in the brain function of mice and humans. Related to this, different species have evolved specific repertoires of adaptive behaviors. It is likely, for example, that sensory modalities used by different species to establish and control behavioral patterns may not be equivalent. For example, social and mating behavior in mice might be driven to a large extent by the olfactory system, and mutations affecting these behaviors may exert their effects through olfactory processing and have little or nothing to do with the complex genetic components underlying human social behavior, which is mainly not olfactory driven. Mice also present a relatively limited spectrum of stereotypic behaviors, whose relation to complex human behaviors is questionable. Nevertheless, researchers have attempted to model certain symptoms of psychiatric disorders in mice based on such stereotypic behaviors. For example, some animal models have focused on increased motor activity and stereotypy as relevant to the positive or negative symptoms of schizophrenia. Amelioration of these behavioral alterations by treatment with antipsychotic drugs used to treat schizophrenia, such as haloperidol or clozapine, has been used as supportive evidence for a correlation between these rodent behaviors and schizophrenia. However, improvement of some of these behavioral alterations after treatment with antipsychotic drugs may simply imply that these behaviors are under the control of the dopaminergic and serotonergic receptors antagonized by these drugs.
3. Constraints imposed by limitations of clinical diagnosis. A great deal of heterogeneity exists among the symptoms of individuals sharing a common diagnosis, and many are not readily classifiable using standard criteria. Uncertainty regarding the clinical features of the human syndromes to be simulated obviously complicates the validation of animal models.

2. MOUSE MODELS OF GENES PREDISPOSING TO PSYCHIATRIC DISORDERS

The impossibility of developing an animal model that captures the totality of any given complex psychiatric disorder argues for a more piecemeal recreation of components of the disorder. In this context, mouse models of "susceptibility genes" identified through forward genetic studies in humans (Fig. 1) hold tremendous promise in understanding the function of a gene in the context of simple cellular pathways, or even at the level of simple neural circuits and behavior. However, several important factors need to be considered for the generation of such models and for the design and interpretation of their analysis.

3. GENERATION OF MOUSE MODELS

Gene-targeting approaches that permit inactivation of genes in mice have revolutionized our understanding of development and, more recently, of learning and memory, and are likely to impact our understanding of human complex psychiatric disorders as well. Almost certainly, this goal will be facilitated by recent efforts generating publicly available collections of mice carrying mutations in all known genes *(3,4)*.

Genetic Mouse Models

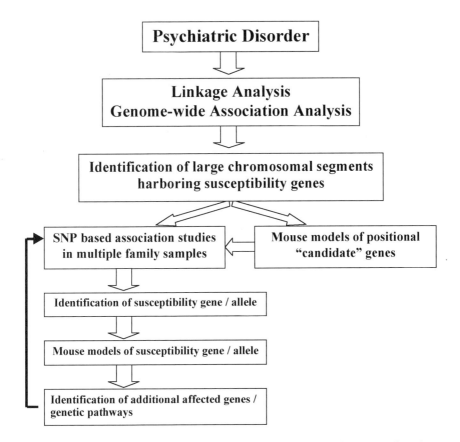

Fig. 1. Use of genetic mouse models in the functional analysis of genes predisposing to psychiatric disorders.

In addition, the recent use of sophisticated tools for the generation of conditional mutants exhibiting gene loss in a temporally and spatially restricted manner is proving particularly useful in the understanding of behavior (5). In general, phenotypic changes in conditional mutants are more specific and subtle than in loss-of-function mutations of the same gene and, therefore, allow better understanding of underlying behavioral, neuropathological, or physiological consequences. In the case of genes that play a role in both development and behavior, conditional and induced mutations permit studies of behavior without any influence from the role that the disrupted gene plays during prenatal or postnatal development (6).

Independent of the approach, several recent findings highlight the importance of defining and controlling both the interactions between genes and between genes and environment in the use of genetic manipulations to develop animal behavioral models of psychiatric disorders. These findings highlight the need for strict control of environmental variables and genetic backgrounds as well as careful phenotyping of animals, using batteries of tasks (7–9).

Unlike studies in development and learning/memory, which can benefit by extreme system perturbations introduced by traditionally generated null mutations, the issue of what is the best gene-targeting strategy to model a psychiatric disease allele in mice has not been addressed extensively. Unlike simple Mendelian disorders caused by highly penetrant mutations in a single gene (that may result in complete abolition of gene function), the genetic component of complex psychiatric disorders is likely to be associated with low penetrance functional variations in a number of susceptibility genes; very few, if any, of these variations are expected to be null alleles. Indeed, it is conceivable that in modeling disease genes in mice, null alleles may introduce unanticipated problems and introduce confusion in the interpretation of the analysis. One consideration is that null alleles may result in a severe phenotype that could mask the disease-relevant phenotypes (for example, a null allele in mice may result in severe epilepsy that would complicate behavioral analysis). In that respect, knockout mice are not necessarily ideal to model psychiatric disease alleles, and the generation of hypomorphic alleles (alleles that retain some residual function) may be the most appropriate approach *(10)*.

4. MOUSE MODELS OF PREDISPOSING ALLELES

One important consideration in developing genetic mouse models of psychiatric disorders has to do with the nature of the susceptibility allele. Unfortunately, even for psychiatric diseases for which strong candidate genes exist (i.e., schizophrenia), the functional implications of the risk alleles are largely unknown, with only a few exceptions *(11,12)*. Obviously, we need to know whether the risk allele is a hypomorph and, therefore, a mouse knockout allele can model it accurately. What if it is a gain-of-function allele that, for example, increases expression of the gene? Alternatively, what if it modulates splicing in such a way that it favors forms that are more active in specific signaling pathways? Even if the risk allele is a hypomorph, it could affect splicing and/or expression in specific brain areas, or specific cellular populations (i.e., subsets of inhibitory neurons or glia), or it could be critical during specific developmental stages. In such instances, it is highly unlikely that a mouse model with a widespread neuronal and nonneuronal decrease in the target gene levels throughout development and adulthood could serve as an accurate model of the disease allele. Indeed, such a model can be misleading. These are extremely important issues, directly relevant to the design and interpretation of any study's findings.

A related consideration has to do with the wide expression and the pleiotropic effect of target genes. Several of the strong candidate susceptibility genes (i.e., *BDNF*) *(13–16)* seem to participate in virtually all aspects of brain development, maturation, and function and to modulate signaling through a large numbers of neurotransmitter receptors. Given such pleiotropy and complexity, one needs to consider carefully when modeling such genes (rather than disease alleles) using mouse knockout approaches, which, of a large number of alternative phenotypes, is a critical link between the actual genetic risk variant and susceptibility to the specific psychiatric disorder. Under these circumstances, considerable work may be needed for any claim to be proven unequivocally.

Therefore, an important goal of human genetics will be to decipher the functional implications of risk haplotypes *(17)*, which can then be modeled in mice as closely as possible (given, of course, the limitations of the available technologies). At present, given our restricted knowledge about the functional impact of human genetic variation, accurate mouse models of risk alleles are currently straightforward for only a small number of susceptibility genes. Nevertheless, to responsibly follow up human genetic studies with mouse models, it is essential that the mouse models generated mimic the risk allele as accurately as possible.

5. MULTILEVEL ANALYSIS OF MOUSE MODELS

Traditionally, animal models of susceptibility genes are analyzed through behavioral assays. Indeed, much of the skepticism toward animal models of psychiatric disorders arises from the unfounded expectation that such animals should convincingly model hallmark features of the human psychiatric disorders, such as grandiosity, delusions, hallucinations, or depression. Mechanistic insights into the nature of contribution of susceptibility genes cannot be gleaned by behavioral observation alone, but rather be obtained from a combined approach that begins at the molecular and cellular levels (which are more likely to be evolutionary conserved) and culminates at the systems and behavioral levels. It is only within this context and by using approaches that are either off-limits or confounded by drug treatment in humans that one can understand the consequences of a mutation on a susceptibility gene.

Molecular-level approaches can be used, for example, to identify changes in gene expression or other posttranslational modifications in response to a specific gene disruption to identify individual genes, gene pathways, and cellular processes that either serve as direct targets or interact with the disrupted pathways to increase susceptibility to a given psychiatric disorder. Cellular- or synaptic-level approaches can be used to identify the cellular or synaptic substrates underlying observed molecular changes. For example, such approaches can be used to distinguish between neuron vs neuropil loss, and to assess gray matter reduction, number of neurons, dendritic and synaptic morphology, neuronal apoptosis, status of the myelinated white matter, and total brain content of selected neurotransmitters. Systems level approaches will likely be indispensable in identifying interacting brain regions contributing to cellular/synaptic abnormalities observed in a given brain region (for example, an observed exaggerated induction of dopamine release in the cortex may be caused by an abnormal inhibition of dopaminergic cell firing in the ventral tegmental area, as suggested previously for pharmacological models of schizophrenia). Behavioral level approaches can be used to dissect complex phenotypes into components, called endophenotypes, which may be more amenable to genetic studies than the fully expressed clinical manifestation of a psychiatric disease. In schizophrenia, for example, abnormalities in attention, sensorimotor gating, olfactory function, and cognitive performance have been suggested as useful endophenotypes *(18–24)*. Therefore, in a mouse model of a schizophrenia susceptibility gene, one can use tests assessing sensorimotor gating and aspects of cognition, i.e., working memory/attention and contextual learning, central processes known to be affected in patients with schizophrenia. It should be noted that a distinct advantage of genetic mouse models is that they allow for the identification of early manifestations of the mutation and the study of the progression of the mutational effects. In that context, other, previously unsuspected behavioral deficits may also be identified and correlated with abnormalities in synaptic physiology, or in communication between different brain regions.

Our argument, thus far, is based on the assumption that generation of genetic mouse models of psychiatric disorders follows unequivocal gene identification via human genetic approaches. Although this can be the case, it will more likely represent the exception rather than the rule. The reason is that risk gene identification in complex psychiatric disorders almost invariably relies on genetic association studies. These studies are indispensable for identifying strong candidate genes for psychiatric disorders, but they suffer from limitations imposed on them by the combined phenotypic and genetic heterogeneity of these disorders and, therefore, are likely to result in inconclusive findings *(25)*. These limitations and the as yet restricted knowledge of the functional impact of human genetic variation are likely to blur the line between human genetic and animal model studies, in the sense that studies in animal models not only will be imperative in understanding the function of susceptibility genes, but will also aid in their identification (Fig. 1) *(26)*. Indeed, in the remaining part of this chapter, we outline recent work on the chromosome 22q11 schizophrenia susceptibility locus that represents one of the first examples of the interplay between human genetic studies and generation of animal models for candidate susceptibility genes.

6. THE 22Q11 SCHIZOPHRENIA SUSCEPTIBILITY LOCUS AS AN EXAMPLE OF THE INTERPLAY BETWEEN HUMAN GENETICS AND ANIMAL MODELS

Schizophrenia affects a staggering 1% of the world's population and is associated with severe functional decline and lifelong disability. It is diagnosed as a clinical syndrome presenting most commonly with psychotic symptoms (delusional ideas, hallucinations, and disordered thinking) and bizarre behavior, accompanied often with a later emergence of "negative" symptoms, including low levels of emotional arousal, mental activity, and social drive. There are also severe cognitive impairments present, particularly in attention, memory, and executive functions. The age of onset of the full set of diagnostic symptoms of schizophrenia is approx 20 yr of age, but there is substantial evidence to suggest that the negative symptoms and neurocognitive impairments are often present in childhood, predating the full phenotypic expression of schizophrenia. The etiology of schizophrenia is unknown. Although genetic epidemiological studies have identified a strong genetic component in the etiology of the disease *(27,28)*, gene identification has been extremely challenging, primarily because the mode of inheritance of the disease is complex and likely involves interaction among multiple genes and environmental factors *(29)*.

During the past 3 yr, however, an increasing number of variants in individual genes were identified convincingly as contributing to schizophrenia susceptibility. However, in every case:

1. The increase of the risk associated with each variant is small (twofold at most).
2. In most cases, the effect of the variant on the expression and function of the gene is unknown.

This immensely complicates efforts to generate disease-related mouse models of these susceptibility genes. An exception to this is the q11 region on the long arm of chromosome 22.

In addition to modest but consistent evidence from linkage studies *(30–32)*, it has been shown that patients with a hemizygous deletion of the q11 region of their chromosome 22 are at very high risk for serious psychiatric illness. Several studies have now established conclusively that the risk to develop schizophrenia for a patient who carries this deletion is approx 30 times the general population risk (three times the risk of a first-degree relative of an individual with schizophrenia) *(33,34)*. Microdeletions of 22q11 account for approx 1–2% of schizophrenia in Caucasian populations *(35–37)*, although mutations and other causative variants in individual genes from this region may account for a higher percentage of the disease in patients without this deletion. Furthermore, recent studies suggest that patients with schizophrenia who carry the 22q11 deletion do not differ from those who do not carry the deletion, in either the core clinical phenotype, or the hallmark neuropsychological and neuroanatomical features of schizophrenia, which include smaller total gray matter volume and larger lateral ventricles than healthy controls *(38,39)*.

22q11 microdeletions represent the first unequivocal association between a well-defined genetic lesion and schizophrenia. The overwhelming majority (approx 87%) of the 22q11 deletions are 3 Mb in size; whereas a smaller percentage (approx 8%) involves the same proximal breakpoint but a different distal breakpoint, resulting in a smaller 1.5 Mb deletion (Fig. 2). All deletions are mediated by low copy repeat (LCR) sequences *(40,41)*. At least one patient with schizophrenia has been reported to carry the smaller 22q11 microdeletion *(35)*, and, therefore, the "schizophrenia critical region" has been defined as 1.5 Mb (LCR-A to LCR-B; Fig. 2) that contains approx 30 genes. The syntenic region of the human 22q11 locus lies on mouse chromosome 16. All human genes (except for one) are represented in the mouse, although the order of the genes is different (Fig. 2) *(42)*.

The robust association between a well-defined genetic lesion and a dramatic increase in the risk to develop schizophrenia, coupled with our precise knowledge of the human and mouse sequences and chromosomal synteny at this locus provide a unique opportunity to use a mouse model to understand the biological basis of the increased psychosis risk associated with this genetic lesion. Because more than one gene from this region seem to contribute to the approx 30-fold increase in risk, a

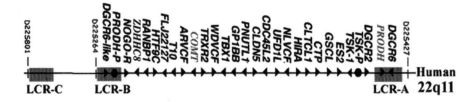

Fig. 2. Map of the 22q11 region. Indicated is the order and transcriptional direction of the known genes between low-copy repeat (LCR)-A and LCR-B, which define the 1.5 Mb "schizophrenia critical region" *(35)*. The region of mouse chromosome 16 syntenic to the human interval LCR-A and LCR-B is shown.

mouse model of the entire schizophrenia-associated microdeletion is likely to capture the interactions among the culprit genes and, therefore, be an accurate model of the disease.

We and others have modeled the human 22q11 deletion in the mouse using gene targeting and chromosomal engineering approaches to introduce an equivalent deletion on the syntenic locus at mouse chromosome 16. (Stark K, Bagchi A, Liu H, Mills A, Karayiorgou M, and Gogos JA, unpublished data) *(43)*. The strategy of chromosomal engineering is based on the techniques of gene targeting in embryonic stem cells and the Cre/loxP system, and is outlined in Mills and Bradley *(44)*. In principle, an appropriate, disease-relevant analysis of this mouse model can be accomplished in two phases. In the first phase, one can test the hypothesis that mutant mice carrying the deletion recapitulate the phenomenology observed in the brains of schizophrenic patients and at least some of the behavioral and neurophysiological deficits of the disease. In the second phase, one can begin dissecting the biological bases underlying any abnormal phenomenology identified in the first phase using higher resolution molecular, cellular, and electrophysiological approaches, depending on the nature of the abnormality and the potentially affected molecular pathways.

As already mentioned, several studies collectively suggest that prominent sensory gating deficits and cognitive impairments, particularly in attention and executive functions, may contribute to schizophrenia susceptibility. These phenotypic indicators (endophenotypes) probably reflect discrete components of pathophysiological processes, mediating between particular sets of predisposing genes and clinical diagnosis; are often present in childhood, predating the full phenotypic expression of the disease; and are evident in clinically unaffected, first-degree relatives of patients with schizophrenia. The most notable examples of schizophrenia-related endophenotypes are sensorimotor gating, working memory, and attention, all of which can be modeled relatively accurately in mice.

Sensorimotor gating refers to the forebrain influence on the automatic brain stem startle response *(45)*. It is usually evaluated from the degree of inhibition of the startle response by an acoustic prepulse. The prepulse tone precedes a much stronger, abrupt acoustic startling stimulus by 100 ms (prepulse inhibition [PPI]). Sensorimotor gating is a major central processing mechanism that is affected in many patients with schizophrenia and some of their first-degree, clinically unaffected relatives. It should be noted that PPI deficits are not specific to schizophrenia, but are present in other disorders characterized symptomatically by a loss of gating in sensory, motor, or cognitive domains (such as obsessive–compulsive disorder, Huntington's disease, nocturnal enuresis and attention deficit disorder, Tourette's syndrome, and nonepileptic seizures [reviewed in ref. *45*]). One largely untested hypothesis is that deficits in sensorimotor gating are related to the inability of schizophrenic patients to automatically filter or "gate" irrelevant thoughts and sensory stimuli from intruding into conscious awareness *(18,19)*. Mice demonstrate a robust and reliable PPI *(46)*. Indeed, one of the most attractive features of PPI is that it is one of a few neuropsychological measures in which humans and rodents can be evaluated in a similar fashion.

Working memory and attention are characteristically impaired in patients with schizophrenia, irrespective of their level of intelligence *(21,23)*. Deficits in these tests have been observed in patients during both active and remitted phases of illness and both before and during treatment with antipsychotic drugs, as well as in a portion of their nonschizophrenic, first-degree relatives *(20,21,47)*. Therefore, problems in these cognitive domains seem to be at the very core of the dysfunction in this disease *(22)*. Working memory can be assessed with at least two different assays that require memory retention over brief delays, the Delayed Alternation Task *(48)* and the Latent Inhibition Assay *(49)*. Disruption of latent inhibition has been proposed as a possible model of the cognitive abnormality that underlies the psychotic symptoms of acute schizophrenia *(24,50)*.

Initial investigation of the long-range deletion mouse model *(43)* assessed different domains of central nervous system functioning using tests such as PPI and the Pavlovian conditioned-fear test, which assesses learning and memory. Additionally, simple tests of sensory and motor function, exploratory activity, anxiety-related traits, and analgesia-related responses were evaluated. Mice carrying long-range deletions showed deficits in sensorimotor gating, as well as in learning and memory. Indeed, the response of these mice in the Pavlovian conditioned-fear test suggests that they have difficulty remembering the types of cues associated with a complex training environment for long (24 h), but not short (1 h) periods. These mice successfully learn and remember that a single auditory cue was paired with a foot shock, suggesting that the conditioned fear impairment is selective and similar in nature to deficits observed in rodents with hippocampal damage *(51,52)*.

The finding of learning and sensorimotor gating deficits is particularly significant because premorbid children with 22q11 microdeletions show similar deficits, including problems with complex forms of learning that require problem solving, planning, and abstract thinking *(53,54)* as well as severe deficits in PPI. Specifically, as compared with sibling controls, children with 22q11 microdeletions showed 20% less PPI; secondary analyses suggested that this decrement did not reflect a developmental delay *(55)*. In addition, comparison of the Attention Network Tests index scores showed that children with 22q11 microdeletions had less efficient scores on measures of executive attention, indicating less efficient processing in uncued conditions *(56)*. In light of these findings, it would be interesting to examine working memory and attention indices in the long-range deletion mouse model.

We and others have also focused significant efforts on identifying susceptibility genes from the 22q11 locus. Two independent systematic approaches, as well as several candidate gene studies designed to identify schizophrenia susceptibility genes from the 22q11 region have been reported. Both systematic approaches, using different methodologies to analyze almost all of the genes in this locus, provided convergent evidence for an involvement of the gene encoding proline dehydrogenase (PRODH). The study of Liu et al. 2002 *(57)* provided evidence that a haplotypic variant of the gene located at the 3' end of the gene is preferentially transmitted in individuals with schizophrenia, a finding recently replicated in two independent large-family studies *(58,59)*, and also identified rare variants of the gene (likely generated through gene conversion from a nearby pseudogene) that affect highly conserved amino acids and are enriched to various degrees in samples of individuals with schizophrenia. Jacquet et al. *(60)* confirmed the existence and partial enrichment of the rare variants and, in addition, identified a small deletion encompassing the *PRODH* gene that co-segregated with schizophrenia in one family. Other genes in the 22q11 region, including the gene encoding for catechol-O-methyltransferase (COMT), have also been implicated by candidate gene approaches *(61,62)*. Taken together, these studies suggest that the 22q11 microdeletion-associated schizophrenia may have the characteristics of a contiguous gene syndrome, in which more than one gene contributes to the dramatic increase in disease risk.

We have followed up these findings by generating mouse models for the identified susceptibility genes to facilitate understanding of the function of these genes and how they may impact on schizophrenia. Specifically, we found a mutation that introduces a premature termination (E453X) and reduces enzymatic activity in the mouse ortholog of the human *PRODH* gene in the Pro/Re hyperprolinemic mouse strain *(63)*. We found that these mice have regional neurochemical alterations in the brain accompanied by a deficit in sensorimotor gating, similar to the one seen in individuals with schizophrenia. To minimize the influence of genetic background on gene expression, we introduced the E453X mutation into the 129/Sv genetic background through backcrossing for 10 generations. Levels of serum L-proline in the homozygous mutant mice range from 300 to 600 μM (compared with less than 100 μM in wild-type littermate control mice). These levels are comparable to those observed in some individuals with the 22q11 microdeletion, or in heterozygous carriers of *PRODH* rare variants, but are well below the levels observed in patients with hyperprolinemia type I, a rare condition often accompanied by epilepsy and mental retardation *(64)*. Therefore, both in terms of the nature of the mutation and the ensuing increases in L-proline levels, this mutant strain represents an accurate model not only of a susceptibility gene but also of a susceptibility allele.

We reasoned that transcriptional profiling *(65–67)* in the brain of a mouse model for a schizophrenia susceptibility allele might provide an unbiased evaluation of the transcriptional programs affected by the disruption of the gene, reflecting either downstream effects of the mutation or compensatory changes. Because functional and structural pathology of the frontal cortex in schizophrenia has been suggested by numerous studies, we started our analysis from this brain region. In control experiments in our mouse model, histological analysis, cell counting, as well as cell death assessment in the developing cortex did not reveal any gross anatomical abnormalities in this region. Transcriptional profiling in the cortex of Prodh-deficient mice identified a few hundred genes as being differentially expressed *(68)*. Most notably, the list of upregulated genes included the *COMT* gene, which, as we already mentioned, also maps within the 22q11 locus and is a candidate schizophrenia susceptibility gene.

Follow-up behavioral analysis of these mice revealed a pronounced hyperresponsivity to dopamine, which is unmasked under challenge with amphetamine and is modulated by the level of COMT activity. In that respect, it is worth noting that dopaminergic hypersensitivity in schizophrenia is well-established, based primarily on the therapeutic effect of dopamine receptor antagonists *(69)*. Further pharmacological analysis revealed a previously unsuspected interaction between *PRODH* and *COMT* that, in principle, can modulate the risk and/or the expression of the 22q11-associated psychiatric phenotypes. If, indeed, COMT upregulation is one of the mechanisms used to control cortical dopaminergic hypersensitivity, then schizophrenic patients with a 22q11 deletion are at a particular disadvantage because they are deficient for both genes and perhaps are unable to compensate efficiently through COMT for the cortical dopaminergic hyperactivity induced by PRODH deficiency. Such synergistic interaction between two physically linked genes that disrupts well-established dopamine–glutamate interactions, could, in principle, lead to the high disease risk associated with this locus and/or modulate the expression of the phenotype. Similar patterns of genetic interactions (impaired synaptic function-impaired compensation) among nonlinked loci may also account for the epistatic component of the genetic risk of psychiatric disorders in general.

Therefore, analysis of a reliable mouse model of a susceptibility allele provided us with a unique insight, as well as with a testable model that can now be explored through further experimentation. A comprehensive research program on the long-range 22q11 deletion and individual gene mouse models could dramatically improve our understanding of disease pathobiology, allow us to specify the function and mode of dysfunction of some of the genes that are responsible for schizophrenia, and also lead to identification of critical disease pathways and of additional candidate genes through sequence homology, or involvement in the same or similar pathways or processes.

REFERENCES

1. Pritchard JK, Cox NJ. The allelic architecture of human disease genes: common disease-common variant...or not? Hum Mol Genet 2002;11:2417–2423.
2. Carroll SB. Genetics and the making of Homo sapiens. Nature 2003;422:849–857.
3. Austin CP, Battey JF, Bradley A, et al. The knockout mouse project. Nat Genet 2004;36:921–924.
4. Auwerx J, Avner P, Baldock R, et al. The European dimension for the mouse genome mutagenesis program. Nat Genet 2004;36:925–927.
5. Lewandoski M. Mouse genomic technologies: conditional control of gene expression in the mouse. Nat Rev Genet 2001;2:743–755.
6. Miyakawa T, Leiter LM, Gerber DJ, et al. Conditional calcineurin knockout mice exhibit multiple abnormal behaviors related to schizophrenia. Proc Natl Acad Sci USA 2003;100:8987–8992.
7. Gerlai R. Gene-targeting studies of mammalian behavior: is it the mutation or the background genotype? Trends Neurosci 1996;19:177–181. Erratum in: Trends Neurosci 1996;19:271.
8. Crabbe JC, Wahlsten D, Dudek BC. Genetics of mouse behavior: interactions with laboratory environment. Science 1999;284:1670–1672.
9. Cabib S, Orsini C, Le Moal M, Piazza PV. Abolition and reversal of strain differences in behavioral responses to drugs of abuse after a brief experience. Science 2000;289:463–465.
10. Mohn AR, Gainetdinov RR, Caron MG, Koller, B.H. Mice with reduced NMDA receptor expression display behaviors related to schizophrenia. Cell 1999;98:427–436.
11. Millar JK, Wilson-Annan JC, Anderson S, et al. Disruption of two novel genes by a translocation co-segregating with schizophrenia. Hum Mol Genet 2000;9:1415–1423.
12. Mukai J, Liu H, Burt RA, et al. Evidence that the gene encoding ZDHHC8 contributes to the risk of schizophrenia. Nat Genet 2004;36:725–731.
13. Sklar P, Gabriel SB, McInnis MG, et al. Family-based association study of 76 candidate genes in bipolar disorder: BDNF is a potential risk locus. Mol Psychiatry 2002;7:579–593.
14. Egan MF, Kojima M, Callicott JH, et al. The BDNF val66met polymorphism affects activity-dependent secretion of BDNF and human memory and hippocampal function. Cell 2003;112: 257–269.
15. Hall D, Dhilla A, Charalambous A, Gogos JA, Karayiorgou M. Sequence variants of the brain-derived neurotrophic factor (BDNF) gene are strongly associated with obsessive–compulsive disorder. Am J Hum Genet 2003;73:370–376.
16. Geller B, Badner JA, Tillman R, Christian SL, Bolhofner K, Cook EH Jr. Linkage disequilibrium of the brain-derived neurotrophic factor Val66Met polymorphism in children with a prepubertal and early adolescent bipolar disorder phenotype. Am J Psychiatry 2004;161:1698–1700.
17. Rebbeck TR, Spitz M, Wu X. Assessing the function of genetic variants in candidate gene association studies. Nat Rev Genet 2004;5:589–597.
18. Braff DL. Information processing and attention dysfunctions in schizophrenia. Schizophr Bull 1993;19:233–259.
19. Perry W, Braff DL. Information-processing deficits and thought disorder in schizophrenia. Am J Psychiat 1994;151:363–367.
20. Cannon TD, Zorrilla LE, Shtasel D, et al. Neuropsychological functioning in siblings discordant for schizophrenia and healthy volunteers. Arch Gen Psychiat 1994;51:651–661.
21. Cannon TD, Huttumen MO, Lonnqvist J, et al. The inheritance of neuropsychological dysfunction in twins discordant for schizophrenia. Am J Hum Genet 2000;67:369–382.
22. Goldman-Rakic PS. The physiological approach: functional architecture of working memory and disordered cognition in schizophrenia. Biol Psychiat 1999;46:650–661.

23. Elvevag B, Goldberg TE. Cognitive impairment in schizophrenia is the core of the disorder. Crit Rev Neurobiol 2000;14:1–21.
24. Martins Serra A, Jones SH, Toone B, Gray JA. Impaired associative learning in chronic schizophrenics and their first-degree relatives: a study of latent inhibition and the Kamin blocking effect. Schizophr Res 2001;48:273–289.
25. Freimer N, Sabatti C. The use of pedigree, sib-pair and association studies of common diseases for genetic mapping and epidemiology. Nat Genet 2004;36:1045–1051.
26. Gerber DJ, Hall D, Miyakawa T, et al. Evidence for association of schizophrenia with genetic variation in the 8p21.3 gene, PPP3CC, encoding the calcineurin gamma subunit. Proc Natl Acad Sci USA 2003;100:8993–8998.
27. Gottesman II, Shields J. Schizophrenia: The Epigenetic Puzzle. Cambridge, UK: Cambridge University Press, 1982.
28. Kety SS, Wender PH, Jacobsen B, et al. Mental illness in the biological and adoptive relatives of schizophrenia adoptees. Replication of the Copenhagen study in the rest of Denmark. Arch Gen Psychiat 1994;51:442–455.
29. Karayiorgou M, Gogos JA. A turning point in schizophrenia genetics. Neuron 1997;19:967–979.
30. Blouin JL, Dombroski BA, Nath SK, et al. Schizophrenia susceptibility loci on chromosomes 13q32 and 8p21. Nat Genet 1998;20:70–73.
31. Shaw SH, Kelly M, Smith AB, et al. A genome-wide search for schizophrenia susceptibility genes. Am J Med Genet 1998;81:364–376.
32. Williams NM, Norton N, Williams H, et al. A systematic genomewide linkage study in 353 sib pairs with schizophrenia. Am J Hum Genet 2003;73:1355–1367.
33. Pulver AE, Nestadt G, Goldberg R, et al. Psychotic illness in patients diagnosed with velo-cardio-facial syndrome and their relatives. J Nerv Ment Dis 1994;182:476–478.
34. Murphy KC, Jones LA, Owen MJ. High rates of schizophrenia in adults with velo-cardio-facial syndrome. Arch Gen Psychiat 1999;56:940–945.
35. Karayiorgou M, Morris MA, Morrow B, et al. Schizophrenia susceptibility associated with interstitial deletions of chromosome 22q11. Proc Natl Acad Sci USA 1995;92:7612–7616.
36. Usiskin SI, Nicolson R, Krasnewich DM, et al. Velocardiofacial syndrome in childhood-onset schizophrenia. J Am Acad Child Adolesc Psychiat 1999;38:1536–1543.
37. Wiehahn GJ, Bosch GP, du Preez RR, Pretorius HW, Karayiorgou M, Roos JL. Assessment of the frequency of the 22q11 deletion in Afrikaner schizophrenic patients. Am J Med Genet 2004;129B:20–22.
38. Chow EWC, Mikulis DJ, Zipursky RB, Scutt LE, Weksberg R, Bassett AS. Qualitative MRI findings in adults with 22q11 deletion syndrome and schizophrenia. Biol Psychiat 1999;46:1436–1442.
39. Chow EWC, Zipursky RB, Mikulis DJ, Basset AS. Structural brain abnormalities in patients with schizophrenia and 22q11 deletion syndrome. Biol Psychiat 2002;51:208–215.
40. Edelmann L, Pandita RK, Morrow BE. Low-copy repeats mediate the common 3-Mb deletion in patients with velo-cardio-facial syndrome. Am J Hum Genet 1999;64:1076–1086.
41. Shaikh TH, Kurahashi H, Saitta SC, et al. Chromosome 22-specific low copy repeats and the 22q11.2 deletion syndrome: genomic organization and deletion endpoint analysis. Hum Mol Genet 2000;9:489–501.
42. Puech A, Saint-Jore B, Funke B, et al. Comparative mapping of the human 22q11 chromosomal region and the orthologous region in mice reveals complex changes in gene organization. Proc Natl Acad Sci USA 1997;94:14,608–14,613.
43. Paylor R, McIlwain KL, McAninch R, et al. Mice deleted for the DGS/VCFS syndrome region show abnormal sensorimotor gating and learning and memory impairments. Hum Mol Genet 2001;10:2645–2650.
44. Mills AA, Bradley A. From mouse to man: generating megabase chromosome rearrangements. Trends Genet 2001;17:331–339.
45. Braff DL, Geyer MA, Swerdlow NR. Human studies of prepulse inhibition of startle: normal subjects, patient groups, and pharmacological studies. Psychopharm 2001;156:234–258.

46. Paylor R, Crawley JN. Inbred strain differences in prepulse inhibition of the mouse startle response. Psychopharmacology (Berl) 1997;132:169–180.
47. Saykin AJ, Shtasel DL, Gur RE, et al. Neuropsychological deficits in neuroleptic naive patients with first-episode schizophrenia. Arch Gen Psychiat 1994;51:124–131.
48. Tanila H, Mustonen K, Sallinen J, Scheinin M, Riekkinen P Jr. Role of alpha2C-adrenoceptor subtype in spatial working memory as revealed by mice with targeted disruption of the alpha2C-adrenoceptor gene. Eur J Neurosci 1999;11:599–603.
49. Weiner I, Feldon J. Facilitation of latent inhibition by haloperidol in rats. Psychopharmacology (Berl) 1987;91:248–253.
50. Joseph MHM, Jones SHS. Latent inhibition and blocking: further consideration of their construct validity as animal models of schizophrenia. Commentary on Ellenbroek and Cools "Animal models with construct validity for schizophrenia." Behav Pharmacol 1991;2:521–526.
51. Abel T, Nguyen PV, Barad M, Deuel TA, Kandel ER, Bourtchouladze R. Genetic demonstration of a role for PKA in the late phase of LTP and in hippocampus-based long-term memory. Cell 1997;88:615–626.
52. Le Pen G, Grottick AJ, Higgins GA, Martin JR, Jenck F, Moreau JL. Spatial and associative learning deficits induced by neonatal excitotoxic hippocampal damage in rats: further evaluation of an animal model of schizophrenia. Behav Pharmacol 2000;11:257–268.
53. Bearden CE, Woodin MF, Wang PP, et al. The neurocognitive phenotype of the 22q11.2 deletion syndrome: selective deficit in visual-spatial memory. J Clin Exp Neuropsychol 2001;23: 447–464.
54. Woodin M, Wang PP, Aleman D, McDonald-McGinn D, Zackai E, Moss E. Neuropsychological profile of children and adolescents with the 22q11.2 microdeletion. Genet Med 2001;3:34–39.
55. Sobin C, Kiley-Brabeck K, Karayiorgou M. Lower pre-pulse inhibition in children with the 22q11 deletion syndrome. Am J Psychiatry 2005;162:1090–1099.
56. Sobin C, Kiley-Brabeck K, Daniels S, Blundell M, Anyane-Yeboa K, Karayiorgou M. Networks of attention in children with the 22q11 deletion syndrome. Devel Neuropsychology 2004; 26:611–626.
57. Liu H, Heath SC, Sobin C, et al. Genetic variation at the 22q11 *PRODH2/DGCR6* locus presents an unusual pattern and increases susceptibility to schizophrenia. Proc Natl Acad Sci USA 2002;99:3717–3722.
58. Liu CM, Liu YL, Lin CY, et al. Significant association evidence of polymorphisms of PRODH/DGCR6 with negative symptoms of schizophrenia. Am J Med Genet 2004;130B, published online Sept. 15. Accessed October, 10, 2004.
59. Li T, Ma X, Sham PC, et al. Evidence for association between novel polymorphisms in the PRODH gene and schizophrenia in a Chinese population. Am J Med Genet 2004;129B:13–15.
60. Jacquet H, Raux G, Thibaut F, et al. PRODH mutations and hyperprolinemia in a subset of schizophrenic patients. Hum Mol Genet 2002;11:2243–2249.
61. Shifman S, Bronstein M, Sternfeld M, et al. A highly significant association between a COMT haplotype and schizophrenia. Am J Hum Genet 2002;71:1296–1302.
62. Chen X, Wang X, O'Neill AF, Walsh D, Kendler KS. Variants in the catechol-o-methyltransferase (COMT) gene are associated with schizophrenia in Irish high-density families. Mol Psychiat 2004;9:962–967.
63. Gogos JA, Santha M, Takacs Z, et al. The gene encoding proline dehydrogenase modulates sensorimotor gating in mice. Nat Genet 1999;21:434–439.
64. Jacquet H, Berthelot J, Bonnemains C, et al. The severe form of type I hyperprolinaemia results from homozygous inactivation of the PRODH gene. J Med Genet 2003;40:e7.
65. Chee M, Yang R, Hubbell E, et al. Accessing genetic information with high-density DNA arrays. Science 1996;274:610–614.
66. Lipshutz RJ, Fodor SP, Gingeras TR, Lockhart DJ. High density synthetic oligonucleotide arrays. Nat Genet 1999;21:20–24.

67. Lockhart DJ, Winzeler EA. Genomics, gene expression and DNA arrays. Nature 2000;405: 827–836.
68. Paterlini M, Zakharenk SS, Lai WS, et al. Transcriptional and behavioral interaction between 22q11.2 orthologs modulates schizophrenia-related phenotypes in mice. Nature Neuroscience, 2005;8:1586–1594.
69. Seeman P. Dopamine receptors and the dopamine hypothesis of schizophrenia. Synapse 1987;1:133–152.

10
Animal Models of Psychosis

Stephen I. Deutsch, Katrice Long, Richard B. Rosse, Yousef Tizabi, Ronit Weizman, Judy Eller, and John Mastropaolo

Summary

Although the thinking and affective and social disturbances of psychosis and schizophrenia may not be easily modeled, if at all, in infrahuman species, animal models can clarify genetic and developmental lesions leading to disruption of some of the key anatomical circuitry involved in their pathophysiology. Increasingly, it is appreciated that patients with schizophrenia manifest symptoms in a variety of discrete domains of psychopathology, including positive (e.g., hallucinations), negative (e.g., affective flattening and social withdrawal), cognitive (e.g., concretization of thought), mood (e.g., anhedonia), and motor (e.g., mannerisms and posturing) symptoms. These symptoms may reflect, in part, the spatially and temporally integrated outputs from these disrupted or faulty circuits. Major goals of current descriptive and pathological research in schizophrenia include the development of sensitive behavioral rating instruments for the assessment of the presence and severity of symptoms in discrete psychopathological domains, elucidation of unique neurotransmitter abnormalities that may underlie each of these discrete domains of psychopathology, and determining the quantitative contribution of each of these discrete domains of psychopathology to the functional disability manifested by patients with schizophrenia and other psychosis. Thus, animal models that reflect nondopaminergic neurotransmitter abnormalities implicated in the pathophysiology of these discrete domains of psychopathology are especially useful. In addition to clarifying aspects of the pathophysiology of these disorders, animal models are crucial for identifying candidate compounds that may be developed as medications; novel medications are especially needed for the negative and cognitive symptom domains of psychopathology, which may be less dependent on abnormalities of dopaminergic neurotransmission. The contributions of dopaminergic abnormalities to the pathophysiology of schizophrenia have been studied most intensively. The focus on dopaminergic abnormalities in schizophrenia was prompted by the complementary observations in humans that psychosis can be elicited by psychostimulant medications such as d-amphetamine, especially when they are abused, whereas the ability to inhibit competitively the binding of dopamine to the D2 type of dopamine receptor is a pharmacological property shared by all of the conventional antipsychotic medications. Psychostimulant medications are either indirect or directly acting dopamine agonists. These pharmacological observations in humans stimulated interest in the quantita-

From: *Contemporary Clinical Neuroscience: Transgenic and Knockout Models of Neuropsychiatric Disorders*
Edited by: G. S. Fisch and J. Flint © Humana Press Inc., Totowa, NJ

tive characterization of a variety of "hardwired" rodent behaviors elicited by dopamine agonists such as apomorphine and d-amphetamine; these behaviors include a variety of stereotypic behaviors (e.g., rearing, grooming, and sniffing), horizontal locomotion and "mouse climbing," among other behaviors. These animal procedures have served as valuable screens for the identification of "dopamine blockers" and medications whose primary pharmacological actions involve modulation (dampening) of dopaminergic neurotransmission, which have proven especially effective in the attenuation of positive symptoms. However, the negative and cognitive symptom domains of psychopathology, which contribute very significantly to the functional disability of schizophrenia and other psychotic disorders, are not dramatically affected by these primarily dopaminergic interventions. Thus, there is also intense interest in animal models that mimic neurodevelopmental abnormalities and/or disruptions of neurotransmitter systems other than dopamine. The existence of neurodevelopmental abnormalities in at least some patients with schizophrenia, as reflected in subtle histopathological abnormalities in cortical lamination, the orientation and alignment of neurons within the hippocampus, and diminished cortical neuropil, has heightened interest in the adult developmental consequences of neonatal lesions of the hippocampus, which, thereby, deprive the developing frontal cortex of afferent inputs from the hippocampus, and genetic models associated with altered cortical lamination, such as the reeler mouse.

This chapter is intended to familiarize the readers with some of these latter activities. Thus, this chapter considers the impact of disrupting N-methyl-D-aspartate (NMDA) receptor-mediated neurotransmission on rodent models of sensorimotor (e.g., prepulse inhibition of the acoustic startle response) and sensory gating, and spatial learning paradigms referable to the hippocampus. These models are very relevant because phencyclidine is a noncompetitive NMDA receptor antagonist and patients with schizophrenia manifest abnormalities of sensorimotor and sensory gating. The pharmacological creation of NMDA receptor hypofunction has also been shown to antagonize electrically precipitated tonic hindlimb extension and elicit irregular episodes of intense jumping behavior in mice, referred to as "popping." These latter behaviors related to pharmacologically induced NMDA receptor hypofunction have proven very useful to early medication discovery efforts. Notably, adult mutant mice deficient in their expression of the NR1 subunit of the NMDA receptor show aberrant social and sexual behavior, which is ameliorated by clozapine administration. Importantly, animal models have been able to highlight important and therapeutically relevant differences between clozapine, the prototypic atypical antipsychotic medication, and conventional antipsychotic medications. This chapter also considers pharmacological and genetic mouse models that support the pathological involvement of a selective subtype of nicotinic acetylcholine receptor in schizophrenia. Patients with schizophrenia show diminished expression of the α_7 nicotinic acetylcholine receptor in hippocampus, frontal cortex and thalamus and deficits in sensory gating, as reflected in a failure to blunt the P50-evoked electrophysiological response to the second of a pair of identical auditory stimuli presented 500 ms apart. The DBA/2 genetically inbred mouse strain is deficient in sensory gating and density of hippocampal α_7 nicotinic acetylcholine receptors; thus, it may serve as a useful genetic mouse model of at least some aspects of the attentional and cognitive abnormalities of schizophrenia.

Ultimately, because of the complexity of schizophrenia and other psychotic disorders, understanding their pathophysiology will necessitate integrating information from many of these seemingly unrelated animal models; further, the process of medication development should be facilitated by evaluating candidate compounds in multiple animal models. The ideal antipsychotic medication or regimen of medications will address multiple domains of psychopathology. Hopefully, attenuation of symptoms in each of the discrete domains will improve functionality and quality of life, allowing our patients with schizophrenia and other psychosis to live independently.

Key Words: Schizophrenia; dopamine; glutomate; acetylcholine; sensorimotor gating; sensory gating; *reeler*; neurodevelopmental disorder.

1. INTRODUCTION

Although psychosis and schizophrenia may be uniquely human conditions, animal models are important because symptomatic manifestations reflect, in many instances, temporally and spatially integrated outputs from dysfunctional circuits connecting different areas of the brain. The clinical presentations of schizophrenia are varied and complex and, because its more dramatic signs and symptoms are uniquely human, cannot be easily modeled in infrahuman animals. Schizophrenia affects abstract and imaginative thinking, language, and socialization, among other uniquely human domains of function.

In the past, animal models were focused on characterizing behavioral consequences of excessive dopaminergic stimulation in striatal and limbic regions *(1)*. The models were understandable derivatives of the dopamine hypothesis of schizophrenia, a pharmacological hypothesis based on the abilities of psychostimulants (indirectly acting dopamine agonists) to precipitate psychosis, and of antipsychotic medications (competitive dopamine antagonists) to attenuate psychosis. These models have proven extremely generative; they have been especially useful for screening candidate antipsychotic medications. This chapter focuses primarily on animal models that may be more relevant to the negative and cognitive symptom domains of psychopathology, etiological hypotheses related to interruption of the normal developmental processes of cortical lamination and synaptic connectivity, as well as neurotransmitter abnormalities involving glutamate and acetylcholine in addition to dopamine.

2. IMPAIRED PREPULSE INHIBITION OF THE ACOUSTIC STARTLE RESPONSE

The startle reflex to a sudden intense stimulus is a defensive response that can be studied across species; the blink reflex component of the startle reflex is frequently studied in humans *(2)*. In rodents, the force of ballistic movements applied to a platform in response to an intense, usually auditory, stimulus is the outcome measure used to quantify the startle response. The startle reflex is a rapid integrated whole-animal response whose intensity varies, reflecting its behavioral plasticity. For example, in the context of fear, the intensity of the startle response is potentiated. Moreover, the intensity of the startle response diminishes with repeated exposures to the intense startling or pulse stimulus, a process referred to as habituation, or by exposure to a significantly less intense, prepulse stimulus that is presented 30 to 500 ms before the intense startling stimulus or pulse. The latter process is referred to as prepulse inhibition (PPI).

Although the basic startle reflex to an acoustic stimulus may involve as few as three synaptic connections linking the eighth cranial nerve to spinal motor neurons

(3), a series of elegant rodent studies has revealed a complex neural circuitry involving such structures as the hippocampus, medial prefrontal cortex, basolateral amygdala, thalamus, striato-pallidal projections, and the pedunculopontine nucleus in the behavioral plasticity of this response, especially to its PPI (for review, *see* ref. *2*). This circuitry is thought to regulate the inhibitory tone in the pons, which is the brain area that ultimately mediates the linking of the inhibited or dampened startle response to the weak prepulse stimulus. Thus, normal PPI of the acoustic startle response depends on the integrated activity of a hard-wired circuit that is regulated by the forebrain and conserved across species. Further, the ability of the weak prepulse to inhibit the full startle response is modulated by neurotransmitters, including dopamine, serotonin, acetylcholine, and glutamate, all of which have been implicated in the pathophysiology of psychosis and schizophrenia. PPI is not a form of learning because it can be demonstrated on the first exposure to a prepulse and is not extinguished over trials. Disruption of the PPI of the acoustic startle response in rodents can be quantified and serves as an animal model of impaired sensorimotor gating.

Patients with schizophrenia show deficient PPI, consistent with the loss of sensorimotor gating and presumed abnormalities of the forebrain circuitry that influences pontine inhibitory tone in this disorder; however, deficient PPI is observed in other disorders, including obsessive–compulsive disorder, suggesting that they too share problems with sensorimotor gating and subtle disruption along this circuit. The clinical consequences of impaired PPI in patients with schizophrenia are thought to include the experiences of sensory flooding and the manifestation of cognitive fragmentation, which may be related to faulty mechanisms of filtering or gating the entrance of irrelevant thoughts and sensory stimuli into conscious awareness. Clearly, the failure of the filtering or gating apparatus would also result in problems with sustained attention and distractibility. The significant correlations between the deficits of PPI in schizophrenia and perseverative errors on the Wisconsin Card-Sorting test, severity of thought disorder, and indices of illness severity (such as the number of hospitalizations, antipsychotic medication dose in chlorpromazine equivalents, and severity of negative symptoms) suggest that this basic pathophysiological abnormality may be primarily involved in mediating many of the signs and symptoms of the illness and contribute to functional disability (for review, *see* ref. *2*). Importantly, the deficit of PPI in schizophrenia is not likely to be an epiphenomenon of the illness, resulting from psychosis, medication, or institutionalization, but may be a fundamental and even necessary condition for illness expression; PPI deficits are observed in schizotypal patients who are neither psychotic nor medicated *(4)*.

Diminished PPI of the acoustic startle response in rodents may be relevant not only to the pathophysiology and psychopathology of schizophrenia, but may also serve as a procedure that may be useful in the identification of candidate compounds with antipsychotic properties, including potential atypical antipsychotic medications. Apomorphine, a nonselective dopamine receptor agonist, disrupts PPI in rats. The clinical potency of both the conventional and atypical antipsychotic medications correlates with their ability to restore disrupted PPI in these animals (for review, see ref. 2). Thus, the paradigm of apomorphine-disrupted PPI in rats is a sensitive one for predicting the clinical potency of antipsychotic medications in general. Medications from a variety of other classes, including buspirone, diazepam, imipramine, naloxone, and propranolol, were ineffective in restoring PPI disrupted by apomorphine. Thus, the paradigm seems to enjoy specificity for identifying antipsychotic medications.

Although effective in identifying antipsychotic medications in general, the apomorphine disruption of PPI of the acoustic startle response does not seem to distinguish between candidate medications with either conventional or atypical properties. Negative and cognitive symptoms contribute more significantly to the poor functional outcomes of patients with schizophrenia than do the positive symptoms. Positive symptoms can be attenuated quite dramatically by compounds sharing the pharmacological property of dopamine type 2-receptor blockade. Although definitions of an atypical medication vary, they are better than conventional medications for treatment-resistant patients with schizophrenia, and are not (as strongly) associated with extrapyramidal side effects, hyperprolactinemia, and increased liability for the emergence of tardive dyskinesia. Thus, atypical medications are associated with a separation of the dose–response curves for therapeutic efficacy and extrapyramidal side effects. Data also suggest that atypical medications may have some minimal to modest therapeutic efficacy, distinguishing them from conventional medications, against the targets of negative and cognitive symptoms. In any event, preclinical screening paradigms are needed that can identify potential antipsychotic medications with efficacy against targets that are thought to mediate poor functional outcomes.

New screening paradigms should be less dependent on perturbations of dopamine systems, especially the creation of hyperdopaminergia. Phencyclidine (PCP) is a noncompetitive N-methyl-D-aspartic acid (NMDA) receptor antagonist that binds to a hydrophobic domain within the NMDA receptor-associated ionophore. Although originally developed as a novel dissociative anesthetic medication, its abuse liability and ability to precipitate a schizophreniform psychosis that resembles naturally occurring schizophrenia in all of the relevant domains of psychopathology led to a withdrawal of its approval as a medication. However, because PCP interferes with glutamatergic neurotransmission and precipitates a schizophreniform psychosis that serves as a good model of human schizophrenia, there is much preclinical interest in characterizing behavioral and pharmacological effects of PCP. PCP disrupts PPI in rats. However, in contrast to the disruption caused by apomorphine, atypical medications such as clozapine and quetiapine are able to restore this disruption, whereas haloperidol, a conventional antipsychotic medication, is unable to restore the PPI disruption. Thus, the disruption of PPI by PCP may serve as a screening paradigm to identify candidate antipsychotic medications with properties that differ from conventional antipsychotic medications.

The disruption of PPI by apomorphine and PCP is an acute response to the perturbations of dopaminergic and glutamatergic transmission, respectively; whereas the sensorimotor gating deficit of schizophrenia is a trait disturbance that is influenced by genetic and developmental interactions. Therefore, the fact that isolation-reared rats show a developmentally specific deficit in PPI compared with group-reared controls that was not seen after isolation of adult animals supports the relevance of this paradigm and outcome measure to schizophrenia. Irrespective of strain (i.e., Lister hooded rats or Sprague-Dawley-derived rats), male rats reared in isolation and tested as adults showed a deficit in PPI, especially when the prepulse stimulus was 8 dB louder than the 65 dB background noise, compared with the socially reared control animals *(5)*. Genetic strain differences in other behavioral consequences (i.e., locomotor activity and startle reactivity) in isolation-reared adult animals were also found. PPI reduction observed in the isolation-reared rats was normalized after treatment with 0.05 mg/kg of raclopride, a dopamine D2 receptor antagonist *(5)*. Isolation rearing of rats is a nonpharmacological method to mimic the deficit in sensorimotor gating observed in patients with schizophrenia *(5)*. In humans, PPI begins to appear between the ages of 5 and 8 yr; thus, abnormalities of development may contribute to the deficits observed in adult patients with schizophrenia. Reduced PPI in isolation-reared rats is reversed by both conventional and atypical antipsychotic medications.

Chemical lesions of the hippocampus in the neonatal rat deprive the developing frontal cortex of an important afferent input. Relative to sham-lesioned controls, postpubertal animals with neonatally lesioned hippocampi show hyperactivity, heightened behavioral sensitivity to dopamine agonists, and regionally selective changes in dopaminergic neurotransmission that suggest that these animals model some aspects of schizophrenia *(6)*. Sham-lesioned and chemically lesioned animals do not differ from each other in these outcome measures before puberty. Thus, rats with neonatally lesioned hippocampi are similar to patients with schizophrenia in terms of manifesting overt signs of this early neurodevelopmental lesion after puberty. After puberty, the rats with neonatal lesions of their hippocampi are more sensitive to the disruptive effects of apomorphine on PPI *(7)*. Increased sensitivity to disruption of PPI in rats by apomorphine can detect a postpubertal consequence of an early neurodevelopmental lesion, supporting the relevance of this paradigm for studying developmental issues in schizophrenia.

Rat strains have been bred and inbred strains have been identified that have heightened sensitivity to the disruptive effects of apomorphine on PPI *(8)*. Similarly, inbred mouse strains have been shown to differ in PPI. Furthermore, PPI inheritance seems to follow an autosomal dominant mode, whereas this was not shown for the amplitude of the startle reflex itself. As noted, a reduction of PPI is thought to reflect problems with the inhibitory mechanisms and circuits, whose integrity are essential for sensorimotor gating. Because of homologies between humans and rodents in PPI and the paradigms used to assess it, and the fact that sensorimotor gating deficits may be fundamental to the pathophysiological disturbance(s) of schizophrenia and other disorders (including schizotypal personality disorder, obsessive–compulsive disorder, and Huntington's disease), there is interest in behavioral genetic strategies with inbred mouse strains to discover genes that are "linked to deficits in sensorimotor gating" *(9)*. Further, the ability of antipsychotic medications to normalize apomorphine-induced disruptions of PPI suggests that this may serve as a screening paradigm for identifying candidate compounds that could serve as new medications. Conceivably, the ability of this procedure to identify new medications will be enhanced by testing candidate compounds in relevant inbred strains.

Data with 12 inbred mouse strains highlighted the fact that genetic strain differences do indeed exist in terms of the percentage of reduction of the startle response after a prepulse stimulus *(9)*. Variables, such as the amplitude of the acoustical prepulse stimulus and whether the startling stimulus was acoustical or tactile, influenced the distribution of strains according to the levels of PPI. For example, when the acoustic prepulse stimuli were 74 and 78 dB, the DBA/2J strain showed dramatic reductions of PPI to both acoustical and tactile startling stimuli. Under these conditions, the amplitude of the startle response was 80% or more of the response observed without any prepulse. However, the DBA/2J strain was intermediate with respect to the other strains when the prepulse intensity was 90 dB. In general, irrespective of the dB intensity of the acoustic prepulse stimulus or modality for delivery of the startling stimulus, the C57BL strains (both C57BL/6J and C57BL/10J) showed low levels of PPI *(9)*.

Another variable that could confound interpretation of these results is hearing impairment, which could affect the results, especially in the lower decibel acoustic prepulse conditions. Age-related hearing impairments have been reported for inbred mouse strains. However, because correlations were not observed between the magnitude of the acoustic startle response and the level of PPI, possible hearing impairment was not thought a significant factor influencing these results. Also, testing was begun on mice aged 80 d or less. In any event, the large distribution of levels of PPI among inbred mouse strains, i.e., the identification of phenotypes with high and low levels of PPI, suggests that this is a good behavioral trait to study using quantitative trait loci (QTL) techniques. The QTL approach could lead to the identification of chromosomal loci influencing sensorimotor gating. Ideally, homologous loci in the human would be relevant to risk for schizophrenia in humans. Also, if some of these loci implicate involvement of a specific neurotransmitter system, they could suggest more precisely targeted pharmacological approaches.

As we noted, reduction of PPI of the startle response may be an endophenotype of schizophrenia, either an alternative phenotypic expression or a necessary condition for manifestation of illness. In addition to their genetic association with schizophrenia, endophenotypes can often be measured quantitatively, as opposed to their dichotomous presence or absence, and may be more amenable to statistical procedures linking them to genetic loci. Recombinant congenic mouse strains derived from the parental C57BL/6J and A/J inbred strains showed that the phenotypic variation of PPI of the acoustic startle response is under the control of several genes as QTL, distributed on several chromosomes. Moreover, some of these genes may interact epistatically with each other *(10)*. Several candidate genes of possible relevance to schizophrenia have emerged that are located in the vicinity of the provisional QTLs for PPI, including genes involved in neurodevelopment (such as ephrins A1, A3, and A4; platelet-activating factor acetylhydrolase Ib; integrin α_3; and neural cell adhesion molecule 2), genes involved in presynaptic function (e.g., SNAP-25), and genes regulating neurotransmission (e.g., G protein-coupled receptor kinase 3 and γ-aminobutyric acid [GABA] transporter 1) *(10)*. Examination of the human homologs of these candidate mouse genes in patients with schizophrenia may prove informative.

Rat strains have also been identified that differ with respect to levels of PPI. Wistar-Kyoto and Brown Norway strains have high, and very low levels of PPI, respectively *(8)*. These strain differences are not accounted for by differences in auditory sensitivity. The existence of these two strains prompted the search for QTLs for PPI, which involved phenotyping a large sample of progeny backcrossed to parental strains and conducting genome-wide scans of polymorphic markers in a subsample *(8)*. Suggestive QTLs were identified on rat chromosomes 2 and 18. Potential candidate genes near the QTL on rat chromosome 2 identified in this study include the GABA-A transporter 1 (*Gabt 1*), and nerve growth factor-β (*Ngfb*). In addition, possible candidate genes located in close proximity to the QTL for PPI on chromosome 18 include the β-2 adrenergic receptor (*Adrb2*), galanin receptor 1 (*Galr1*), neuropeptide Y receptor Y6 (*Npy6r*), and gastrin-releasing peptide (*Grp*).

Unfortunately, despite the promising nature of this approach, especially the quantifiable nature of PPI, there are inconsistencies across studies and failures to replicate (for discussion, *see* ref. *8*). Nonetheless, this approach should lead to the identification of candidate genes in humans that regulate PPI and may confer susceptibility to psychosis in general and schizophrenia in particular.

3. IMPAIRED INHIBITORY GATING OF AUDITORY EVOKED POTENTIALS

Patients with schizophrenia have impaired sensory processing. They have difficulty discriminating and responding to specific stimuli, and they are unable to inhibit responding to irrelevant stimuli that do not require their attention *(11,12)*.

The physiological substrate of sensory processing is sensory inhibition or inhibitory gating, which can be assessed formally in humans with a paradigm that measures suppression of the auditory evoked P50 potential response *(11)*. The P50 wave is a small amplitude positive wave occurring about 50 ms after an auditory stimulus. In the absence of pathology, the amplitude of the P50 response to the test stimulus is approx 20% of the amplitude of the response to the conditioning stimulus, reflecting normal sensory inhibition or inhibitory gating mechanisms. In patients with schizophrenia, the amplitude of the P50 auditory evoked potential to the test stimulus is approx 80% of the amplitude of the response to the conditioning stimulus *(11)*.

An abnormally inhibited evoked P50 potential response is also observed in approx 50% of the first-degree biological relatives of index patients with schizophrenia. Most of these relatives are unaffected by the disorder. Thus, the pattern of inheritance of this P50 abnormality of sensory inhibition in schizophrenia is consistent with an autosomal dominant mode, and this pathophysiological deficit may be a necessary, but not sufficient, condition for the phenotypic expression of overt schizophrenia (13–15). The P50 auditory evoked potential abnormality is a trait (referred to as an endophenotype) that segregates among patients with schizophrenia and discriminates between patients and normal controls (16). In a study of 36 unrelated patients with schizophrenia, 91% had TC ratios more than or equal to 0.50, compared with 6% of 43 unrelated normal controls. The TC ratio is the ratio of the amplitude of test and conditioning responses.

GABAergic inhibitory interneurons in the hippocampus that contain α_7 nicotinic acetylcholine receptors on their surface may mediate the normal inhibitory gating of the P50 evoked response to the test stimulus. The stimulation of these α_7 nicotinic acetylcholine receptors leads to the release of GABA, an inhibitory amino acid neurotransmitter, at terminals synapsing on pyramidal cell neurons. In patients with schizophrenia, the expression of the α_7 nicotinic acetylcholine receptor is reduced in several critical regions of the brain, including the hippocampus (the CA3 region and dentate hilus), frontal cortex, and reticular nucleus of the thalamus (17–19). Given the postsynaptic location of the α_7 nicotinic acetylcholine receptor on GABAergic inhibitory interneurons, the decreased density of these receptors in schizophrenia could be the basis of the impaired inhibitory gating mechanisms in this disorder (17). There have been conflicting, but provocative, data suggesting linkage between informative markers of the genetic locus for the α_7 nicotinic acetylcholine receptor polypeptide subunit on chromosome 15 in the q13-q14 region, the P50 sensory gating abnormality, and schizophrenia (16).

Ultimately, the decreased expression of the α_7 nicotinic acetylcholine receptor in selected regions of brain in patients with schizophrenia (e.g., hippocampus) may be caused by the inheritance of abnormal promoter variants, regulating expression of the gene for the α_7 nicotinic acetylcholine receptor polypeptide subunit located on chromosome 15 in the q13-q14 region (20). Specifically, several functional polymorphisms of the core promoter region of the α_7 nicotinic acetylcholine receptor polypeptide subunit gene located on chromosome 15 have been demonstrated; the core promoter region is a 231-basepair fragment located proximally to the ATG translation start site for the gene for the α_7 nicotinic acetylcholine receptor polypeptide subunit. The core promoter region contains consensus-binding sites for several important transcription factors. Functionally, using in vitro procedures whereby some of these promoter variants were attached to a luciferase reporter gene, functional polymorphisms of the core promoter region were shown to cause decreased transcription. Promoter variants causing decreased transcription of the luciferase reporter gene may be associated with, or responsible for, decreased sensory processing (20). Conceivably, abnormal promoter variants may mediate the decreased expression of functionally normal α_7 nicotinic acetylcholine receptors in selected brain regions in schizophrenia.

In mice and rats, sensory inhibition can be assessed directly by electrophysiological recordings of hippocampal CA3 pyramidal cell neurons; recordings from these neurons show a maximally negative deflection (N40) occurring between 20 and 60 ms after an auditory stimulus. The N40 wave is preceded by a positive deflection referred to as P20. Ordinarily, the auditory evoked P20-N40 complex to the second of a pair of identical auditory stimuli presented 500 ms apart is blunted *(21)*. As in humans, the magnitude of this sensory inhibition is defined as the ratio of the amplitude of the response to the second or test stimulus vs the amplitude of the response to the conditioned or first stimulus of the pair. Lesions of the fimbria–fornix, which sever the septal–hippocampal cholinergic projection, disrupt inhibition of the evoked P20-N40 complex to the second stimulus of the pair. Thus, normal sensory inhibition is dependent on the intactness of the cholinergic influence in the hippocampus. Moreover, when the density of α_7 nicotinic acetylcholine receptors in hippocampus was measured in nine inbred mouse strains, a relation was found between the density of these receptors and the magnitude of sensory inhibition *(21)*. As in humans, genetic factors influence sensory inhibition and expression of the α_7 nicotinic acetylcholine receptor in hippocampus. The DBA/2 mouse strain showed a profound deficit in sensory inhibition and a marked reduction in the density of hippocampal α_7 nicotinic acetylcholine receptors. In view of this relation, the DBA/2 mouse strain may serve as a valuable heuristic model of the deficit of sensory inhibition seen in patients with schizophrenia. The DBA/2 inbred mouse's phenotype of deficient sensory inhibition lends itself to the screening of candidate medications that target this fundamental deficit of sensory processing in schizophrenia, as well as studies of the mechanism of action of novel drugs for the treatment of psychosis *(22,23)*.

After weaning at d 21, male Sprague-Dawley rat pups reared in social isolation, as opposed to group housing at three animals per cage, showed a persistent deficit in inhibitory auditory gating, as reflected in high TC ratios *(24)*. Isolation-reared weanling rats were tested after 7 wk of social isolation. Although they were housed individually, they could hear, see, and smell other animals. Isolation-reared animals had lower amplitudes of their evoked responses to conditioned stimuli than the socially reared controls. Amplitudes to test stimuli in isolation-reared animals were higher over 10 baseline recording sessions. Further, latencies of evoked responses to conditioning stimuli in isolation-reared animals were longer. Thus, isolation rearing can affect the early sensory processing of an auditory stimulus, and is a nonpharmacological strategy for causing impaired inhibitory gating of auditory stimuli *(24)*. Nicotine was able to transiently normalize impaired auditory gating in the isolation-reared rats, whereas haloperidol was not able to do so. Thus, socialization and nicotinic interventions may influence the early processing of sensory information *(24)*.

Derivatives and analogs of anabaseine, a naturally occurring nicotinic acetylcholine receptor agonist and toxin extracted from nemertine worms and ants, have been synthesized with selectivity as partial agonists for the α_7 nicotinic acetylcholine receptor. These compounds include GTS-21 (DMXB), DMAB-anabaseine (DMAB), and DMAC-anabaseine *(22)* and were shown to restore sensory inhibition in DBA/2 mice. GTS-21 restored normal sensory inhibition within a period of

10 min and persisted for approx 15 min. Normalization was reflected in a significant decrease in the ratio of the test-to-conditioning amplitude over time at all three of the doses. There was no significant effect of GTS-21 on the amplitude of the conditioning stimulus. The effect seemed specific to the inhibitory gating process. The effect of GTS-21 was antagonized by α-bungarotoxin, a selective $α_7$ nicotinic acetylcholine receptor antagonist *(22)*. The major problem with nicotinic acetylcholine receptor agonist interventions of any type is the rapid desensitization of this general class of receptors. Normalization of deficits in sensory inhibition by $α_7$ nicotinic acetylcholine receptor agonists would, therefore, be expected to be transient.

Unlike conventional antipsychotic medications (such as haloperidol), clozapine, an atypical antipsychotic medication, is able to normalize deficient inhibitory gating of auditory stimuli in patients with schizophrenia and the DBA/2 inbred mouse strain *(23)*. Clozapine improved inhibitory gating, as reflected in a decrease of the TC ratio of the amplitude of the P20-N40 auditory evoked potential. Decreased TC ratios in response to clozapine were largely a result of selective decreases in the amplitude of the evoked response to the test stimulus. The normalizing effect of clozapine was mediated by a selective cholinergic influence. It could be antagonized by the intracerebroventricular administration of α-bungarotoxin, a selective $α_7$ nicotinic acetylcholine receptor competitive antagonist, but not by dihydro-β-erythroidine, an $α_4β_2$ nicotinic acetylcholine receptor antagonist *(23)*. Clozapine is not thought to interact directly with $α_7$ nicotinic acetylcholine receptors. Rather, clozapine may normalize deficient sensory inhibition by enhancing cholinergic influences in the hippocampus. The latter could include elicitation of the release of acetylcholine *(23)*.

4. HETEROZYGOUS REELER MOUSE MODEL AND DIMINISHED REELIN EXPRESSION

The heterozygous reeler mouse expresses approx 50% of the amount of reelin (RELN) found in the wild-type mouse. RELN is a 3461-amino acid glycoprotein that shares structural analogies with extracellular matrix proteins *(25,26)*. In various cortical areas, during the development of the brain and throughout life, a select population of GABAergic interneurons secretes RELN into the extracellular matrix, where it binds to integrin receptors located on the apical and basal dendritic spines of cortical glutamatergic pyramidal neurons and other cells, specifically $α_3$ integrin receptor subunits. The secretion of RELN into the extracellular matrix is a constitutive process. The RELN–integrin receptor interaction links the extracellular matrix with an intracellular transduction process that may ultimately lead to the translocation of soluble tyrosine kinases within the cell. Translocation of soluble tyrosine kinases to the nucleus may lead to the transcription of genes involved in synaptogenesis. Further, the phosphorylation of relevant substrate proteins including the adaptor protein, mouse disabled 1 (mDab1), may positively influence the ribosomal translation of cytoskeletal messenger RNA (mRNA) in the localized regions of spines and dendrites; this is a proposed mechanism of RELN-stimulated changes associated with maturation and learning *(25–27)*.

In the adult, the expression of RELN is thought to regulate dendritic spine density in an experience-dependent fashion; its deficiency is associated with a decreased density of dendritic spines. RELN may be less involved with the regulation of dendritic length and thickness, but more involved with regulating the shape, size, and number of dendritic spines and filopodia *(26)*. The heterozygous reeler mouse may serve as a valuable heuristic animal model of both the neurodevelopmental defect and diminished plasticity, as reflected in the less experience-dependent change in the shape, size, and density of dendritic spines of psychosis in general and schizophrenia in particular *(26,27)*.

During development of the fetal brain, RELN is an important regulator of cell migration and the formation of appropriate connections between cells. These processes are hypothesized to be faulty in schizophrenia. In fact, data show that developing neurons do not migrate normally in schizophrenia, and the disorder has been referred to as a neurodevelopmental disorder of abnormal synaptic connectivity. Decreased synaptic connectivity could contribute to the increased packing density of cells in the cerebral cortex and the decrease in cortical neuropil that is observed in schizophrenia *(25–27)*. The reduced secretion of RELN may also be related to reports of diminished numbers of GABAergic inhibitory interneurons in selected areas of brain in schizophrenia and may serve as the source of RELN.

Alternatively, the diminished transduction of the RELN signal may lead to fewer targets for GABAergic inhibitory neurons. Thus, the reduction of GAD_{67} may be a consequence of the RELN deficiency and decreased density of synaptic spines. There are compelling data suggesting that psychosis, irrespective of whether the diagnosis is schizophrenia or bipolar disorder, is associated with reduced levels of RELN and GAD67 mRNAs and expression of GAD_{67} and RELN proteins in the prefrontal cortex (Brodmann area 9) *(25)*. GAD_{67} is a specific biochemical marker of GABAergic inhibitory interneurons, which are the secretory source of RELN. There was also the suggestion that GAD_{67} protein expression was reduced in the cerebellum from patients with schizophrenia and bipolar disorder with psychosis, and levels of *RELN* mRNA were significantly reduced in the cerebellum of patients with schizophrenia. Reductions of levels of *RELN* mRNA were observed in the cerebellum of patients with bipolar disorder with psychosis, but these reductions were not statistically significant. The downregulation of GAD_{67} and RELN is independent of treatment with neuroleptic medications. Psychosis-related reductions do not seem to be artifactual because of sex or left–right hemispheric differences *(25)*.

As discussed, the haploinsufficient heterozygous reeler mouse expresses approx 50% RELN mRNA in brain compared with the wild-type mouse. It is emerging as a valuable animal model of anatomic and, possibly, behavioral abnormalities of psychosis, especially schizophrenia *(26)*. In this mouse model, RELN and GAD67 mRNA levels in the frontoparietal cortex are approx 50% of the levels observed in the wild-type, without any significant change in the expression of GAD65 mRNA. Further, the stereological counts of RELN-immunoreactive neurons were significantly reduced in five of six cortical layers in the motor frontoparietal cortex of the heterozygous reeler mouse. The RELN-immunoreactive neurons are scattered throughout all layers of the cortex. However, compared with the wild-type animal, their density is significantly reduced. Similarly, the density of GAD_{67}-immunoreactive neurons is reduced significantly in the M2 and Cg1 regions of the frontoparietal cortex of the heterozygous reeler mouse, especially in the superficial layers I–IV. Neuronal packing density of the frontoparietal cortex of the heterozygous reeler mouse is also increased. This is especially striking in cortical layers III–VI. Moreover, the cortical thickness is decreased, consistent with the increased neuronal packing density. There are marked changes in the morphology and density of dendritic spines in the frontoparietal cortex, hippocampus, and cerebellum of the heterozygous reeler mouse, compared with the wild-type. Density of apical and basal dendritic spines in branches of layer III pyramidal neurons in the frontoparietal cortex of the heterozygous reeler mouse is reduced. In addition, the spines themselves are smaller and have shorter necks, and filopodia are difficult to detect. Basal dendritic spine density of CA1 hippocampal pyramidal neurons and the density of dendritic spines on the distal branches of cerebellar Purkinje neurons are reduced significantly in the heterozygous reeler mouse. However, the length and thickness of the dendritic branches do not differ between the heterozygous and wild-type animals.

Mice heterozygous for expression of GAD_{67}, expressing approx 50% of the levels of GAD67 mRNA in frontoparietal cortex, express normal amounts of RELN and have normal density of their dendritic spines in this area. These data suggest that the reduction of GAD67 expression in the heterozygous reeler mouse is secondary to the loss of synaptic targets for the presynaptic GABAergic axon terminals, which are ordinarily located on dendritic spines. Dendritic spines are known to express receptors for GABA, glutamate, and monoamines. Perhaps a reduction in the density of these spines leads to a pruning of axon terminals that use these neurotransmitters for neurotransmission. The behavioral phenotype of the heterozygous reeler mouse includes disruption of PPI, problems with working memory (as reflected in performance of the eight-arm radial-maze task), and problems with socialization and social memory (i.e., abnormal responses to an intruder mouse) *(26)*. In summary, a condition of RELN deficiency may be causative and/or associated with impaired synaptic connectivity and diminished experience-dependent plasticity of dendritic spine dynamics *(26,27)*.

Diminished RELN expression and secretion, which result in abnormalities of lamination of the brain and synaptic connectivity, may be a final common pathway for viral and immunological pathogenic mechanisms of schizophrenia *(28)*. Epidemiological data suggest an association between heightened risk for schizophrenia and maternal infection with the human influenza virus during the second trimester of pregnancy. Strains of the human influenza virus are neurotropic. However, in addition to direct infection of the developing embryonic or fetal brain, the consequences to the developing brain of maternal infection may also result from the stimulation of production of either autoantibodies to a brain-specific protein, or the production of maternal antibodies to an epitope that is shared by the fetal brain and infectious viral particle. In these latter two instances, therefore, evidence of fetal brain abnormalities may occur without recovery of viral cytopathic activity or viral specific antigenicity from the brains of affected neonates *(28)*.

In any event, because of the epidemiological association between schizophrenia and second trimester maternal infection with the human influenza virus, the effects of intranasal instillation and infection of d 9 pregnant C57BL/6 mice with a human influenza virus strain (Influenza A/NWS/33 [H1N1]) on morphological, synaptic changes and RELN production by Cajal–Retzius cells in the brains of d 0 neonatal mice were studied *(28)*. The dose of infectious viral particles inoculated by intranasal instillation of pregnant d 9 mice was not lethal, but did result in reliable infection as shown by lung consolidation, increased lung weight, and recovery of measurable titers of cytopathic virus from lung homogenate. RELN mRNA production is first detected in embryonic mouse brain on d 9.5 and is first produced by pioneer Cajal-Retzius neurons, which are transient neurons that act as pathfinders in the temporally highly orchestrated process of brain lamination. In the reeler mutant mouse, RELN is absent, and because of abnormalities of neuronal migration, more recently differentiated neurons fail to bypass neurons that differentiated earlier in development. Thus, the normal inside-out pattern of lamination of the cerebral cortex is not seen and nests of ectopic neurons are found. Patients with schizophrenia show patterns of abnormal neuronal migration that are similar to the reeler phenotype; as noted, levels of mRNA for *RELN* are decreased by 40 to 50% in the brains of patients with schizophrenia.

The prenatally infected d 0 neonatal mice showed significant reductions in the number of RELN-immunoreactive cells and RELN-immunoreactive cell density in neocortex and hippocampus, compared with sham-infected controls *(28)*. Cajal-Retzius cells were identified in cortical layer 1 of prenatally infected d 0 neonatal mice and were similar in number to sham-infected controls. However, they produced less RELN. In addition, brains from the prenatally infected mice showed global reductions in the thickness of the neocortex, hippocampus, and brain hemisphere. Although a neurotropic human strain of influenza virus was used in this study, viral cytopathic activity or viral specific antigens could not be detected in the brains of prenatally infected d 0 neonates *(28)*. Therefore, immunological mechanisms may mediate the reduced RELN production by Cajal-Retzius and other RELN-immunoreactive neurons in cerebral cortex and hippocampus and its consequences, rather than direct viral infection. Diminished production of RELN as a result of genetic, viral, and/or immunological factors, and their interactions, may be a final common factor responsible for the abnormal corticogenesis, lamination, and synaptic connectivity observed in schizophrenia *(27,28)*.

5. NEONATAL LESION OF THE VENTRAL HIPPOCAMPUS

Preclinical data support the hypothesis that abnormal development of cortical systems, especially those related to prefrontal and temporolimbic cortices and their connectivity to each other, is involved in, or even responsible for, the pathogenesis of schizophrenia *(6)*. Moreover, abnormal development, which may be reflected in alterations of nondopaminergic neurotransmitter systems in the cerebral cortex, may secondarily affect dopamine systems, including subcortical ones. Abnormal cortical development in schizophrenia is subtle and may be related to disruption of the normal developmental process of lamination of the cortex caused by the migratory failures of some young neurons to reach their appropriate target sites *(6)*. A failure of neurons to reach appropriate targets would result in the formation of anomalous connections, affecting how anatomically distributed regions that are linked to each other by neural circuits function in a coordinated manner. These anomalies of normal lamination and connectivity are thought to occur during the second trimester of fetal development, are present at the onset of illness, and may be static, as opposed to other possibly progressive neuroimaging abnormalities that may reflect neurotoxic processes and ongoing excitotoxic damage *(29)*. Histopathological data on postmortem brain from patients with schizophrenia, especially cytoarchitectural disorganization of the entorhinal cortex *(30,31)*, support the neurodevelopmental disruption of normal communication between frontal cortex and hippocampus *(6)*.

The fact that symptoms of schizophrenia in the cognitive and negative symptom domains of psychopathology respond minimally, if at all, to antipsychotic medications support the hypothesis that abnormal cortical development is primary, whereas the disturbance of dopamine systems is secondary. The severity of cognitive and negative symptoms seem to correlate more with structural abnormalities, whereas the response of positive symptoms, such as agitation, hallucinations, and delusional and disorganized thinking, correlate more with competitive blockade of the dopamine type 2 receptor, a pharmacological property shared by conventional and atypical antipsychotic medications *(6)*. A functional downstream consequence of abnormal cortical development is disruption, including subcortical disruption, of dopaminergic neurotransmission. Further, *in situ* gene expression data show that antipsychotic medications have important actions in the region of the nucleus accumbens that receives inputs from prefrontal and temporolimbic cortices. Thus, it is conceivable that antipsychotic medications may have an important therapeutic role to play in normalizing dysfunctional communication between prefrontal and temporolimbic cortices, although this role is likely to be a participatory or cooperative one with other interventions.

The evaluation of neuropsychological test data in schizophrenia suggests that patients are impaired in prefrontal cortical functions that require communication between prefrontal and temporolimbic cortices *(6)*. For example, working memory tasks require this type of communication between prefrontal and temporolimbic cortices, and performance on these tasks is impaired in patients with schizophrenia. Moreover, regional cerebral blood flow studies of both monozygotic twins discordant for schizophrenia and singleton schizophrenia patients support a complex connection and interdependence between the prefrontal cortex and hippocampus, especially while performing working memory tasks and other cognitively effortful tasks, such that diminished activity may be seen in prefrontal cortex, whereas increased activity is observed in the hippocampus *(6)*. Thus, in contrast to patients with discrete anatomical injury of the prefrontal cortex, the impaired function of patients with schizophrenia (as reflected in neuropsychological test performance and regional cerebral blood flow) may be caused by disconnection between circuits linking prefrontal and temporolimbic cortices. It is not the disruption or severing of these circuits in the normally developed adult brain that is thought to account for the symptoms and functional impairment of schizophrenia, rather the lesion in schizophrenia may involve interference with the development of normal circuits, in addition to allowing structures, e.g., prefrontal cortex, to develop deprived of the benefit of normal afferent inputs from other critical developing structures, e.g., hippocampus. Moreover, in addition to the symptoms of schizophrenia, especially the cognitive and negative symptoms, an important consequence of the developmental dysconnectivity of the prefrontal and temporolimbic cortices is the disruption of dopaminergic activity, including subcortical dopaminergic activity.

This conceptualization of the pathogenesis of schizophrenia involving an early developmental disconnection between hippocampus and prefrontal cortex has been modeled successfully in rodents *(6,32,33)*. From a neurodevelopmental perspective, the neurodevelopment of the cerebral cortex of the rat during the first postnatal week corresponds in the primate to the second trimester of gestation. The model is a lesion of the ventral hippocampus of 7-d-old rat pups made with infusion of ibotenic acid, a glutamate analog that is an excitotoxin that destroys intrinsic and projection neurons but spares fibers of passage. The lesion essentially causes a deafferentation of normally occurring hippocampal inputs into the developing medial prefrontal cortex and other cortical and subcortical sites. This lesion is associated with important behavioral consequences related to dopaminergic neurotransmission that do not become manifest until after a considerable developmental delay. Ibotenic acid-lesioned and sham-lesioned animals do not differ from each other before puberty on such measures as haloperidol-induced catalepsy and apomorphine-induced stereotypic behaviors *(32)*. This work was done with Sprague-Dawley rats, an outbred strain.

On postnatal d 56, neonatally lesioned animals show a reduction of catalepsy after an injection of haloperidol and an increased amount of stereotypic activity after an injection of apomorphine. The reduction in haloperidol-induced catalepsy may reflect increased dopaminergic tone in the dorsal striatum. Further, the increased sensitivity to apomorphine-induced stereotypies in the neonatal rats with lesions of the ventral hippocampus is consistent with heightened postsynaptic dopaminergic activity in the dorsal striatum. In addition, after a developmental delay, neonatally lesioned animals show increased locomotor activity in a novel environment and hyperresponsiveness to stress. The injection of saline is a minor stressor; the neonatally lesioned rats show an increased locomotor response to a saline injection at postnatal d 56, but not at postnatal d 35 *(32,34)*. These and other data showing the postpubertal emergence of increased spontaneous locomotion and increased locomotion in response to d-amphetamine in rats with neonatal lesions of the ventral hippocampus suggest that the deafferentation of normal ventral hippocampal inputs into the developing medial prefrontal cortex affects the normal development of this latter area.

Some of the changes that emerge in neonatally lesioned postpubertal animals are reminiscent of similar behavioral changes observed in adult rats with lesions of the medial prefrontal cortex *(33)*. The medial prefrontal cortex of the rat continues to develop into adulthood and is involved in the regulation of subcortical dopaminergic neurotransmission. Its projections innervate the nucleus accumbens and anteromedial caudate–putamen. Thus, the neonatal lesion may result in a primary disruption of cortical development and a secondary effect on dopaminergic neurotransmission in the nucleus accumbens and caudate–putamen. There are, however, projections from the ventral hippocampus directly to the nucleus accumbens. The disruption of these projections with the neonatal lesion of the ventral hippocampus may contribute to the increased behavioral sensitivity to apomorphine, independently of any disruptive effect on the development of the medial prefrontal cortex *(32,34)*. As discussed below, many of the behaviors linked to dopaminergic function in the mesolimbic and nigrostriatal areas are influenced by or modulated by stress.

In addition to the regulation exerted by the medial prefrontal cortex on the response to stress, the prepubertal/postpubertal differences observed in rats may relate to the maturation of the hypothalamic–pituitary–adrenal axis, which is immature before puberty, and disruptive effects on this axis resulting from the faulty development of the medial prefrontal cortex as a result of the neonatal lesion. In any event, this neonatal lesion of the ventral hippocampus in the rat is emerging as a valuable animal model of schizophrenia. The postpubertal animal displays aberrant behaviors related to abnormalities of dopaminergic neurotransmission and these seem to occur secondary to, or relate to, the cytoarchitectural defects in the hippocampal formation and the structural and functional abnormalities of the prefrontal cortex. As discussed below, these neonatally lesioned animals also have problems with sensorimotor gating. They show a postpubertal emergence of diminished PPI of the acoustic startle response *(7)*.

The use of inbred rat strains and varying the size of the excitotoxic damage by adjusting the dose of ibotenic acid showed that genetic factors interact with the extent of the lesion to influence both prepubertal and postpubertal behavioral outcomes, specifically novelty- and amphetamine-induced hyperlocomotion *(33)*. Two different doses of ibotenic acid were used to create small and large ventral hippocampal lesions, respectively, in postnatal d 7 rat pups from the outbred Sprague-Dawley strain and from the inbred Fischer 344 and Lewis strains. The small ventral hippocampal lesion was more restricted to the site of injections, and involved neuronal loss in the hippocampal CA1 and CA2 subregions, no damage in the dentate gyrus or subiculum, and no cavitation. The large ventral hippocampal lesion was characterized by neuronal loss throughout the hippocampal formation, including the dentate gyrus and subiculum, and cavitation at the site of injection. Significant main effects and interactions of age of the animal, extent of the lesion, and genetic strain differences with the behavioral outcome measures were shown *(33)*.

For the outbred Sprague-Dawley strain, behavioral differences in terms of novelty- and amphetamine-induced hyperlocomotion between sham-lesioned and ibotenic acid-lesioned animals emerged after puberty (postnatal d 56) in rats with the large ventral hippocampus lesion only. In contrast to the Sprague-Dawley outbred strain, a small ventral hippocampal lesion of the Fischer 344 strain showed the delayed emergence of hyperlocomotion in response to both novelty and amphetamine. Further, in the Fischer 344 strain, prepubertal animals with a large ventral hippocampal lesion showed hyperlocomotion in response to novelty and amphetamine. The responses of the Fischer 344 rats with the large lesions were mildly exaggerated at postnatal d 56. Whereas the inbred Fischer 344 strain was more sensitive to the effects of the lesion, the inbred Lewis strain was more resistant to the effects of the lesion. With respect to the Lewis rats with large ventral hippocampal lesions, hyperlocomotion in a novel environment and in response to amphetamine did not distinguish between the chemically lesioned and sham-lesioned animals at any age. Thus, there are clear genetic differences with respect to vulnerability to the behavioral effects of an ibotenic acid lesion of the ventral hippocampal formation in neonatal rats. These differences may relate to reported differences between strains in tyrosine hydroxylase in mesolimbic dopaminergic pathways and differences in levels of expression of the serotonin (5-HT) 1A receptor in frontal cortex and hippocampus *(33)*.

The neonatal lesion of the ventral hippocampus of the rat disrupts the architectural integrity of the late-developing medial prefrontal cortex; the latter serves as a neural substrate for the delayed emergence of aberrant behaviors and dopaminergic abnormalities *(35)*. This lesion is also associated with decreased dendritic-spine density and a decreased dendritic length of pyramidal neurons in the medial prefrontal cortex, which could contribute to the dysconnectivity of cortico–cortical associational pathways. The medial prefrontal cortex of the rat is the analog of the dorsolateral prefrontal cortex in primates, a region whose integrity is essential for specialized cognitive tasks, such as working memory. Deficits in working memory have been posited as contributing to the psychopathology of schizophrenia. Thus, it is interesting that working memory deficits were demonstrated in adult rats with neonatal ibotenic acid-lesions of the ventral hippocampus. These tasks were not confounded by a dependence on spatial recognition for their performance *(35)*. Working memory tasks were delayed alternation tasks conducted with food-deprived rats in a T-maze apparatus. Animals were required to remember which arm of the T-maze led to a food reward in a previous trial and select the alternate arm for a similar reward during a subsequent trial. In addition to difficulty making the correct choices, the performance of the neonatally lesioned rats worsened as a function of the duration of the delay between trials. In many respects, the performance of the neonatally lesioned adult rats resembled that of adult animals with lesions of the medial prefrontal cortex *(35)*. Rats with adult lesions of the ventral hippocampus did not show deficits of working memory in these tasks. These data are consistent with hypotheses that an early developmental lesion of the ventral hippocampus disrupts associative functions of the late developing prefrontal cortex, including nonspatially dependent working memory.

6. BEHAVIORAL, MORPHOLOGICAL, AND BIOCHEMICAL EFFECTS OF NONCOMPETITIVE NMDA RECEPTOR ANTAGONISM

PCP is a noncompetitive antagonist of the NMDA receptor complex, a class of glutamate-gated ion channel. PCP binds to a hydrophobic domain within the NMDA receptor-associated ionophore. Thus, PCP is referred to as an open-channel blocker. For PCP to be effective, the channel must assume an open configuration. PCP was originally developed as a dissociative anesthetic agent. However, because of its abuse liability and psychotomimetic properties, it was withdrawn from clinical medicine. From a descriptive perspective, the psychosis precipitated by PCP in normal persons resembles naturally occurring schizophrenia, manifesting positive, negative, cognitive, and motor symptoms. PCP causes exacerbation in patients with schizophrenia. In view of the close resemblance of the PCP-induced psychosis to schizophrenia, and PCP's pharmacological action as a noncompetitive NMDA receptor antagonist, a glutamatergic deficiency or NMDA receptor hypofunction (NRH) has been hypothesized as a possible pathophysiological mechanism of schizophrenia.

The binding of glycine to a site on the NMDA receptor that is distinct from the agonist recognition site for glutamate increases the likelihood that glutamate will be effective in promoting channel opening. In fact, glycine may function as an obligatory co-agonist with glutamate. The existence of a strychnine-insensitive glycine binding site on the NMDA receptor complex has stimulated interest in the development of glycinergic interventions, such as milacemide, an acylated prodrug of glycine, D-cycloserine, and D-serine (glycine agonists), and sarcosine, a glycine reuptake inhibitor, as possible therapeutic interventions for clinical disorders associated with NRH, such as schizophrenia.

Presumptive disturbance of NMDA receptor-mediated neurotransmission in patients with schizophrenia has led to the preclinical characterization of behavioral effects of administering noncompetitive NMDA receptor antagonists, such as PCP and MK-801 (dizocilpine), to rodents and monkeys. Quantitative behavioral effects that have been characterized include locomotor activity, stereotypic behaviors, ataxia, and social interactions. Using operant drug discrimination procedures, food-deprived rats have been trained to discriminate between an injection of PCP and saline, using food as a reinforcer *(36)*.

The NMDA receptor plays an important role in the development and maturation of the fetal brain; thus, the delayed effects of neonatal PCP/MK-801 administration on morphology, behavior, and biochemistry are being actively investigated. These studies are relevant to the neurodevelopmental hypothesis and hypotheses suggesting that schizophrenia results from disruptions of normal synaptic connectivity. Our laboratory has used two behavioral procedures with mice to evaluate the influences of stress, pharmacological interventions, and genetic strain differences on NMDA receptor-mediated neurotransmission in the intact adult animal: the ability of MK-801, an analog of PCP that binds with high-affinity to the same hydrophobic channel domain, to antagonize electrically precipitated tonic hindlimb extension and elicit irregular episodes of intense jumping, referred to as popping.

In one social interaction paradigm, the social interactions of two male Wistar rats placed in an open arena were studied after acute, subchronic, and chronic administration of varying dose levels of PCP *(37)*. In general, PCP caused dose-dependent increases in stereotyped behaviors and ataxia and decreases in exploratory and social behaviors. Ordinarily, pairs of vehicle-treated animals spend substantial amounts of time in close proximity to each other, exploring the unfamiliar arena together. PCP decreases this active social interaction in a dose-dependent fashion. The analysis of the time-course of the behavioral effects of PCP after injection showed that the decrease in active social interaction persisted for up to 3 h, relative to the more short-lived effects of PCP on stereotyped behaviors and ataxia. Further, tolerance did not develop to the effect of PCP on active social interaction even after 21 d of continuous infusion. This contrasted with the effects of PCP on stereotyped behaviors and ataxia, showing that effects on social behavior occurred independently of effects on these other behaviors. The effects of subacute and chronic administration of haloperidol on the behavioral effects of PCP were complex. Nonetheless, in general, haloperidol suppressed stereotyped behaviors and ataxia, but did not have a significant effect on social interaction. In contrast, after 21 d of its administration, clozapine partially normalized the active social interaction of pairs of male Wistar rats treated with PCP *(37)*. Chronic administration of clozapine also positively affected PCP-elicited stereotyped behaviors, but did not interact significantly with PCP-induced ataxia.

Even after 21 d of continuous infusion of PCP, tolerance did not develop to its significant reductions in social behavior, whereas tolerance did develop to the elicitation of stereotyped behaviors and induction of ataxia *(37)*. Stereotyped behaviors were thought to be isomorphic with positive symptoms of schizophrenia, whereas reductions in social interactions are proposed as the modeling of negative symptoms. These data suggest the independence of the so-called positive and negative symptoms in this paradigm. Moreover, the persistence and lack of the development of tolerance to PCP's ability to reduce social interactions implicate NRH in the mediation of the negative symptom domain of psychopathology. This is important because negative symptoms may be more enduring than positive symptoms and contribute more significantly to the functional disability associated with this disorder. Further, social interaction as an outcome measure affected by PCP seems to be able to distinguish between an atypical (clozapine) and conventional (haloperidol) antipsychotic medication *(37)*. The effect of clozapine to partially normalize both the PCP reduction in social interactions and the PCP increase in stereotyped behaviors emerged after its chronic administration. This may have some correlate with the clinical observation that therapeutic effects of clozapine in schizophrenia may emerge as long as 6 mo after its initiation.

In the early development of the central nervous system, stimulation of NMDA receptors may be critical for neuronal cell survival and differentiation, as well as for the establishment of neural networks resulting from experience-dependent plasticity. However, the trophic influence of NMDA receptor stimulation may be present only during a certain critical period of development. Several studies have shown disruptive effects of noncompetitive NMDA receptor antagonists on normal neurobehavioral development *(38)*. These studies have heightened concern about the therapeutic administration of noncompetitive NMDA receptor antagonists, such as MK-801, as neuroprotective and anticonvulsant agents to pregnant women, neonates, infants, and young children. Further, this research has important public health implications because PCP, a frequently abused drug, may disrupt brain development *in utero* when abused by pregnant women. However, the research on the adult consequences of neonatal blockade of the NMDA receptor complex in animals has led to useful models of human neurodevelopmental disorders. These models mimic relevant neurodevelopmental aspects of at least some forms of schizophrenia, especially the early developmental disconnection of circuits between the hippocampus and frontal cortex.

Mice lacking any expression of the NR1 receptor subunit, a common subunit of the heteromeric NMDA receptor complex, die in the perinatal period. Unlike the NR1-null mouse, a mouse line surviving into adulthood has been developed using homologous recombination in embryonic stem cells that express 5–10% of normal levels of NR1 *(39)*. The mRNA for NR1 in these mutant mice showed no alterations of the primary sequence. Thus, what occurs in the brains of these mice is a severe reduction in the expression of the normal transcript compared with wild-type animals. This reduced expression was reflected in a reduction of the maximal density for the binding (B_{MAX}) of ^{3}H-MK-801 to membrane homogenates prepared from the prefrontal cortex of the homozygous mutant mice with diminished expression of NR1, compared with the B_{MAX} observed with homogenates prepared from wild-type animals (0.12 pmol/mg and 1.2 pmol/mg, respectively). There was no apparent change in the affinity of the binding.

Compared with the adult wild-type animals, the adult mutant mice expressing 5–10% of NR1 showed increased locomotor activity and stereotypic behaviors during a 2-h habituation period in a digital activity monitor. The intensity of these behaviors resembled that seen with wild-type adult animals injected intraperitoneally with either 3 mg/kg of PCP or 0.2 mg/kg of MK-801. Furthermore, the increased locomotor activity and stereotypy of the adult mutant mice were attenuated with haloperidol and clozapine. Adult NR1-deficient mice were very sensitive to these behavioral effects of clozapine. Clozapine attenuated abnormal behaviors of the adult NR1-deficient mice at a dose that did not affect the motor behavior of the wild-type littermates. Thus, a nonspecific effect on locomotion or a primary effect on sedation did not confound the attenuation of stereotypic behaviors in the NR1-deficient mice by clozapine. Finally, the NR1-deficient mice were insensitive to the stimulatory effects of PCP or MK-801, consistent with their known pharmacological actions as noncompetitive NMDA receptor antagonists.

NR1-deficient mice also displayed aberrant social and sexual behavior. Compared with wild-type resident male mice, these mice were more likely to socially withdraw or engage in escape behaviors, as opposed to social investigation and fighting, when an intruder male mouse was placed in the home cage of the resident NR1-deficient mutant mouse. Moreover, the NR1-deficient mutant mice were less successful in engaging in mating behaviors with superovulated female mice than the wild-type male mice. Clozapine ameliorated the deficits observed in the social and sexual behaviors of the NR1-deficient mutant male mice *(39)*. These data show that mice with persistence of NRH since conception show behaviors as adults that model some important aspects of schizophrenia. Several of the behaviors were responsive, at least partially, to treatment with clozapine, the prototypic atypical antipsychotic medication.

The transient blockade of NMDA receptors, when they are hypersensitive during a specific stage in ontogenesis, results in apoptotic neurodegeneration *(40)*. Specifically, compared with vehicle-injected neonates, the intraperitoneal injection of [+]MK-801 on postnatal d 7 at 0, 8, and 16 h resulted in a 3- to 39-fold enrichment of the density and number of degenerating neurons in widespread areas of the brain on postnatal d 8, including: CA1 hippocampus; dentate gyrus; laterodorsal, mediodorsal, and ventral thalamus; ventromedial hypothalamus; frontal cortex; parietal cortex; cingulate cortex; and retrosplenial cortex. The proapoptotic effect of MK-801 in neonatal mice was confined to neurons. There was no evidence of a histological process resembling programmed cell death affecting nonneuronal cells. The apoptotic effect of MK-801 was dose dependent. Moreover, after a single dose of MK-801 administered on postnatal d 7, the apoptotic response, as reflected in the increasing density of degenerating neurons, progressed until 24 h after injection, declining dramatically by 48 h. This apoptotic effect of MK-801 seemed to be mediated by NMDA receptors, because it was mimicked by PCP and ketamine (noncompetitive NMDA receptor antagonists) and carboxypiperazin-4-yl-propyl-1-phosphonic acid (a competitive NMDA receptor antagonist). Also, significant apoptosis was not observed after the neonatal injection of 6-nitro-7-sulfamoyl-benzo[f]quinoxaline-2,3-dione (a non-NMDA glutamate receptor antagonist), scopolamine hydrobromide (a muscarinic cholinergic receptor antagonist), haloperidol (a dopamine receptor antagonist), or nimodipine and nifedipine (voltage-gated calcium ion channel blockers).

Thus, antagonism of NMDA receptors may mediate a relatively specific neurotransmitter receptor-mediated role in apoptosis. The spontaneous occurrence of apoptosis occurs normally during development, showing regional selectivity. The apoptotic effect of MK-801 in developing rat forebrain was observable during a critical period of brain development. Apoptosis in the cerebral cortex was not detected on postnatal d 0 in either vehicle-treated or MK-801-treated animals. However, on postnatal d 7, the spontaneous rate of apoptotic neurodegeneration declined to modest levels in all regions. MK-801 treatment resulted in a 14.6-fold increase of apoptotic neurodegeneration over the spontaneous rate in the frontal cortex. On postnatal d 21, there was a relative insensitivity to the proapoptotic effects of MK-801. Thus, the NMDA receptor may play an important developmentally dependent role in the production of apoptosis. Developing rat forebrain was most sensitive to apoptosis on postnatal d 7. The first postnatal week of brain development in the rat corresponds to developmental events occurring during the third trimester of human pregnancy, a period characterized by the densest expression of NMDA receptors and the beginning of the brain growth spurt. These data have implications for the *in utero* exposure of developing fetuses during the third trimester of pregnancy to PCP, ketamine, and ethanol (an NMDA receptor antagonist) in the context of substance abuse *(40)*.

The administration of PCP to female rat pups on postnatal d 7, 9, and 11 resulted in behavioral changes in the postweanling and adult animals, consistent with the modeling of schizophrenia *(41)*. Moreover, these behavioral changes may be related to apoptotic neurodegenerative changes in layers II and III of the dorsolateral frontal cortex and alterations in the expression of NMDA receptors, detectable on postnatal d 12. Biochemical effects of neonatal treatment were prominent in anterior forebrain, including dorsolateral frontal cortex. Apoptotic neurodegeneration was assessed quantitatively with a specialized histological technique, terminal deoxynucleotide transferase-mediated dUTP nick-end labeling (TUNEL), which stains nucleosomal DNA fragmentation. The technique uses the enzyme deoxynucleotidyl transferase, a template-independent polymerase, to incorporate biotinylated nucleotides at sites of DNA breaks. The incorporation of the biotinylated nucleotides into the fragmented DNA associated with apoptosis is amplified and visualized for light microscopy by an avidin–biotin–peroxidase reaction. TUNEL-positive cells increased significantly after PCP treatment in the piriform and dorsolateral frontal cortices. However, there were no significant increases of TUNEL-positive cells in the anterior cingulate cortex, hippocampus, or cerebellum. TUNEL-positive cells increased 3.5-fold, relative to vehicle-treated controls, in dorsolateral frontal cortex.

Neonatal PCP treatment also seemed to be associated with astrocytes in the dorsolateral frontal cortex and the migration of astrocytes into deeper cortical layers, as reflected in the immunohistochemical staining of glial fibrillary acidic protein. The neonatal PCP-treated rat pups showed an alteration in the concentrations of two cytoplasmic proteins in the dorsolateral frontal cortex, whose ratio to each other may determine the likelihood of occurrence of apoptosis: Bax (a proapoptotic protein), and Bcl-X_L (an antiapoptotic protein). Specifically, the treatment resulted in a significant reduction of the ratio of Bcl-X_L to Bax of more than 50%. PCP treatment increased the nuclear translocation of a specific transcription factor, nuclear factor-κB, whose translocation is sensitive to the redox status of the cell and may be involved in the mediation of apoptosis. Moreover, as shown by *in situ* molecular hybridization with a radiolabeled oligonucleotide probe complementary to the mRNA for all of the splice variants of the *NR1* subunit, the expression of *NR1* mRNA was increased throughout the anterior forebrain, including the frontal cortex, striatum, nucleus accumbens, and olfactory tubercle. The amount of *NR1* mRNA was more than twofold higher in the dorsolateral frontal cortex, compared with appropriately neonatally injected controls. In contrast to the anterior forebrain, there were no changes in the level of *NR1* mRNA in posterior sections of hippocampus or cerebellum. Moreover, the increased expression of *NR1* mRNA in the neonatal pups treated with PCP was associated with an actual increase in the amount of NR1 subunit protein that was translated in dorsolateral frontal cortex. Based on the work with the *NR1*-deficient mutant mice *(39)*, it is possible that these increases in *NR1* expression reflect some type of compensatory response.

There was also a variety of enduring behavioral deficits associated with the neonatal administration of PCP, which may have some relation to the biochemical and histological findings that were evident on postnatal d 12, especially in frontal cortex. Rats treated neonatally with PCP were significantly impaired in their speed of acquisition of a delayed spatial alternation task, a task that may reflect working memory and is sensitive to dopaminergic- and NMDA receptor-targeted manipulations. Between postnatal d 38 and 49, the improvement of accuracy of the neonatal PCP-treated rats in the delayed spatial alternation task was significantly slower than the saline-treated controls. This deficit in a task relevant to working memory persisted in the neonatal PCP-treated rats until they reached postnatal d 50 to 70. Neonatal treatment with PCP significantly reduced the PPI of the acoustic startle response in postweanling rats on postnatal d 24 and 25. On postnatal d 42, the neonatal PCP-treated rats showed an exaggerated locomotor response to a PCP challenge, compared with saline-injected controls, when assessed for 100 min in an open-field postchallenge.

In summary, these data show clearly that perinatal PCP administration is associated with dysconnectivity, apoptotic changes, and compensatory changes in NR1 expression, consistent with alterations of NMDA receptor-mediated neurotransmission, especially in frontal cortex *(41)*. Moreover, there are very provocative data suggesting that the disruptive effects of PCP administration on the developing brain, including the behavioral effects, can be antagonized by pretreatment with olanzapine, an atypical antipsychotic medication, during the perinatal period *(41)*. These latter data complement a growing body of exciting recent reports suggesting that atypical antipsychotic medications are distinguished from conventional antipsychotic medications by their ability to prevent progressive pathomorphological changes in first-episode patients with schizophrenia *(42)*. Further, the unique therapeutic actions of the atypical antipsychotic medications may involve expression of brain-derived neurotrophic factor. Quetiapine was reported to antagonize immobilization stress-induced reduction of brain-derived neurotrophic factor in hippocampus *(43)*. Finally, antipsychotic medications, including the atypical ones, were shown to antagonize acute neurotoxic effects of noncompetitive NMDA receptor antagonists in adult rodents *(44)*.

Because of our interest in the PCP model of schizophrenia and the existence of the strychnine-insensitive glycine-binding site on the NMDA receptor, which could serve as a site for the development of medications to treat disorders associated with NRH, our group developed two behavioral procedures to assess NMDA receptor-mediated neurotransmission in the intact mouse *(45–47)*. Not surprisingly, we demonstrated a dose–response relation for the ability of MK-801 to raise the threshold voltage for the precipitation of tonic hindlimb extension. However, quite unexpectedly, we found that if animals are stressed by forcing them to swim for up to 10 min in cold water 24 h before testing the ability of MK-801 to antagonize seizures, there is a dramatic reduction in antiseizure efficacy of MK-801 in our paradigm. Thus, a single session of forced swimming in cold water is a nonpharmacological strategy for altering NMDA receptor-mediated neurotransmission *(46,48)*. A reduction in the antiseizure efficacy of MK-801 is consistent with NRH. For MK-801 to be effective, the NMDA receptor-associated channels must be activated or in the open configuration. Further, we showed that a glycinergic intervention (specifically milacemide, an acylated prodrug of glycine) can up-shift and left-shift the dose–response relation between dose of MK-801 and threshold voltage that is required for the elicitation of seizures. Presumably, this reflects a milacemide-associated increase in the concentration of glycine neurotransmitter within the synaptic cleft of the NMDA receptor, resulting in a higher percentage of channels transiently assuming the open configuration. Milacemide normalized the stress-induced reduction of the antiseizure efficacy of MK-801. These data provided compelling support for the development of glycinergic interventions to treat disorders associated with NRH, especially schizophrenia. Genetically inbred mouse strains were also shown to differ from each other regarding their sensitivity to the antiseizure efficacy of MK-801. The BALB/c strain showed high sensitivity to this effect *(49)*.

Our laboratory has also characterized the dose–dependent elicitation of irregular episodes of intense jumping behavior (popping), in mice by MK-801 *(45)*. Popping behavior was attenuated by haloperidol and clozapine. Further, inbred genetic mouse strains differ in their sensitivity to the elicitation of this behavior. Again, the BALB/c strain was shown to be a sensitive strain *(50)*. To elicit popping behavior by MK-801, the NMDA receptor-associated channel must be in the open configuration. Pretreatment with an NMDA receptor competitive antagonist blocked the ability of MK-801 to elicit popping. Presumably, pretreatment with the competitive NMDA receptor antagonist decreased the proportion of NMDA receptor-associated channels in the open configuration and, thereby, prevented MK-801 from gaining access to its binding site within the hydrophobic domain of the channel. These data highlighted the fact that the elicitation of popping by MK-801 is linked to its specific binding to the unique hydrophobic channel domain. In addition, a dissociation of behavioral effects from MK-801 (the noncompetitive receptor antagonist) and carboxypiperazin-4-yl-propyl-1-phosphonic acid (the competitive receptor antagonist) was shown with this paradigm. The paradigm has been used to show complex interactions between stress, inbred mouse strains, and glycinergic interventions on the elicitation of popping *(51,52)*. Moreover, the paradigm has been used to screen for novel approaches to the attenuation of NRH, such as with inhibitors of nitric oxide synthase, galantamine (a positive allosteric effector of nicotinic receptors), anabasine (a relatively selective α_7 nicotinic acetylcholine receptor agonist), and topiramate (an anticonvulsant with GABA-potentiating and excitatory amino acid antagonist properties), which may be relevant to the future pharmacotherapy of schizophrenia *(53–55)*.

In summary, the PCP model has encouraged intense examination of the consequences of inducing NRH with PCP and MK-801 in neonatal and adult animals.

These models offer a broader spectrum of information, which may be related to the negative and cognitive symptom domains of psychopathology, than was obtained from models focused exclusively on the behavioral and neurochemical consequences of dopaminergic agonist interventions.

ACKNOWLEDGMENT

This work is supported by the Mental Illness Research, Education and Clinical Center (MIRECC) in VISN 5 (Alan S. Bellack, PhD, Director).

REFERENCES

1. Creese I, Iversen SD. The pharmacological and anatomical substrates of the amphetamine response in the rat. Brain Res 1975;83:419–436.
2. Swerdlow NR, Geyer MA. Using an animal model of deficient sensorimotor gating to study the pathophysiology and new treatments of schizophrenia. Schizophrenia Bulletin 1998;24(2): 285–301.
3. Davis M, Gendelman D, Tischler M, Gendelman P. A primary acoustic startle circuit; lesion and stimulation studies. J Neurosci 1982;2:791–805.
4. Cadenhead KS, Geyer MA, Braff DL. Impaired startle prepulse inhibition and habituation in schizotypal patients. Am J Psychiatry 1993;150:1862–1867.
5. Geyer MA, Wilkinson LS, Humby T, Robbins TW. Isolation rearing of rats produces a deficit in prepulse inhibition of acoustic startle similar to that in schizophrenia. Biol Psychiatry 1993;34:361–372.
6. Weinberger DR, Lipska BK. Cortical maldevelopment, anti-psychotic drugs, and schizophrenia: a search for common ground. Schizophr Res 1995;16:87–110.
7. Lipska BK, Swerdlow NR, Geyer MA, Jaskiw GE, Braff DL, Weinberger DR. Neonatal excitotoxic hippocampal damage disrupts sensorimotor gating in postpubertal rats. Psychopharmacology 1995;122:35–43.
8. Palmer AA, Breen LL, Flodman P, Conti LH, Spence MA, Printz MP. Identification of quantitative trait loci for prepulse inhibition in rats. Psychopharmacology 2003;165:270–279.
9. Paylor R, Crawley JN. Inbred strain differences in prepulse inhibition of the mouse startle response. Psychopharmacology 1997;132:169–180.
10. Joober R, Zarate J-M, Rouleau G-A, Skamene E, Boksa P. Provisional mapping of quantitative trait loci modulating the acoustic startle response and prepulse inhibition of acoustic startle. Neuropsychopharmacology 2002;27:765–781.
11. Freedman R, Adler LE, Myles-Worsley M, et al. Inhibitory gating of an evoked response to repeated auditory stimuli in schizophrenic and normal subjects. Arch Gen Psychiatry 1996;53:1114–1121.
12. Venables P. Input dysfunction in schizophrenia. Prog Exp Pers Psychyopathol Res 1964;1:1–47.
13. Freedman R, Adler LE, Bickford P, et al. Schizophrenia and nicotinic receptors. Harv Rev Psychiatry 1994;2:179–192.
14. Leonard S, Adams C, Breese CR, et al. Nicotine receptor function in schizophrenia. Schizophr Bull 1996;22(3):421–445.
15. Adler LE, Olincy A, Waldo M, et al. Schizophrenia, sensory gating, and nicotine receptors. Schizophr Bull 1998;24(2):189–202.
16. Freedman R, Coon H, Myles-Worsley M, et al. Linkage of a neurophysiological deficit in schizophrenia to a chromosome 15 locus. Proc Natl Acad Sci USA 1997;94:587–592.
17. Freedman R, Hall M, Adler LE, Leonard S. Evidence in postmortem brain tissue for decreased numbers of hippocampal nicotinic receptors in schizophrenia. Biol Psychiatry 1995;38(1):22–33.
18. Court J, Spurden D, Lloyd S, et al. Neuronal nicotinic receptors in dementia with Lewy bodies and schizophrenia: α-bingarotoxin and nicotine binding in thalamus. J Neurochem 1999;73: 1590–1597.

19. Guan Z-Z, Zhang X, Blennow K, Nordberg A. Decreased protein level of nicotinic receptor α7 subunit in the frontal cortex from schizophrenic brain. NeuroReport 1999;10:1779–1782.
20. Leonard S, Gault J, Hopkins J, et al. Association of promoter variants in the α7 nicotinic acetylcholine receptor subunit gene with an inhibitory deficit found in schizophrenia. Arch Gen Psychiatry 2002;59:1085–1096.
21. Stevens KE, Freedman R, Collins AC, et al. Genetic correlation of hippocampal auditory evoked response and α-bungarotoxin binding in inbred mouse strains. Neuropsychopharmacology 1996;15:152–162.
22. Stevens KE, Kem WR, Mahnir VM, Freedman R. Selective α7-nicotinic agonists normalize inhibition of auditory response in DBA mice. Psychopharmacology 1998;136:320–327.
23. Simosky JK, Stevens KE, Adler LE, Freedman R. Clozapine improves deficient inhibitory auditory processing in DBA/2 mice, via a nicotinic cholinergic mechanism. Psychyopharmacology 2003;165:386–396.
24. Stevens KE, Johnson RG, Rose GM. Rats reared in social isolation show schizophrenia-like changes in auditory gating. Pharmacol Biochem Behav 1997;58(4):1031–1036.
25. Guidotti A, Auta J, Davis JM, et al. Decrease in reelin and glutamic acid decarboxylase67 (GAD67) expression in schizophrenia and bipolar disorder. A postmortem brain study. Arch Gen Psychiatry 2000;57:1061–1069.
26. Liu WS, Pesold C, Rodriguez MA, et al. Down-regulation of dendritic spine and glutamic acid decarboxylase 67 expressions in the reelin haploinsufficient heterozygous reeler mouse. Proc Natl Acad Sci USA 2001;98(6):3477–3482.
27. Costa E, Chen Y, Davis J, et al. Reelin and schizophrenia: a disease at the interface of the genome and epigenome. Mol Interv 2002;2:47–57.
28. Fatemi SH, Emamian ES, Kist D, et al. Defective corticogenesis and reduction in Reelin immunoreactivity in cortex and hippocampus of prenatally infected neonatal mice. Mol Psychiatry 1999;4:145–154.
29. Deutsch SI, Rosse RB, Schwartz BL, Mastropaolo J. A revised excitotoxic hypothesis of schizophrenia: therapeutic implications. Clin Neuropharmacol 2001;24(1):43–49.
30. Jakob H, Beckman H. Prenatal developmental disturbances in the limbic allocortex in schizophrenics. J Neural Trans 1986;65:303–326.
31. Arnold SE, Hyman BT, van Hoesen GW, Damasio AR. Some cytoarchitectural abnormalities of the entorhinal cortex in schizophrenia. Arch Gen Psychiatry 1991;48:625–632.
32. Lipska BK, Weinberger DR. Delayed effects of neonatal hippocampal damage on haloperidol-induced catalepsy and apomorphine-induced stereotypic behaviors in the rat. Dev Brain Res 1993;75:213–222.
33. Lipska BK, Weinberger DR. Genetic variation in vulnerability to the behavioral effects of neonatal hippocampal damage in rats. Proc Natl Acad Sci USA 1995;92:8906–8910.
34. Lipska BK, Jaskiw GE, Weinberger DR. Postpubertal emergence of hyperresponsiveness to stress and to amphetamine after neonatal excitotoxic hippocampal damage: a potential animal model of schizophrenia. Neuropsychopharmacology 1993;9:67–75.
35. Lipska BK, Aultman JM, Verma A, Weinberger DR, Moghaddam B. Neonatal damage of the ventral hippocampus impairs working memory in the rat. Neuropsychopharmacology 2002;27:47–54.
36. Brady K, Balster R. Discriminative stimulus properties of ketamine stereoisomers in phencyclidine-trained rats. Pharmacol Biochem Behav 1982;17:291–295.
37. Sams-Dodd F. Phencyclidine-induced stereotyped behaviour and social isolation in rats: a possible animal model of schizophrenia. Behav Pharmacol 1996;7:3–23.
38. Deutsch SI, Mastropaolo J, Rosse RB. Neurodevelopmental consequences of early exposure to phencyclidine and related drugs. Clin Neuropharmacol 1998;21(6):320–332.
39. Mohn AR, Gainetdinov RR, Caron MG, Koller BH. Mice with reduced NMDA receptor expression display behaviors related to schizophrenia. Cell 1999;98:427–436.
40. Ikonomidou C, Bosch F, Miksa M, et al. Blockade of NMDA receptors and apoptotic neurodegeneration in the developing brain. Science 1999;283(5398):70–74.

41. Wang C, McInnis J, Ross-Sanchez M, Shinnick-Gallagher P, Wiley JL, Johnson KM. Long-term behavioral and neurodegenerative effects of perinatal phencyclidine administration: implications for schizophrenia. Neuroscience 2001;107(4):535–550.
42. Lieberman J, Charles C, Sharma T, et al., for the HGDH Research Group. Antipsychotic treatment effects on progression of brain pathomorphology in first episode schizophrenia. ACNP 41st Annual Meeting, December 8–12, 2002, San Juan, Puerto Rico.
43. Li X-M, Xu H, Qing H, Lu W, Keegan D, Richardson JS. Quetiapine attenuates the immobilization-induced decrease of brain-derived neurotrophic factor in hippocampus. ACNP 41st Annual Meeting, December 8–12, 2002, San Juan, Puerto Rico.
44. Farber NB, Foster J, Duhan NL, Olney JW. Olanzapine and fluperlapine mimic clozapine in preventing MK-801 neurotoxicity. Schizophr Res 1996;21:33–37.
45. Deutsch SI, Hitri A. Measurement of an explosive behavior in the mouse, induced by MK-801, a PCP analogue. Clin Neuropharmacol 1993;16:251–257.
46. Norris DO, Mastropaolo J, O'Connor DA, Novitzki M, Deutsch SI. Glycinergic interventions potentiate the ability of MK-801 to raise the threshold voltage for tonic hindlimb extension in mice. Pharmacol Biochem Behav 1992;43, 609–612.
47. Deutsch SI, Rosse RB, Paul SM, Riggs RL, Mastroapolo J. Inbred mouse strains differ in sensitivity to popping behavior elicited by MK-801. Pharmacol Biochem Behav 1997;57:315–317.
48. Norris DO, Mastropaolo J, O'Connor DA, Cohen JM, Deutsch SI. A glycinergic intervention potentiates the antiseizure efficacies of MK-801, flurazepam, and carbamazepine. Neurochem Res 1994;19:161–165.
49. Deutsch SI, Mastropaolo J, Powell DG, Rosse RB, Bachus SE. Inbred mouse strains differ in their sensitivity to an antiseizure effect of MK-801. Clin Neuropharmacol 1998;21(4):255–257.
50. Deutsch SI, Rosse RB, Mastropaolo J. Behavioral approaches to the functional assessment of NMDA-mediated neural transmission in intact mice. Clin Neuropharmacol 1997;20(5):375–384.
51. Deutsch SI, Rosse RB, Schwartz BL, Powell DG, Mastropaolo J. Stress and a glycinergic intervention interact in the modulation of MK-801-elicited mouse popping behavior. Pharmacol Biochem Behav 1999;62(2):395–398.
52. Billingslea EN, Mastropaolo J, Rosse RB, Bellack AS, Deutsch SI. Interaction of stress and strain on glutamatergic neurotransmission: relevance to schizophrenia. Pharmacol Biochem Behav 2003;74:351–356.
53. Deutsch SI, Rosse RB, Paul SM, et al. 7-Nitroindazole and methylene blue, inhibitors of neuronal nitric oxide synthase and NO-stimulated guanylate cyclase, block MK-801 elicited behaviors in mice. Neuropsychopharmacology 1996;15:37–43.
54. Deutsch SI, Rosse RB, Billingslea EN, Bellack AS, Mastropaolo J. Topiramate antagonizes MK-801 in an animal model of schizophrenia. Eur J Pharmacol 2002;449:121–125.
55. Deutsch SI, Rosse RB, Billingslea EN, Bellack AS, Mastropaolo J. Modulation of MK-801-elicited mouse popping behavior by galantamine is complex and dose-dependent. Life Sci 2003;73:2355–2361.

11
Animal Models of Anxiety
The Ethological Perspective

Robert Gerlai, Robert Blanchard, and Caroline Blanchard

Summary

Anxiety is a complex neurobehavioral disorder that often includes co-morbid features, e.g., obsessive–compulsive behavior. Animal models can be generated by several techniques, one of which includes transgenic and knockout technologies. Although anxiety can be assessed in various environments, we examine our animals in a naturalistic or ethological setting in what we call the mouse defense test battery (MDTB), in which predator model-induced anxiety and fear responses are studied. Additionally, we use several pharmacological agents known to induce or reduce fear and anxiety in humans. The pharmacological data gathered from the MDTB demonstrated that panic-modulating agents specifically and appropriately affect the flight responses of mice. Using the principles of ethology, we also examined motor and posture patterns of Alzheimer's disease mouse mutants (PD-APP), and found abnormalities not associated with learning and memory but suggestive of alterations in fear responses and anxiety. Animal models such as the PD-APP mouse may, thus, allow investigators to screen existing and novel drugs for their ability to alter noncognitive traits, including anxiety, in a preclinical setting.

Key Words: Anxiety; fear; knockout mice; transgenic mice; mouse defense test battery; anxiolytic drugs.

1. INTRODUCTION: ANXIETY IN ANIMALS?

Anxiety is a complex and diverse human disorder that may manifest in numerous ways. From a clinical standpoint, anxiety includes, for example, generalized anxiety disorder, panic, obsessive–compulsive disorder, posttraumatic stress disorder, and numerous phobias, to mention but a few symptom clusters. The severity of these disorders also varies and occasionally the definition and the borderline among the different disorders is somewhat blurred. Given the large prevalence of anxiety disorders in humans, extensive efforts have been focused on the understanding of the biological underpinnings of these diseases. A potential fruitful method to pursue this question is to use animal models in which certain features of human anxiety may be more easily tested in well-controlled laboratory settings. With the use of animals,

From: *Contemporary Clinical Neuroscience: Transgenic and Knockout Models of Neuropsychiatric Disorders*
Edited by: G. S. Fisch and J. Flint © Humana Press Inc., Totowa, NJ

however, arise the questions of how one can define a human problem such as anxiety in animals, and how one can model this complex disease. Another problem is whether anxiety is one single disease with differing severity of symptoms or a set of unrelated disorders. The critical answer to this question will lie in the discovery of the mechanisms that underlie these diseases. Animal models may facilitate the discovery of such mechanisms. We propose that the most fruitful way to investigate anxiety-like characteristics in animals is to take the ethological approach, i.e., to try to understand natural behavioral responses, those that are relevant from an ecological viewpoint, a point we return to shortly.

Animal models of anxiety, like other models of human psychiatric diseases, may be grouped into five main approaches according to the way the alteration of brain function is elicited or manipulated. First, there are behavioral models in which anxiety is induced by presenting particular environmental stimuli (e.g., predators or stimuli-characterizing predators, or pain-inducing stimulation, and so on). The second category concerns the models of anxiety in which the behavioral change is elicited by systemic administration of compounds ranging from pharmacological agents (e.g., anxiogenic drugs) to hormones. Novel recombinant DNA techniques, including viral vector-mediated delivery of certain genes (gene therapy) or antisense oligonucleotide-based knock-down studies, including RNA interference techniques, leading to temporary reduction of gene expression of particular target genes also fall in this category. The third group is genetic models, for example, transgenic approaches, in which particular genes are overexpressed or underexpressed. The fourth approach investigates the effect of spontaneously occurring mutations or differences among inbred strains. The fifth category is that in which electrical stimulation of particular brain areas and other more invasive surgical techniques are used.

The general problem with all anxiety models is that anxiety is a complex human behavioral phenomenon that is influenced by a large number of environmental and genetic factors. The conundrum scientists using animal models have to face is that without knowing the exact mechanisms of the human disease it is difficult to know how to model the disease properly, but without appropriate models it is difficult to unravel the mechanisms of the human disease, a situation analogous to a "catch 22." Nevertheless, because science is always an iterative process, numerous human and animal studies have been conducted providing useful feedback to one another. The increasing amount of information gained suggests that unraveling the biology of anxiety, albeit difficult, will not be impossible.

Among the numerous potentially useful approaches, we propose that naturalistic methods have one of the best chances to move the field forward. This follows from a view that anxiety, and fear in general, is based fundamentally on particular circuitry or functional characteristics of the brain that have been shaped by natural selection. What we observe as abnormally anxious behavior in the 21st-century human is based on a neurobiology that evolved in response to the demands of the environment in which *Homo sapiens* and its ancestors lived. On the one hand, we argue that the general layout of the mammalian brain is identical across a broad range of species and that there are numerous genetic and physiological similarities among mammals, thus, it is reasonable to assume that some of the neurobiological characteristics of the mechanisms underlying anxiety are evolutionarily conserved. On the other hand, we also appreciate that each species is adapted to its own environment and, thus, anxiety may be induced, and may manifest, differently, from species to species. That is, the stimuli to which a species responds may differ, and the manner in which the particular species responds may also reflect its own abilities uniquely shaped by its evolutionary past. Thus, we propose that for one to understand the common underlying biological mechanisms of anxiety, one needs to design experiments that take into account the ecological and ethological constraints and specificities of the model organism used *(1)*.

This chapter focuses on these ethological approaches. Instead of presenting a comprehensive review of all anxiety models, we will give two main examples to illuminate the use of the naturalistic approach. In the first example, we present a thorough discussion on the use of a naturalistic experimental set up, the mouse defense test battery (MDTB) in which a predator model induces anxiety, and fear responses are studied. The concept of this test battery has been validated for multiple species and its characteristics have been investigated using behavioral and pharmacological methods in a detailed manner, allowing numerous comparisons between the ethological and other animal psychological approaches.

In the second part of this chapter, the ethological method of recording motor and posture patterns is highlighted. This example is presented to show how the investigators of a particular study *(2)* could discover anxiety-related alterations in an animal model of Alzheimer's disease (AD) when they used the ethological method.

2. DEFENSIVE BEHAVIORS AS MODELS OF ANXIETY: FROM EVOLUTIONARY CONCEPTUALIZATION TO PHARMACOLOGICAL FOCUS

According to the Academic Press *Dictionary of Science and Technology* the term *fear* refers to "the state in which the perception of danger, aversive stimuli, or pain produces defensive or escape behavior" or to "a strong and unpleasant emotional state brought about by the threat of danger, pain, or suffering. "Anxiety" is defined as "a strong and unpleasant feeling of nervousness or distress in response to a feared situation." Clearly, the two definitions do not differ substantially. In this chapter, we also use these terms somewhat interchangeably, noting, however, that although fear may be an adaptive response, anxiety has a pathological component (i.e., its manifestation may be abnormal fear responses). Furthermore, we also note that, as defined, anxiety is a human condition and, thus, we prefer to use the term *anxiety-like* responses when we discuss behavior of animals.

Having discussed these definitions, we turn to the question of how fear may be elicited. In contrast to the classic view of fear as a learned response to stimuli associated with aversive (painful) events, ethological studies have clearly demonstrated that a number of unconditioned, nonpainful, stimuli are also capable of eliciting a wide range of specific defensive behaviors related to fear. These factors, in addition to some unusually specific patterns of behaviors associated with particular defenses, have contributed to an increasing emphasis on defensive behaviors in response to unconditioned threat stimuli in animal models of anxiety. The behaviors described in these ethological models are coming to be incorporated into other aversive situations, including the elevated plus maze, which are widely used to evaluate anxiety *(3)*.

The ethological approach suggests that careful attention to behavior may make it possible to identify manipulation-induced changes in defensive behaviors in a variety of situations and models, including those with a primary focus on psychopathologies other than anxiety. Thus, in addition to providing new models of anxiety, defensive behaviors may supplement information on primary variables (e.g., cognitive abilities in the case of AD) and contribute to a more complete understanding of the effects of these paradigms and the conditions they model.

3. DEFENSIVE BEHAVIORS IN REPONSE TO UNCONDITIONED THREAT STIMULI

The unconditioned defensive behaviors of rodents seem to consist of at least the following: flight, hiding, freezing, defensive threat, defensive attack, and risk assessment. These defensive behaviors are species typical (i.e., typically expressed by individuals of those species under appropriate circumstances), but not species-specific. They occur in much the same form across a variety of mammalian species *(4)*. For instance, the most primitive mammals, monotremes, display immobility and hiding (rolling into a ball, digging holes and hiding there; echidnas) as well as flight and defensive threat and attack (platypuses). Marsupials also show flight, freezing, and defensive threat and attack, under circumstances that seem to be similar to those that control these same behaviors in placental mammals *(5)*. Indeed, a host of nonmammalian species also show many defensive behaviors similar to those of mammals, although nonmammalian species tend to have additional structural defenses, such as crypsis and protective scales or spines, which are less common in mammals.

The statement that these responses are unconditioned reflects the fact that each behavior can be elicited in wild rats or wild mice and—sometimes—in laboratory or domesticated strains of rats and mice, without previous relevant experience (*see* ref. *6* for review of behavioral analyses in wild vs domesticated rats and mice). Predator odors, as well as live predators are commonly used threat stimuli in such ethological tests, but salient and unexpected stimuli, such as loud noises, sudden motion, novel situations, high places, or moving substrate *(7–9)* all can elicit defense.

A characteristic of defensive behavior that contributes greatly to the ease of specifically eliciting particular defenses is its dependence on features of the threat stimulus and the situation in which the threat is encountered. Manifest and tangible threat stimuli, such as living predators, elicit high-magnitude defenses, the forms of which are strongly dependent on the situation in which the predator is encountered. Predators tend to elicit flight when an escape route is available, hiding if there is a place of protection or concealment; and freezing when neither of these features is present. A predator closely approaching and contacting the mouse or rat subject first elicits defensive threat (e.g., sonic vocalizations) and then defensive attack. The tangibility of the threat stimulus is particularly important to invoke a defensive attack. Although it could be argued that defensive attack may simply be hard to identify in the absence of a tangible threat, we have never seen anything that looked remotely like an attack (e.g., lunging or jaw snapping) in thousands of observations of rats and mice in the presence of intangible threat stimuli.

When the predator threat is less clearly manifest, the dominant form of defense is often risk assessment, a category that includes a number of activities involved in gathering information about a threat source *(10)*. These may include orientation to the threat source; sensory scanning (sniffing and auditory and visual scanning, marked by side-to-side head sweeps); and approach/investigation. A particularly useful marker for risk assessment in rats and mice (and, potentially, in other species) is a low-back, long-body stretched posture or movement that seems to permit investigation of the threat source while minimizing the probability that the animal will be detected by the threat. Movement during the stretch-approach phase of risk assessment is typically marked by periods of motionlessness interspersed with rapid, stealthy, movement.

The concept of a less manifest threat may involve partial predator stimuli, such as odors. However, complete predator stimuli, i.e., the presence of a living predator, may also constitute an ambiguous threat. Although defensive behaviors are generally very resistant to habituation *(11)*, predators that do not attack or are inactive seem to induce uncertainty regarding their immediate dangerousness to potential prey. Even a chasing predator elicits occasional stopping and looking back (orientation) in a fleeing mouse, presumably because of uncertainty about whether the predator is still chasing. Escaping unnecessarily is costly from an evolutionary perspective.

4. THE MDTB: DEVELOPMENT AND BEHAVIOR ANALYSIS

The MDTB was largely developed by Guy Griebel, who was a postdoctoral fellow in the Blanchard laboratory at the time. Its development was facilitated by a number of previous studies of defensive behaviors and their response to drugs in laboratory and wild rats (summarized in ref. *6*). The MDTB was designed specifically to elicit each of these behaviors in as precise a manner as possible, within the confines of a short (10- to 12-min long) test. It is run in an oval runway with a longitudinal central wall not extending to either end; an arrangement that permits the mouse to run along the alley on one side of the wall, around the end and on, without hindrance. It uses a deeply anesthetized rat, moved by hand, as the stimulus set representing a predator. The designation of the rat as predator is not arbitrary, because rats are indeed predators of mice, and mice are clearly stressed by their proximity and are behaviorally responsive to rats *(12,13)*. Lastly, anesthetizing the rat allows the experimenter to standardize the movement of the predator and, thus, reduce experimental error variation. The MDTB comprises several subtests, including a prethreat period in which activity is measured; a period of being chased by the oncoming rat in the chase–flight test; a slower approach by the rat when the mouse subject is trapped in a straight alley formed by blocking off one side of the oval runway; contact by the rat in the same situation; and a postthreat period after the rat is removed. The behaviors evaluated in the MDTB consist of:

1. Flight: in the chase–flight test, measured as avoidance frequency and the distance between the threat stimulus and the subject when avoidance occurred; also average and maximal flight speed (measured over a straight section of the alley).
2. Freezing: in the straight-alley test.
3. Defensive threat (vocalizations) in response to approach and contact in the straight-alley test.
4. Defensive attack: (jump attacks or biting) in the straight-alley test.
5. Risk assessment: measured as stops, orientations, and reversals in the chase–flight (oval runway), and approaches to and withdrawals from the rat in the straight alley.
6. Contextual defense: attempts to escape the alley after the rat has been removed.

The contextual-defense measure is compared with escape attempts during an initial period of identical length at the beginning of the test, before introduction of the rat, to provide an index of enhanced defensiveness to the apparatus (context) in which the rat was presented, after rat confrontation. The same preperiod also provides a measure of activity (line crossings) before rat exposure.

Because the MDTB was developed on the basis of a great deal of information on the ability of specific conjunctions of situational/stimulus characteristics to elicit particular defensive behaviors (albeit in rats) it was no surprise to find that these situational/stimulus configurations elicited similar responses in mice. However, the organization of defense in the MDTB was subsequently confirmed *(14)* by factor analyses identifying four main independent factors. Factor 1 included cognitive aspects of defensive behaviors that seem to be related to the process of acquiring and analyzing information in the presence of threatening stimuli (i.e., risk assessment). Factor 2 included a number of measures of flight. Factor 3 included several defensive threat/attack reactions (i.e., upright postures and biting), i.e., more affective-oriented defense reactions. Factor 4 included escape attempts in the period after removal of the rat, or contextual defensiveness. The independence of these factors suggests that the threat-related behaviors of mice exposed to a predator may involve underlying neural systems that may differ for particular behaviors; a view that has led to a great deal of work involving drug effects on the behaviors measured in the MDTB.

5. DRUGS, ANXIETY-LIKE RESPONSES, AND DEFENSIVE BEHAVIOR

To date, more than 70 drug studies using the MDTB have been reported. The results of these studies have recently been summarized (*see* ref. *4*, a review emphasizing the effects of antipanic and propanic drugs in the MDTB; and ref. *15*, reviewing all published MDTB drug to that date).

Because of earlier work on defensive behaviors in rats, drug studies in the MDTB started out with the expectation that anxiolytic drugs would decrease defensive threat/attack and have some type of effect on risk assessment, altering flight and freezing much less consistently. In fact, benzodiazepine full agonists with known antianxiety efficacy, such as diazepam, chlordiazepoxide, and clobazam, do reduce defensive threat/attack and risk assessment while the mouse is being chased (reviewed in ref. *15*). Serotonergic anxiolytics, such as the serotonin 1A agonist, buspirone, and selective serotonin reuptake inhibitors, such as fluoxetine (given on a chronic basis), also consistently reduce defensive threat/attack, as does chronic administration of the tricyclic drug, imipramine. Consistent with a common finding that initial exposure to either fluoxetine or imipramine tends to exacerbate anxiety; acute administration of these drugs actually enhances defensive threat/attack behaviors *(16)*. In general, drugs of known efficacy against generalized anxiety disorder reduce both defensive threat/attack and risk assessment, with the interesting exception that serotonin 1A agonists seem to largely impact the former measure. A number of benzodiazepine partial agonists have had more variable effects than the classic, full agonist, anxiolytics. This is consonant with the lack of reports of anxiolytic efficacy for these drugs, some of which have been used in clinical trials *(17)*.

The consistent reduction of risk assessment in the chase–flight test by drugs effective against generalized anxiety is particularly felicitous, in that this consistent reduction avoids a problem of bidirectional effects of anxiolytics on risk assessment, as had been previously seen in rats *(18)*. In fact, there seems to be a genuine difference between mice and rats regarding the expression of risk assessment to threat stimuli. Studies of rats *(10)* or mice *(19)* living in habitats with tunnels and burrows as well as an open area suggest that the mice, but not rats, show immediate risk assessment after presentation of a predator. Specifically, mice rapidly return to the entrance of a burrow to which they had fled and peer out at a cat in the open area, whereas rats only return to the entrance some hours later. Thus, risk assessment in rats, but not mice, seems to be greatly inhibited by high-level defensiveness, such that risk assessment appears only as such high-level defensiveness declines, or in situations in which it is relatively low to begin with. This likely underlies the bidirectional effect of anxiolytics on this measure in tasks varying in threat magnitude and immediacy, with anxiolytics from a baseline of high defensiveness, but reducing it when baseline levels of risk assessment are already high *(18)*. In contrast, mice show risk assessment even when confronted by high-level threat stimuli, and, in the midst of expressing high-level defenses, such as flight. This is obviously not a problem when the difference between the two species is recognized, but it emphasizes the necessity for tests or models developed in one species to be thoroughly described and validated again when applied to other species.

6. FLIGHT AND PANIC: THE MDTB AND OTHER MEASURES RELATED TO FLIGHT

Just before the development of the MDTB, Deakin, Graeff, and colleagues suggested that the desire to flee may be a core component of panic *(20)*. This view was largely based on clinical observations, but additionally suggested that panic may be caused by the spontaneous activation of hypothalamic–periaqueductal grey (PAG) flight mechanisms. Because the MDTB includes several direct measures of flight, it provided an appropriate situation for evaluation of this hypothesis in an animal model.

The results for propanic, antipanic, and panic-neutral drugs in the MDTB are summarized in *(4)*. Briefly, two high-potency benzodiazepines (alprazolam and clonazepam) and chronic treatments with various antidepressants (imipramine, fluoxetine, moclobemide, and phenelzine) all effective against panic attacks, also reduced flight behavior in the MDTB *(14,21–23)*. A number of drugs known to trigger or potentiate human panic responses, such as acute imipramine, fluoxetine, yohimbine, cocaine, or the benzodiazepine receptor antagonist, flumazenil *(24–27)*, were found to increase flight *(14,28,29)*. Findings that the classic benzodiazepine full agonist, chlordiazepoxide, the serotonin 1A receptor agonist, buspirone, and the serotonin 2 receptor antagonist, mianserin, did not modify flight behavior in this test *(30,31)* are also in agreement with clinical data. Mixed findings with reference to diazepam are consonant with reports that this benzodiazepine full agonist impacts panic, but only at higher doses *(32,33)*.

This view that flight mechanisms are important in panic has resulted in the development of other procedures with a focus on flight. One of these measures is locomotion away from an unconditioned threat source (an open, elevated arm) in an elevated T-maze *(34,35)*, whereas another has a focus on operant behaviors that allow the animal to interrupt aversive PAG stimulation and escape from it *(36,37)*. The results of pharmacological studies using these measures have been mixed, but generally supportive of the view of a relationship between flight and panic.

Recent studies using a protocol for the elevated T-maze procedure in which animals are habituated to the open arms of the maze before drug testing, a procedure that decreases baseline escape latencies compared with the original nonhabituated version, have indicated that chronic, but not acute, administration of clomipramine and fluoxetine increases escape latency, as does chronic administration of imipramine *(38,39)*. These findings are consonant with the effects of chronic administration of all of these drugs on panic, although acute administration of imipramine failed to show the expected escape enhancing (i.e., reduced latency) effect. However, contrary to expectations that it might enhance flight, 1-(m-chlorophenyl) piperazine (mCPP) increased open-arm escape latencies in both habituated and nonhabituated forms of the elevated T-maze *(39)*. Buspirone, used as a negative control, had no effect when given on either an acute or a chronic basis *(38)*. Previous studies using a nonhabituated version of the test produced more variable results (compare Table 1 in this chapter to Table 2 in ref. *4*, which includes the earlier results), indicating that the more rapid and, thus, more flight-like respond-

Table 1
Flight-Based Models of Panic: Effects of Some Antipanic and Propanic Drugs on Flight in the MDTB, Self-Interruption Task, and (Habituated) Elevated T-Maze[a]

	MDTB	Self-interruption task	Elevated T-maze
Clinically effective antipanic treatments			
Alprazolam	+	+	
Clobazam	+		
Clomipramine [a]			+
Clonazepam	+	+	
Clorazepate	+	+	
Diazepam	+		
Fluoxetine [c]	+	+	+
Imipramine [c]	+		+
Moclobemide [c]	+		
Plenelzine [c]	+		
Panic-provoking challenges			
Caffeine	—		
CCK$_4$		0	
Clomipramine [a]			—
Cocaine	—		
Flesinoxan		—	
Flumazenil	—		
Fluoxetine [a]	—	+	—
Fluvoxamine [a]		+	
Imipramine [a]	—	0	0
mCPP		+	+
Yohimbine	—	—	

[a] + indicates a panicolytic response; —, a panicogenic response; 0, no response. MDBT, mouse defense test battery; mCPP, 1-(m-chlorophenyl) piperazine; CCK4, cholecystokinin 4.

ing seen in previously habituated animals enhances the sensitivity of the elevated T-maze model to panic-altering drugs.

In the self-interruption task, alprazolam and clonazepam, both of which acutely ameliorate panic, reduce the aversive response to dorsal PAG stimulation, whereas yohimbine, caffeine, and flesinoxan, which precipitate panic attacks, potentiate this response *(40,41)*; findings supporting a view that PAG stimulation-induced aversion is appropriately responsive to antipanic and propanic drugs. However, some panic-provoking or potentiating drug treatments, such as mCPP, or acute administration of fluoxetine and fluvoxamine, tended to reduce the effects of dorsal PAG stimulation in this model *(37,42,43)*. These effects were also reduced by some compounds (haloperidol, or the cholecystokinin receptor B antagonist, L-365,260) *(36,44)* that have no clinical effect on panic.

Taken together, these pharmacological data from the MDTB demonstrate that panic-modulating agents specifically and appropriately affect the flight responses of mice, when these are evaluated directly. The considerable success of some additional tasks designed to tap flight- or escape-related responses found in additional situations involved altered behaviors in response to antipanic and propanic drugs, and is consonant with this view. That these protocols become more responsive to such drugs when fine-tuned to produce a clear flight response also suggests the importance of using measures that provide a clear and sole focus on the behavior of interest. Although this focus is clearly important in any animal model, models of defense have the great advantage of using previously identified, unconditioned behaviors of known cross-species validity. The use of ethologically relevant environmental stimuli, such as exposure to a natural predator, should facilitate the analysis of fear and anxiety-like responses in animals. Test paradigms using this approach may shed light on alterations in anxiety-like responses induced by a variety of methods, including genetic and pharmacological manipulations. The two main aspects of the naturalistic approach are the use of ecologically relevant stimuli and the quantification of species-typical motor and posture patterns. The latter quantification method is rarely used in the era of computer-automated recording. Nevertheless, as the following example demonstrates, the method may reveal previously undiscovered findings.

7. THE EXAMPLE OF THE PD-APP MOUSE: HINTS OF PREVIOUSLY MISSED CHARACTERISTICS

Numerous transgenic mouse models of AD have been generated and all of them have been thoroughly investigated regarding their cognitive or mnemonic characteristics. Focusing on cognitive abilities of the transgenic mice is understandable, given that AD is associated with progressive deterioration of learning and memory. However, this narrow focus represents a potential problem because it may result in failure to identify the possibly broad range of phenotypical changes that a mutation generates. The present experimental example *(2)* shows that analysis of motor and posture patterns, a method often used by ethologists, can reveal numerous abnormalities that may have nothing to do with learning and memory *per se*; some of these suggest alterations in fear responses and anxiety. The mouse model of this example was one of the first AD mouse models generated, the APP^{V717F} mouse, also known as the PD-APP mouse *(45)*.

7.1. PD-APP Mice Were Designed to Model Pathological and Behavioral Characteristics of Human AD

PD-APP mice express the human amyloid precursor protein *(APP)* gene carrying a mutation leading to substitution of valine by phenylalanine at amino acid residue 717, the first mutation found in humans that leads to an early onset familial AD *(46)*. The expression of the APP transgene was driven by the platelet-derived growth factor (PD) promoter in these transgenic mice. AD-associated neuropathological changes, including the appearance of neurofibrillary tangles and neuritic plaques containing the Aβ peptide have been thoroughly investigated in the PD-APP mice and these mice were found to exhibit only Aβ neuritic plaques, not neurofibrillary tangles. In addition, the mice showed astrocytosis and microgliosis *(45,47)*, pathological alterations observed in human AD. At the behavioral level, PD-APP mice exhibit impaired performance in spatial learning paradigms in the water maze *(48)* or radial maze *(47)*, as well as in object-recognition tasks *(47,48)*.

It is notable, however, that although impaired cognitive abilities are not the only concern in AD patients, the noncognitive alterations are often ignored in preclinical (animal) studies. Complex disturbances of emotion and behavior have been recognized as part of the syndrome of AD since Dr. Alzheimer reported his original observation in 1907 (for review, *see* ref. *49*). The behavioral disturbances include psychosis, depression, aphasia, agnosia, apraxia, and personality disorders *(50–53)*. Analyses of the PD-APP mice previously ignored the possibility that changes other than cognitive alterations may also occur in mouse models of AD.

8. FEAR CONDITIONING WITH EVENT RECORDING: A TOOL FOR BEHAVIORAL PROFILING

To address this problem, PD-APP mice were tested in a learning paradigm, context- and cue-dependent fear conditioning *(54)*, which allowed the investigators to study not only memory performance but also several other aspects of behavior, including behavioral posture patterns associated with motor activity and numerous responses associated with fear *(2)*. The study included comparative quantification of motor and posture patterns that were exhibited by the PD-APP mice and their wild-type control counterparts in fear conditioning. The principle of this event-recording approach is that the behavioral patterns, just like gestures or facial expressions in human communication, reveal fundamental characteristics of the psychological/behavioral status of the experimental animal. The results of event-recording analysis in the fear-conditioning paradigm showed that although memory performance of PD-APP mice seemed impaired, this impairment was associated with alterations in fear responses and modified exploratory behavior, changes that were not known before.

Briefly, fear conditioning entailed placing the mice into a chamber in which they received three electric shocks during a short (6 min) training session. The shocks were paired with a tone cue. In subsequent tests, the mice were studied for their responses to the tone cue (the associative stimulus) and to the shock chamber itself (the so-called context). Traditionally, the behavioral response to pain or fear in rodents is tested by measuring immobility, or freezing *(54)*. This reaction is a natural behavior that is elicited by predators but can also be observed under artificial laboratory conditions *(55)*. In addition to this behavior, numerous other behavioral patterns were tested in the paradigm (for detailed definitions of the patterns, *see* refs. *2* and *56*). What was found was somewhat surprising. PD-APP transgenic mice exhibited significant alterations in freezing in the cue test only, which, at first sight, may suggest impaired elemental learning, i.e., reduced ability to make association between two stimuli. However, the reduction of freezing in the cue test could be seen even before the tone cue was presented. Thus, it could not have been related to elemental learning but perhaps was caused by a generalized alteration of fear. The pattern of changes across all three phases of the fear-conditioning paradigm also suggested consistent, albeit not always significant, reduction of freezing, a result that is indeed compatible with altered fear.

Nevertheless, because fear cannot be measured *per se*, one could also argue that reduction of freezing was perhaps caused by altered motor performance, e.g., increased general activity. Quantification of locomotion, however, revealed that PD-APP mice did not exhibit more activity compared with their wild-type counterparts during training. In fact, these mice showed a modest reduction of locomotor activity. Similarly, locomotion of PD-APP mice was not increased in the context test compared with control. Thus, elevation of general activity could be ruled out. The increased locomotion of PD-APP mice found in the cue test may be caused by factors other than general activity levels. A potential factor may be emotionality or fear.

Several behavioral elements have been previously shown to correlate with fear (*see*, e.g., refs. *57* and *58*). One of them, grooming, has been shown to decrease in response to electric shocks or presentation of stimuli that were previously associated with shocks. Grooming was decreased in PD-APP mice during the training phase of fear conditioning, and a similar trend was observed in the context test, suggesting that the transgenic mice were more responsive (fearful) to the training environment and the shock. This conclusion was supported by the analysis of another behavioral element, long-body posture (or stretch-attend posture, as first described in the literature by the Blanchards, e.g., ref. *18*). Increased amount of long-body posture was found in PD-APP mice compared with control during training and during the context test, and a similar trend was also seen during the cue test.

Long-body posture has been observed under aversive conditions in mice (and rats) when cues associated with natural predators are present *(18)*. This behavior is also evoked in mice by other fear-inducing stimuli, including electric shocks or the context in which the shocks were delivered *(57,58)*. Thus, long-body posture has been interpreted as a sign of fear or anxiety *(18,57,58)*. Accordingly, increased long-body posture suggested increased fear in PD-APP mice. It is also notable that long-body posture may represent a type of fear that is qualitatively different from the behavioral state associated with freezing. Although the latter behavior can be seen in the case of clear presence of danger or pain *(55)*, long-body posture is usually observed when a rodent is ambivalent regarding the nature or degree of danger *(18)*.

The experimental analysis of PD-APP mice uncovered yet another previously not described alteration. Analysis of spontaneous activity of PD-APP mice before shocks were delivered in the training session showed that these mice exhibited significantly reduced frequency of leaning, a behavior characterized by upright posture during which the front paws are touching the wall of the apparatus. This behavior is thought to represent exploratory activity *(59–61)*. It has been suggested to be associated with gathering information about the surroundings of the experimental subjects. The robust reduction of leaning frequency seen in PD-APP mice compared with wild type control implies that these mice exhibited reduced exploratory behavior, which, in turn, might have led to an impaired ability of these mice to gather information about their environment.

This hypothesis is in accordance with previous findings showing impaired object recognition *(47)* and spatial learning performance *(48)* in PD-APP mice, and it may also explain why PD-APP mice behaved like experimental subjects with incomplete information about the contextual cues of the fear-conditioning chamber.

In summary, the profile of behavioral abnormalities of PD-APP mice revealed by the quantification of motor and posture patterns suggests that these mice may suffer from alterations in fear and exploratory behaviors. What was previously thought to be impaired cognition *per se* may actually be caused, at least in part, by alterations in these characteristics, a working hypothesis whose validity will be investigated by tests specifically developed for the analysis of fear or anxiety and exploratory behavior.

Proper characterization and discovery of changes in noncognitive traits in animal models may have crucial relevance for drug development and, thus, for the human clinic. Disturbances including affective and emotional behavioral abnormalities and sleep-related problems are common among AD patients (for review, *see* ref. *49*) and these alterations are often more debilitating than memory loss itself. It is also important to note that numerous drugs have already been, or are being, developed for psychiatric diseases whose symptoms include some of the aforementioned disturbances. Animal models such as the PD-APP mouse may thus allow investigators to screen existing and novel drugs for their ability to alter noncognitive traits, including anxiety, in a preclinical setting, which will significantly accelerate the development of pharmacotherapies for the patient.

9. CONCLUSION

The data from genetic models of AD and from studies with a focus on defensive behaviors in response to a variety of stimuli and situations all contribute to a view that an ethological emphasis on natural behavioral patterns can add substantially to our understanding of the consequences of important organismic and situational manipulations. They suggest the importance of defense systems under natural as well as pathological conditions. In particular, systematic differences in responsivity of different defensive behaviors to drugs effective against particular classes of anxiety disorders raise the possibility that differences in human symptomatology may, in part, reflect abnormal functioning of separate neural systems associated with these particular behaviors. If so, then defensive behaviors may provide an especially useful means of approaching the biology of anxiety disorders.

REFERENCES

1. Gerlai R, Clayton NS. Analysing hippocampal function in transgenic mice: An ethological perspective. Trends Neurosci 1999;22:47–51.
2. Gerlai R, Fitch T, Bales K, Gitter B. Behavioral impairment of APPV717F mice in fear conditioning: Is it only cognition? Behav Brain Res 2002;136:503–509.
3. Rodgers RJ. Animal models of 'anxiety': where next? Behav Pharmacol 1997;8:477–496.

4. Blanchard DC, Griebel G, Blanchard RJ. Mouse defensive behaviors: pharmacological and behavioral assays for anxiety and panic. Neurosci Biobehav Rev 2001;25:205–218.
5. Eisenberg JF. The Mammalian Radiations: An Analysis of Trends in Evolution, Adaptation, and Behavior. Chicago: University of Chicago Press, 1981.
6. Blanchard DC. Stimulus and Environmental Control of Defensive Behaviors. In: Bouton M, Fanselow M, eds. The Functional Behaviorism of Robert C. Bolles: Learning, Motivation and Cognition. Washington, DC: American Psychological Association, 1997, pp. 283–305.
7. Blanchard DC, Lee EMC, Williams G, Blanchard RJ. Taming of Rattus norvegicus by lesions of the mesencephalic central gray. Physiological Psychology 1981;9:157–163.
8. Endler JA. Defense against predators. In: Feder ME, Lauder GV, eds. Predator-Prey Relationships. Chicago: University of Chicago Press, 1986, pp. 109–134.
9. King SM. Escape-related behaviours in an unstable elevated and exposed environment. I. A new behavioural model of extreme anxiety. Behav Brain Res 1999;98:113–126.
10. Blanchard RJ, Blanchard DC. Anti-predator defensive behaviors in a visible burrow system. J Comp Psychol 1989;103:70–82.
11. Blanchard RJ, Nikulina JN, Sakai RR, McKittrick C, McEwen BS, Blanchard DC. Behavioral and endocrine change following chronic predatory stress. Physiol Behav 1998;63:561–569.
12. de Catanzaro D. Effect of predator exposure upon early pregnancy in mice. Physiol Behav 1988;43(6):691–696.
13. Nikulina EM. Neural control of predatory aggression in wild and domesticated animals. Neurosci Biobehav Rev Winter 1991;15:545–547.
14. Griebel G, Blanchard DC, Blanchard RJ. Predator-elicited flight responses in Swiss-Webster mice: an experimental model of panic attacks. Prog Neuro-Psych Biol Psych 1996;20:185–205.
15. Blanchard DC, Griebel G, Blanchard RJ. The mouse defense test battery: pharmacological and behavioral assays for anxiety and panic. Eur J Psychol 2003;463:97–116.
16. Griebel G, Blanchard DC, Jung A, Lee J, Masuda CK, Blanchard RJ. Further evidence that the mouse defense test battery is useful for screening anxiolytic and panicolytic drugs: effects of acute and chronic treatment with alprazolam. Neuropharmacology 1995;34:1625–1633.
17. Haefely W, Martin JR, Schoch P. Novel anxiolytics that act as partial agonists at benzodiazepine receptors. Trends Pharmacol Sci 1990;11:452–456.
18. Blanchard DC, Blanchard RJ, Rodgers RJ. Risk Assessment and Animal Models of Anxiety. In: Olivier B, Mos J, Slangen JL. Animal Models in Psychopharmacology. Basel, Switzerland: Birkhauser Verlag AG, 1991, pp. 117–134.
19. Blanchard RJ, Parmigiani S, Bjornson C, Masuda C, Weiss SM, Blanchard DC. Antipredator behavior of Swiss-Webster mice in a visible burrow system. Aggress Behav 1995;21:123–136.
20. Deakin JFW, Graeff FG. 5-HT and mechanisms of defense. J Psychopharmacol 1991;5:305–315.
21. Blanchard RJ, Griebel G, Henrie JA, Blanchard DC. Differentiation of anxiolytic and panicolytic drugs by effects on rat and mouse defense test batteries. Neurosci Biobehav Rev 1997;21: 783–789.
22. Griebel G, Perrault G, Sanger DJ. Behavioural profiles of the reversible monoamine-oxidase-A inhibitors befloxatone and moclobemide in an experimental model for screening anxiolytic and anti-panic drugs. Psychopharmacology 1997;131:180–186.
23. Griebel G, Curet O, Perrault G, Sanger DJ. Behavioral effects of phenelzine in an experimental model for screening anxiolytic and anti-panic drugs: correlation with changes in monoamine-oxidase activity and monoamine levels. Neuropharm 1998;37:927–935.
24. Darragh A, Lambe R, O'Boyle C, Kenny M, Brick I. Absence of central effects in man of the benzodiazepine antagonist Ro 15-1788. Psychopharmacology 1983;80:192–195.
25. Harrisonread PE, Tyrer P, Lawson C, Lack S, Fernandes C, File SE. Flumazenil-precipitated panic and dysphoria in patients dependent on benzodiazepines: a possible aid to abstinence. J Psychopharmacol 1996;10:89–97.
26. Higgitt AC, Lader MH, Fonagy P. The effects of the benzodiazepine antagonist Ro 15-1788 on psychophysiological performance and subjective measures in normal subjects. Psychopharmacology 1986;89:395–403.

27. Schopf J, Laurian S, Le PK, Gaillard JM. Intrinsic activity of the benzodiazepine antagonist Ro 15-1788 in man: an electrophysiological investigation. Pharmacopsychiatry 1984;17:79–83.
28. Blanchard RJ, Kaawaloa JN, Hebert MA, Blanchard DC. Cocaine produces panic-like flight responses in mice in the mouse defense test battery. Pharmacol Biochem Behav 1999;64: 523–528.
29. Blanchard RJ, Taukulis HK, Rodgers RJ, Magee LK, Blanchard DC. Yohimbine potentiates active defensive responses to threatening stimuli in Swiss-Webster mice. Pharmacol Biochem Behav 1993;44:673–681.
30. Griebel G, Perrault G, Sanger DJ. A comparative study of the effects of selective and non-selective 5-HT2 receptor subtype antagonists in rat and mouse models of anxiety. Neuropharm 1997;36:793–802.
31. Griebel G, Perrault G, Sanger DJ. Characterization of the behavioral profile of the non peptide CRF receptor antagonist CP-154,526 in anxiety models in rodents—comparison with diazepam and buspirone. Psychopharmacology 1998;138:55–66.
32. Ballenger JC. Efficacy of benzodiazepines in panic disorder and agoraphobia. J Psychiatr Res 1990;24(Suppl 2):15–25.
33. Denboer JA, Slaap BR. Review of current treatment in panic disorder. Int Clin Psychopharmacol 1998;13:S25–S30.
34. Graeff FG, Netto CF, Zangrossi H. The elevated T-maze as an experimental model of anxiety. Neurosci Biobehav Rev 1998;23:237–246.
35. Graeff FG, Viana MB, Mora PO. Opposed regulation by dorsal raphe nucleus 5-HT pathways of two types of fear in the elevated T-maze. Pharmacol Biochem Behav 1996;53:171–177.
36. Jenck F, Broekkamp CL, Van Delft AM. Effects of serotonin receptor antagonists on PAG stimulation induced aversion: different contributions of 5HT1, 5HT2 and 5HT3 receptors. Psychopharmacology 1989;97:489–495.
37. Jenck F, Broekkamp CL, Van Delft AM. Opposite control mediated by central 5-HT1A and non-5-HT1A (5- HT1B or 5-HT1C) receptors on periaqueductal gray aversion. Eur J Pharmacol 1989;161:219–221.
38. Poltronieri SC, Zangrossi H Jr, de Barros Viana M. Antipanic-like effect of serotonin reuptake inhibitors in the elevated T-maze. Behav Brain Res 2003;147:185–192.
39. Zangrossi H Jr, Viana MB, Zanoveli J, Bueno C, Nogueira RL, Graeff FG. Serotonergic regulation of inhibitory avoidance and one-way escape in the rat elevated T-maze. Neurosci Biobehav Rev 2001;25:637–645.
40. Jenck F, Moreau JL, Martin JR. Dorsal periaqueductal gray-induced aversion as a simulation of panic anxiety: Elements of face and predictive validity. Psychiatry Res 1995;57:181–191.
41. Jenck F, Martin JR, Moreau JL. The 5-HT1A receptor agonist flesinoxan increases aversion in a model of panic-like anxiety in rats. J Psychopharmacol 1999;13:166–170.
42. Jenck F, Broekkamp CL, Van Delft AM. 5-HT-sub(1C) receptors in the serotonergic control of periaqueductal gray induced aversion in rats. Psychopharmacology 1990;100:372–376.
43. Jenck F, Moreau JL, Berendsen HHG, et al. Antiaversive effects of 5HT(2C) receptor agonists and fluoxetine in a model of panic-like anxiety in rats. Eur Neuropsychopharmacol 1998;8: 161–168.
44. Jenck F, Martin JR, Moreau JL. Behavioral effects of CCKB receptor ligands in a validated simulation of panic anxiety in rats. Eur Neuropsychopharmacol 1996;6:291–298.
45. Games D, Adams D, Alessandrini R, et al. Alzheimer-type neuropathology in transgenic mice overexpressing V717F beta-amyloid precursor protein. Nature 1995;373:523–527.
46. Goate A, Chartier-Harlin MC, Mullan M, et al. Segregation of a missense mutation in the amyloid precursor protein gene with familial Alzheimer's disease. Nature 1991;349:704–706.
47. Dodart JC, Meziane H, Mathis C, Bales KR, Paul SM, Ungerer A. Behavioral disturbances in transgenic mice overexpressing the V717F beta-amyloid precursor protein. Behavioral Neuroscience 1999;113:982–990.
48. Chen G, Chen KS, Knox J, et al. A learning deficit related to age and beta-amyloid plaques in a mouse model of Alzheimer's disease. Nature 2000;408:975–979.

49. Gilley DW. Behavioral and affective disturbances in Alzheimer's Disease. In: Parks RW, Zec RF, Wilson RS, eds. Neuropsychology of Alzheimer's Disease and Other Dementias. Oxford, UK: Oxford University Press, 1993, pp. 112–137.
50. Ferris SH, Kluger A. Assessing cognition in Alzheimer disease research. Alzheimer Dis Assoc Disord 1997;11:45–49.
51. Perry RJ, Hodges JR. Attention and executive deficits in Alzheimer's disease. A critical review. Brain 1999;122:383–404.
52. Collie A, Maruff P. The neuropsychology of preclinical Alzheimer's disease and mild cognitive impairment. Neurosci Biobehav Rev 2000;24:365–374.
53. Cummings JL. Cognitive and behavioral heterogeneity in Alzheimer's disease: seeking the neurobiological basis. Neurobiol Aging 2000;21:845–861.
54. Gerlai R. Behavioral tests of hippocampal function: simple paradigms, complex problems. Behav Brain Res 2001;125:269–277.
55. Blanchard RJ, Blanchard DC. Crouching as an index of fear. J Comp Physiol Psychol 1969;67:370–375.
56. Gerlai R, Friend W, Becker L, O'Hanlon R, Marks A, Roder J. Female transgenic mice carrying the human gene for S100β are hyperactive. Behavioural Brain Research 1993;55:51–55.
57. Fitch T, Adams B, Chaney S, Gerlai R. Force transducer based movement detection in fear conditioning in mice: a comparative analysis. Hippocampus 2002;12:4–17.
58. Gerlai R, Shinsky N, Shih A, et al. Regulation of learning by EphA receptors: a protein targeting study. J Neurosci 1999;19:9538–9549.
59. Cabib S, Algeri S, Perego C, Puglisi-Allegra S. Behavioral and biochemical changes monitored in two inbred strains of mice during exploration of an unfamiliar environment. Physiol Behav 1990;47:749–753.
60. van Abeelen JH, van den Heuvel CM. Behavioural responses to novelty in two inbred mouse strains after intrahippocampal naloxone and morphine. Behav Brain Res 1982;5:199–207.
61. van Daal JH, Herbergs PJ, Crusio WE, et al. A genetic-correlational study of hippocampal structural variation and variation in exploratory activities of mice. Behav Brain Res 1991;43:57–64.

12
Modeling Human Anxiety and Depression in Mutant Mice

Andrew Holmes and John F. Cryan

Summary

Mood and anxiety disorders represent some of the most common and proliferating health problems worldwide, but are inadequately treated by existing therapeutic interventions. Valid animal models of anxiety and depression have a critical role to play in identifying novel therapeutic targets for these debilitating conditions. The emergence of techniques that allow genetic manipulation of specific molecules in mice has added a valuable new dimension to this field of research. In this chapter, we discuss some of the conceptual issues surrounding the use of mutant mice to study anxiety and depression, the behavioral tasks commonly used for assessment, and important caveats associated with the use of mutant mice. Anxiety-related behavior is most commonly assayed in mice using tests based on exploratory approach/avoid conflict (e.g., elevated plus maze, open field, light–dark exploration, and hyponeophagia), although a variety of alternatives exist (e.g., stress-induced hyperthermia, mouse defense test battery, Vogel conflict, shock-probe burying, four-plate test, marble burying, and separation-induced pup ultrasonic vocalizations). Mouse tests for depression-related behaviors include models based on "behavioral despair" (forced-swim test, tail-suspension test, and learned helplessness), as well as chronic mild stress, olfactory bulbectomy, and psychostimulant withdrawal. Various factors can confound performance and complicate interpretation of the behavior of mutant mice on anxiety- and depression-related tasks, including genetic background, abnormal motor and sensory phenotypes, previous test history, and variability in early life environment and parental behavior. In addition, lifelong constitutive mutations can recruit compensatory changes that occlude the normal function of a molecule in neural circuits mediating emotion-related behaviors. Mutant mice provide a particularly valuable approach to the study of anxiety disorders and depression and their treatment when used in conjunction with other techniques to generate converging lines of evidence regarding the role of a molecule in these circuits. When viewed as such, we believe that mutant mice will continue to foster the study of anxiety and depression.

Key Words: Anxiety; depression; approach/avoidance behavior; fear conditioning; learned helplessness; stress tests; behavioral despair; mutant mice; genetic strains.

1. INTRODUCTION

1.1. Opening Remarks

Mood and anxiety disorders represent some of the most common and proliferating health problems worldwide *(1,2)*. The seriousness of these conditions cannot be overestimated. In sufferers, they can interfere considerably with normal everyday functioning throughout the lifespan, and are associated with significant mortality in the form of greater risk for a variety of medical complaints as well as suicide *(3–5)*. Mood and anxiety disorders are also a major financial burden; the cost of depression to the US economy alone has recently been estimated to be in excess of $80 billion per year *(6)*. Against this emerging pandemic, current treatments for depression and anxiety disorders are of limited efficacy in a significant proportion of patients and are associated with a troublesome side-effect burden in many others *(7)*. Clearly, a better understanding of the pathophysiology of these disorders and the development of novel, improved therapeutic treatments would fill a major unmet medical need *(8)*.

For many years, research in infrahumans, primarily rodents, has provided the foundation for efforts to understand the neural systems subserving anxiety and depression, how these systems dysfunction under pathological conditions and how they can be therapeutically modulated. In clinical practice, many tools have been developed and validated to better diagnose anxiety disorders and depression and to assess the efficacy of treatment strategies in humans. These range from the Diagnostic and Statistical Manual, 4th Edition (DSM-IV) of the American Psychiatric Association *(3)* to the various rating scales, such as the Hamilton rating scales *(9)*. Further, clinicians also rely on self-reports from patients for diagnosis in depression, and it goes without saying that no such tools are available when we shift to animal models. Numerous attempts have been made to create rodent models of anxiety and depression, or at least of the symptoms of these disorders. Although evaluating the usefulness of these models is less than straightforward, general criteria have been proposed (*see* refs. *10–13*). Some of the most widely cited criteria were developed by McKinney and Bunny *(14)*, who proposed that the minimum requirements for an animal model of a psychiatric disease (in this case, depression) are that it:

1. Be "reasonably analogous" to the human disorder in its manifestations or symptomatology.
2. Causes a behavioral change that can be monitored objectively.
3. Produces behavioral changes that are reversed by the same treatment modalities that are effective in humans.
4. Should be reproducible between investigators.

These principles provide a foundation for animal models and apply across various approaches to the study of anxiety and depression.

Research on brain systems mediating anxiety and depression in rodents has traditionally used a combination of pharmacological, lesional, and electrophysiological approaches. The emergence of techniques that allow targeted manipulation of specific molecules at the genetic level has added a valuable new dimension to this and many other fields of neuroscience research (15). Many gene knockout, knock-in, and transgenic mice have been assessed as part of studies aimed at understanding the neural basis of anxiety, stress, and depression. The use of these mutant mice has led to some important advances in our understanding of anxiety- and depression-related behaviors in rodents. In some instances, findings have reinforced and refined existing hypotheses regarding the importance of certain systems in mediating emotion (e.g., corticotropin-releasing factor [CRF], serotonin). In other cases, phenotypic analysis of mutant mice has helped nurture novel avenues of research that are more difficult to tackle with more traditional methods (e.g., the role of specific intracellular signaling molecules, such as cyclic adenosine monophosphate response element-binding protein).

Although research on genetically mutated mice has provided some major advances in this field of research, there are, as with any research tool, certain important caveats associated with the use of mutant mice. In this chapter, we discuss some of the conceptual issues surrounding the use of mutant mice to study anxiety and depression and the behavioral tasks commonly used to assess these processes. A description of the many abnormal anxiety-like and depression-related phenotypes in various mutant mice is not provided; the reader is referred to several recent reviews for this information (16–20).

2. ANXIETY AND FEAR

Anxiety is an appropriate, adaptive response to impending danger that is integral to an organism's preparations to deal with or avoid a potential environmental threat. In this sense, anxiety differs from depression in that, although people will increasingly describe themselves as being "depressed" when experiencing low mood, depression is still most appropriately used as a term to describe a pathological condition (parenthetically, some investigators do conceive of depression as an adaptive response to insurmountable environment demands, e.g., see ref. 21). Anxiety can also be differentiated from fear (Table 1). Fear responses occur in response to explicit, imminent threats and are usually short lived, evoking intense escape and avoidance of the threat responses. In contrast, anxiety responses occur in reaction to less explicit, more generalized threats, and promote preparedness by increasing arousal and risk assessment. Anxiety responses tend to be more sustained and longer lasting than fear responses. Although being highly adaptive responses to threat, abnormal fear and anxiety responses can manifest as serious pathological conditions requiring clinical intervention. Individuals suffering from anxiety disorders (of which there are various subclassifications) repeatedly experience fear and/or anxiety responses that are out of proportion to the level of tangible threat to the point that everyday functioning is compromised (3).

Table 1
Differentiating Fear and Anxiety in Humans and Mice

	Nature of threat	Response evoked	Clinical manifestations	How could it be modeled in mice?
Anxiety	• Potential • Ambiguous • Distant	• Long-lasting/sustained • Hyperarousal, increased cognitive risk assessment	• Generalized anxiety disorder • Agoraphobia	• Tests based on approach/avoid conflict (e.g., elevated plus maze) • Exposure to ambiguous/partial cue associated with fear
Fear	• Explicit • Proximal-imminent • Unambiguous	• Short-lived/phasic • Intense motivation to avoid and escape	• Panic attack • Specific phobia	• Exposure to predator (e.g., mouse defense battery) • Exposure to explicit cue associated with fear

2.1. Mouse Tests for Anxiety-Like Behavior Based on Approach/Avoidance Conflict

There is a rich behavioral pharmacological literature in the rat that has informed the study of anxiety-like behavior in mutant mice. Many behavioral tasks that were developed and validated on the basis of their ability to predict the effects of anxiolytics in rats have been adapted for use in mice. However, mice are not little rats, and rat tests have translated into mouse versions with varying success. The social interaction *(22)* and the fear-potentiated startle tests are two examples that have proven useful in rats studies, but have, thus far, been less widely validated in mice *(23)*. On the other hand, there are tasks that were explicitly developed for use in mice, including the light–dark exploration *(24)* and stress-induced hyperthermia *(25)* tests. These and many other commonly used mouse tests for anxiety-like behavior are ethologically based, exploiting what is known about the natural behavioral patterns of the species *(26)*. Perhaps most popular among these are the exploration-based tasks.

Mice have an innate aversion to open and brightly lit spaces (presumably an adaptive response to reduce the risk of predation), but are also a naturally foraging, exploratory species. Exploration-based tasks exploit these conflicting tendencies in the form of a test apparatus in which the mouse's drive to *approach* is in conflict with the *avoidance* of potential threat. The aversive/threatening area can take various manifestations: an open, elevated arm (elevated plus maze); open, elevated quadrant (elevated zero maze); light compartment/arena (light–dark-exploration test, dark/light-emergence test); mirrored arena (mirrored-chamber test); a staircase (staircase test); or central area of a novel or brightly lit open field (open-field test) (for a more complete description of these tests, *see* refs. *18* and *27*). Over a typical test session, nonmutant, nondrug-treated mice are expected to exhibit a clear avoidance of these aversive areas relative to the more protected zones in the apparatus.

Although the concept of approach/avoid conflict provides these tests with some face validity, they also, critically, demonstrate predictive validity. That is, drugs with anxiolytic action in humans produce decreases in avoidance of aversive/threatening areas in mice. Conversely, drugs with proanxiety effects in humans often potentiate the anxiety-like avoidance response in these tests. The same logic is extended to the interpretation of phenotypic abnormalities in mutant mice on these tests. Thus, a decreased level of avoidance of the threatening areas of the apparatus (e.g., elevated plus maze open arms) in a mutant mouse, relative to a nonmutant wild-type control, is interpreted as a decreased level of anxiety-like behavior or an "anxiolytic-like phenotype." Contrariwise, increased avoidance behavior in a mutant relative to wild-type controls is interpreted as evidence of heightened anxiety-like behavior, an "anxiogenic-like phenotype."

Although the exploration-based tests have undoubtedly demonstrated true positive anxiety-related phenotypes in various mutant mice *(18–20,27)*, there are some important caveats associated with their use. One conceptual issue stems from the

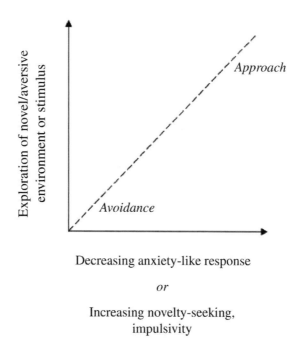

Fig. 1. Anxious or impulsive? Mouse tests for anxiety-like behavior generate a conflict between a natural tendency to approach an aversive and *novel* environment or stimulus vs an innate avoidance of such situations because of the potential for encountering danger. Therefore, increased exploration of aversive, novel stimuli in a mutant mouse may indicate abnormally high novelty seeking or impulsivity, rather than reduced anxiety-like behavior, and vice versa. Specific tests for novelty seeking or impulsivity may be required to fully resolve the true nature of an abnormal mutant phenotype.

fact that, as discussed, these tests are based on the interplay between two parallel processes, approach and avoidance. Therefore, it is conceivable that a decrease in avoidance behavior does not always indicate a decrease in anxiety-like behavior, but rather reflects an increase in approach (Fig. 1). Increased approach could derive from abnormally high novelty seeking or impulsivity in a mutant. The potential for such ambiguity is illustrated by studies of dopamine D4 receptor knockout mice that show behavioral changes that would be consistent with either an increase in anxiety-like behavior or a decrease in novelty seeking *(28,29)*. In such cases, further testing for anxiety- and emotion-related behaviors, or assessment of novelty seeking or impulsivity may be required to fully resolve the true nature of an abnormal mutant phenotype.

Exploration-based tests are explicitly dependent on intact sensory and motor function. It is obvious that a mutation that causes abnormal tactile or visual acuity could alter performance on anxiety tests in which the absence of thigmotactic cues and high levels of illumination drive avoidance behaviors. Furthermore, abnormalities in locomotor function have the potential to cause a false-positive anxiety-related phenotype in these tests. Quantification of general measures of locomotor activity in a test, e.g., distance traveled in the light–dark test; closed-arm entries in the elevated plus maze, can increase confidence in the anxiety-related specificity of an abnormal behavioral phenotype in a mutant mouse. Alternatively, mutants can be assessed on formal tests of locomotor activity (e.g., open field) and motor function (e.g., rotarod coordination); however, caution should be taken to minimize the stressful nature of such tasks to avoid cross-contamination with anxiety! Lastly, measuring certain forms of ethological, defensive behavior that may be less reliant on motor function (e.g., such as risk assessment/scanning of the elevated plus maze open arms), has also proven to be a valuable adjunct to mouse anxiety tests *(26,45)*.

2.2. Nonlocomotor Activity-Based Mouse Tests for Anxiety-Related Behavior

When a locomotor deficit is evident in a mutant mouse, alternatives to exploration-based tasks can be used to measure anxiety-like behavior. The stress-induced hyperthermia paradigm *(25)* and mouse defense test battery *(30)* are two tasks that have been well validated for use in mice. Other available tasks include the Vogel conflict test *(31)*, hyponeophagia (novelty suppressed feeding) *(32)*, shock-probe burying *(33)*, four-plate test *(34)*, marble burying *(35)*, and separation-induced pup ultrasonic vocalizations *(36)*. Of course, although such tests may be less influenced by motor deficits in mutants, they come with their own unique sensorimotor requirements and, therefore, potential confounds. For instance, altered pain perception would complicate assessment of an anxiolytic-related phenotype on the Vogel conflict, shock-probe burying, or four-plate tests. Assessment of sensory, neurological, and motor functions may help substantiate the specificity of anxiety-related phenotypes *(37,38)*. A more far-ranging approach to assay anxiety-related phenotypes in mutant mice is to examine adjunct physiological responses to anxiety-provoking stimuli in combination with behavior (Fig. 2). The potential value of this strategy is borne out by recent studies. For example, Salas and colleagues *(39)* were able to demonstrate that decreased anxiety-like behavior on the elevated plus maze and staircase tests in α4 nicotine receptor knockout mice was concomitant with reduced sympathetic arousal during testing (as measured by radiotelemetry).

Even when there are no overt sensory or motor abnormalities that could confound the measurement of an anxiety-like behavior in a given test, compelling reasons remain for using multiple behavioral tests. Foremost among these is the possibility that different mouse tests measure different forms of anxiety-like behavior. This is supported by a number of lines of evidence. First, factor analyses of mouse behavior on multiple anxiety-like tasks (*see* Section 4.1.) suggests qualitative differences between tests *(40)*. Second, the pattern of mouse strain differences in anxiety-like behavior is dependent on the task used *(41)*. Third, there are a number of examples of mutant mice that show anxiety-related abnormalities on some tasks, but not oth-

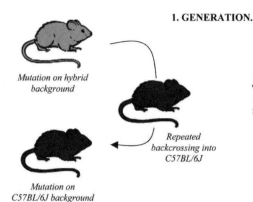

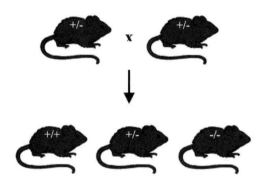

Fig. 2. Optimal generation, breeding, and testing scheme for phenotyping anxiety- and depression-related phenotypes in mutant mice.

ers (e.g., CRF receptor-2 knockout mice; refs. *42–44*). Fourth, although some quantitative trait loci for anxiety-like behavior are general to multiple tests, others are task-specific *(45)*. Taken together, these findings suggest that anxiety-like behaviors measured by different tests likely recruit overlapping, but partly distinct, neural systems, molecules, and genes, and, therefore, strongly advocate the use of multiple tests to phenotype a mutant (Fig. 2).

2.3. Modeling Anxiety-Related Cognitive Abnormalities in Mice

There is significant potential for developing novel approaches to measure mouse anxiety-like behavior. One promising avenue for future research involves modeling the cognitive symptoms of anxiety disorders. Cognitive disturbances, such as misappraisal and overattention to threatening stimuli are cardinal features of emotional disorders *(3,46)*. To date, insights into emotional cognition in rodents has centered on certain well-established paradigms, notably those based on Pavlovian fear conditioning *(47)*. Although fear conditioning tasks have principally been used as models of learning and memory, rather than emotion *per se*, the behaviors measured in these tasks recruit the same neural circuits mediating cognition in human fear and anxiety, including the amygdala, hippocampus, and prefrontal cortex *(48)*. Therefore, further modification of these paradigms could provide additional insights into the role of these structures and other neural systems in anxiety disorders. For example, measuring conditioned emotional responses to stimuli that are partial, incomplete representations of the learned stimulus (e.g., an auditory tone of different frequency to the conditioned tone) may provide a means to assess responses to partial or ambiguous threat cues. This modified approach could have significant clinical relevance, given the abnormally high reactivity of individuals with anxiety disorders, such as posttraumatic stress disorder, to overgeneralized threat cues that inappropriately evoke memories of stimuli previously associated with traumatic events *(49,50)*. This approach has been successfully used in recent studies of mutant mice, and more work along these lines is warranted *(51)*. Failure to extinguish learned fear responses is another prepotent feature of posttraumatic stress disorder, phobias, and other anxiety disorders that is readily studied in rodents, but has, thus far, been understudied in mutant mouse models of anxiety *(52,53)*. Further emphasis on the study of cognitive aspects of emotional behavior, especially when incorporated into studies using the more well-established tests of anxiety-like behavior discussed in Subheadings 2.1. and 2.2., could greatly add to research in this field.

3. ASSESSING DEPRESSION-RELATED BEHAVIOR IN MICE

Similar to anxiety disorders, depression is a heterogeneous disorder with symptoms manifested at the psychological, behavioral, and physiological levels, which leads to difficulty in attempting to mimic the disorder in the laboratory (Table 2). Indeed, many of the human symptoms of depression as described in the DSM-IV (such as recurring thoughts of death or suicide or having excessive thoughts of guilt) are impossible to model in mice. The evolution of a greatly elaborated cere-

Table 2
Modeling DSM-IV-Based Endophenotypes of Depression in Mice

Depression symptom	How could it be modeled in mice?	References
Markedly diminished interest or pleasure in all or most activities most of the day (anhedonia)	Intracranial self-stimulation	86,156
	Progressive ratio responding for positive reward (e.g., sucrose)	193
	Social withdrawal	194
Large changes in appetite or weight gain	Body weight loss after exposure to chronic stressors	195,196
Insomnia or excessive sleeping	Sleep architecture using electroencephalogy (EEG)	197–199
Psychomotor agitation or slowness of movement	Ease of handling	167
	Various measures of locomotor activity and motor function are available	200,201
Fatigue or loss of energy	Energy expenditure	202
	Treadmill/running wheel	203
	Nesting building	204
	Active waking EEG	199,205
Indecisiveness or diminished ability to think or concentrate	Working memory	206
	Spatial memory	207
	Attention	208
Recurrent thoughts of death or of suicide	Not applicable	
Feelings of worthlessness or excessive or inappropriate guilt	Not applicable	

DSM, Diagnostic and Statistical Manual, 4th edition.

bral cortex in humans permits the processing of complex psychological concepts, such as self-esteem and the ability to perceive and plan future events, that seem to be affected in depression but are absent in mice. The question of whether we can ever know whether a mouse is "depressed," therefore, remains unanswerable (*see* Cryan and Mombereau, ref. *16* for further discussion). Nonetheless, there are also many fundamental physiological and behavioral responses that have been evolutionarily conserved between species. Largely through inference, therefore, we can exploit these responses to elucidate phenotypes in genetically mutated mice relevant to emotional behaviors. Depression, and many psychiatric disorders, can be viewed as a manifestation of an inability to cope with various lifetime stressors. This coping deficit may be caused by previous exposure to acute or chronic stressors and is strongly influenced by genetic factors *(54–56)*. Consequently, there are a large number of physiological and neurochemical alterations seen in majorly depressed patients that provide clues to the causation, or are at least are coincident with, some of the behavioral manifestations that are delineated in DSM-IV criteria of depression *(57,58)*. Unfortunately, it is extremely difficult to ascertain whether such hallmarks are the root cause of the disease or are consequences of suffering from depression. Further, it is becoming clear that a more useful strategy may be to model single endophenotypic differences, i.e., one clear-cut behavioral, physiological, or neurochemical endpoint, relevant to the disease state rather than an entire syndrome *(59,60)*. In parallel, similar urgings have been made for dissecting the DSM-IV criteria of various psychiatric disorders into specific endophenotypes and using this modified classification as a basis for investigating novel treatments and underlying genes for specific endophenotypes *(61–64)*.

3.1. Models Based on Behavioral Response to Acute Inescapable Stress

The forced-swim test (FST) (a.k.a. Porsolt's test or the "behavioral despair" test) is probably the most widely and most frequently used experimental paradigm for detecting antidepressant activity in rodents, largely because of its relative reliability across laboratories and its ability to detect the activity of a broad spectrum of clinically effective antidepressants *(17)*. The FST is the most widely used paradigm to assess depression and antidepressant-related phenotypes in mutant mice *(17,65,66)*. The test is based on the observation that rodents, after initial escape-oriented movements, develop an immobile posture in an inescapable cylinder filled with water. If antidepressant treatments are given before the test, the subjects will actively persist in engaging in escape-directed behaviors for longer periods of time than after vehicle treatment. A related test is the tail-suspension test (TST), in which animals are subjected to the inescapable stress of being hung by their tail and, as in FST, they become immobile in an antidepressant-reversible manner *(67,68)*. Although both the FST and TST are similar in the constructs that they purport to assess they are probably different in terms of the biological substrates that underlie the observed behavior, and often offer converging data on a potential antidepressant *(65,69,70)*. The TST avoids any possible confounds induced by hypothermic exposure that may be problematic in the FST, especially if a targeted gene is involved in thermoregulatory processes. Indeed, the TST, similar to many acute stressors, induces a hyperthermic response in animals *(71)*. The TST also circumvents the need of the mouse to swim, which may be relevant for examining the effects of certain genetically modified animals in which motor coordination may be compromised. A relevant example of this is the 129 inbred mouse, which has problems keeping afloat when tested in the FST after treatment with selective serotonin reuptake inhibitors (SSRIs) *(72)*, but are suitable for testing in the TST. However, the TST itself is also dependent on a motor readout, and animals with severe motoric phenotypes may give misleading information in the test. Perhaps one of the most important differences between the FST and TST is the response to drugs in both tests. The mouse FST has traditionally been seen as not the most consistently sensitive model for detecting activity of SSRIs (reviewed in ref. *73*), whereas these antidepressants are generally reported as active in the TST. More recently, it is becoming clear that procedural alterations and better mouse strain selection may increase the sensitivity to detect action of SSRIs in both the FST and TST *(17,72)*. We have recently shown that, whereas antagonist or genetic knockout of the γ-aminobutyric acid B receptor results in an antidepressant-like effect in the FST, no effect was seen in the TST *(74)*.

The induction of an immobile posture in the FST was originally termed "behavioral despair" by Porsolt, largely based on the assumption that the animals have given up hope of escaping, i.e., that there is a failure of persistence in escape-directed behavior. Other investigators have contended that the behavioral responses comprise an evolutionarily preserved coping strategy *(75)*, in which, immobility behaviors represent the psychological concept of entrapment described in clinical depression *(76–78)*. Thus, the development of passive behavior (immobility) disengages the animal from active forms of coping with stressful stimuli *(78)*. Further, as Willner *(13)* correctly points out, the immobility in the test is caused by inability or reluctance to maintain effort rather than by a generalized hypoactivity. This provides an interesting correlate with clinical findings demonstrating that depressed patients show their most pronounced psychomotor impairments in those tests requiring sustained expenditure of effort *(79)*. In further support of the notion that immobility is actually a beneficial behavioral posture to adopt, Nishimura and colleagues *(80)* demonstrated that animals with high immobility in the initial minutes of the swim test are protected against sinking after prolonged exposure to the inescapable cylinder.

Although we caution overextrapolation of the behavioral readout in the FST and TST, it is noteworthy that immobility in these tests is affected by many factors that either influence or are altered by depression in humans. These include genetic predisposition *(81)*, exposure to stress *(82–84)*, changes in food intake *(85)*, alterations in sleep architecture *(82)*, and anhedonia *(86,87)*, in addition to clinically efficacious pharmacological and nonpharmacological antidepressant treatments *(88,89)*. However, we caution that there are also many nondepression-related factors that can influence immobility in these tasks. For example, M_1 muscarinic acetylcholine receptor knockout animals are hyperactive and, correspondingly, have an artifactual antidepressant-like phenotype in the FST *(90)*.

3.2. Models Based on Behavioral Response to Chronic Inescapable Stress

3.2.1. Learned-Helplessness Model

This paradigm was originally developed following the observations that dogs subjected to repeated inescapable, uncontrollable (but not dogs subjected to controllable) shocks demonstrate subsequent escape deficits *(91,92)*. The model was later translated to the rat *(93)* and subsequently to the mouse *(94,95)*. The rodent studies revealed that the behavioral deficits are sensitive to a broad spectrum of antidepressants usually after short-term treatment *(96–98)*. The major drawbacks of the learned-helplessness model are twofold. First, most of the depression-like symptomatology does not persist beyond 2–3 d after cessation of the uncontrollable shock *(99)*. Second, only a certain percentage (estimates vary between 10 and 80%) of animals develop helplessness behavior *(95,100,101)*. Vollmayr and Henn *(102)* have recently proposed key factors that can be manipulated to enhance both the usability and reliability of the learned-helplessness paradigm in the rat, many of which are, in principle, translatable to the mouse. These include the use of a larger testing apparatus, a mild shock presentation, a relatively difficult shock-avoidance task, and taking into account animals that are artifactually avoiding shock because of their position in the apparatus.

As in many other mouse emotional paradigms, major strain differences are observable in the learned-helplessness test *(100,103–105)* (*see also* Section 4.1.). In addition to baseline differences, Shanks and Anisman showed differential responsivity to three different antidepressants across four strains of mouse tested *(104)*. Chronic treatment with desipramine prevented the escape deficits in A/J mice, but did not affect the performance in BALB/cByJ, C57BL/6J, or CD-1 mice; repeated treatment with bupropion, in contrast, had only a modest effect in CD-1 mice. Unlike these compounds, amitriptyline was found to influence escape performance irrespective of whether the drug was acutely or chronically applied. Caldarone and colleagues *(100)* suggest that C57BL/6 mice are suitable for use in learned-helplessness studies, but caution that strains such as the 129 and the hybrid B6129F1 strain may be inappropriate because nonshocked control mice of these strains show poor escape performance *per se*. Marked sexual dimorphic effects have also been observed in the mouse learned-helplessness paradigm, with female mice being less disrupted by the effects of inescapable shock than male mice *(100)*. One important caveat that must be considered with the learned-helplessness paradigm is that alterations in pain sensitivity caused by pharmacological or genetic manipulation will influence performance. This is relevant when discussing strain effects, because prominent differences in pain sensitivities among inbred mouse strains have been described *(106)*. One example in which such caution is pertinent, is the recent analysis of mice with heterozygous expression of brain-derived neurotrophic factor *(107)*. These mice display antidepressant-like behavior in the learned-helplessness paradigm, but the authors point out that the mice also show a reduced sensitivity to pain, which complicates the dissociation of helplessness from hypoalgesia. As with many behavioral models, motor performance can also affect mouse behavior in this test.

3.2.2. Chronic Mild Stress

The chronic mild stress (CMS) model has long been championed by Willner and colleagues as a realistic model of depression *(108–110)*. As the name suggests, this paradigm consists of exposing rodents to several series of mild unpredictable stressors over a prolonged period (usually >2 wk). This stress regimen induces many long-term behavioral, neurochemical, neuroimmune, and neuroendocrine alterations resembling those dysfunctions observed in depressed patients *(110)*. Primarily, there have been two major antidepressant-sensitive readouts characterized in the CMS model:

1. CMS depresses the consumption and preference for sucrose solution.
2. CMS decreases brain reward function, as assessed using intracranial self-stimulation (ICSS).

Both measures are correlated with anhedonia, one of the core symptoms of depression as defined in DSM-IV. These anhedonia-like behaviors have generally been shown to be reversed by chronic, but not acute, treatment with several classes of antidepressants *(110,111)*.

Although the paradigm has been described as a model with a high predictive, construct, and etiological validity *(13,108–110)*, two major facts limit its widespread use in assessing depression-related behavior in mutant mice. First, this model (at least in rats, in which it has been for the most part characterized) has very poor reliability and could not be reproduced in many laboratories *(112–121)*. Second, there are relatively few studies to date that have used mice *(122)*. A interesting modification of the CMS model in mice has been proposed recently by Belzung and colleagues *(123–125)* which uses a 1- to 3-point scale for the assessment of the physical state of the animals fur. Animals subjected to chronic stress do not groom themselves and subsequently exhibit degradation in the condition of their fur. There is certainly some analogy between this stress-induced state and the observations that depressed patients have a reduced efficiency with which even the smallest tasks are accomplished, leading to the inability to maintain minimal personal hygiene. Further, it has been shown that chronic treatment with the antidepressant fluoxetine and novel antidepressants (e.g., CRF R1 receptor and vasopressin V1B receptor antagonists) improved the physical state index of the mice exposed to chronic stress *(124,125)*. Although this readout seems promising, further studies are required. At present, it remains to be determined whether the CMS model can be applied successfully and more reliably in mice than has been the experience in rats.

3.3. Lesion-Based Model: Olfactory Bulbectomy

The bilateral removal of the olfactory bulbs of rodents results in a complex constellation of behavioral, neurochemical, neuroendocrine, and neuroimmune alterations, many of which are comparable to changes seen in patients with depression *(126–129)*. This model has been best characterized in the rat, with only a handful of studies conducted in mice *(130–137)*. The most consistent behavioral change of bulbectomy is a hyperactive response in a novel brightly lit open-field apparatus, which is reversed almost exclusively by chronic, but not acute, antidepressant treatment, and at doses that do not compromise the performance of sham-lesioned control animals *(127,138,139)*. Recent studies have shown that this hyperactivity might be related to increases in defensive behavior *(140)* or alterations in aversively motivated behavior in the rat *(141)*. Furthermore, it has been shown that antidepressant compounds predominantly improve habituation to novelty in the bulbectomized rat *(142)*. Concurrent with these studies, other groups have focused on neurochemical alterations that might account for these antidepressant-sensitive behavioral alterations. In addition to an increased locomotor response to stress, bulbectomized rats have a heightened acoustic startle responsivity to stress *(143)*, marked deficits in circadian rhythms (also observed in mice) *(133)*, cognitive deficits in Morris water maze and passive avoidance learning *(137,144)*, elevated corticosterone *(145,146)*, and anhedonia-like behaviors, such as decreased sucrose preference and sexual behavior *(128,147)*. In addition, it has been shown that bulbectomized rats also exhibit increased levels of amphetamine self-administration compared with sham animals *(148)*, and bulbectomized mice have elevated alcohol consumption *(149)* compared with sham animals. Several lines of evidence suggest that the behavioral sequelae induced by bulbectomy in the rat is not just a consequence of loss of smell, because peripheral anosmia fails to mimic the syndrome *(142,150,151)*. The olfactory bulbectomy behavioral syndrome is largely thought to be brought about by compensatory neuronal reorganization, changes in synaptic strength and/or loss of spine density in various subcortical limbic regions, such as the amygdala and hippocampus *(127,152,153)*. Recently, the first studies investigating the effects of bulbectomy in mutant mice were published. Mice lacking the gene for tachykinin 1, which encodes for substance P and neurokinin A, seemed to be resistant to the hyperactivity-inducing effects of bulbectomy and also showed an antidepressant-like profile in the FST *(136)*.

3.4. Drug Withdrawal As a Model of Depression

Reward deficits associated with withdrawal from drugs of abuse has been posited as a model for the depression-related anhedonia *(17,59,86,87,154)*. Amphetamine withdrawal has been proposed as a suitable substrate for inducing depression in rodents *(87,154,155)*. Recent rat studies showed that withdrawal from chronic amphetamine treatment is characterized by decreased breaking points under a progressive ratio schedule for a sucrose solution reinforcer, and decrements in anticipatory and motivational measures for sexual reinforcement *(119,154)*. Moreover, amphetamine withdrawal increases immobility in both the rat and mouse FST *(86,87)* and mouse TST *(87)* and also induces escape deficits in the mouse learned-helplessness paradigm *(86)*.

The use of the ICSS paradigm has provided investigators with a reliable and quantifiable behavioral measure of reduced brain reward function after withdrawal from a variety of drugs of abuse *(86,120)*. Many investigators have adapted the ICSS procedures for use with mice *(86,156,157)*. It will be of interest to assess the effects of drug withdrawal-induced anhedonia across various strains using ICSS, including those that have been genetically modified.

4. GENERAL ISSUES OF STUDY DESIGN AND METHODOLOGY

4.1. Are So Many Mutants Really Unhappy? Avoiding False-Positives

Abnormalities in anxiety- and depression-related behaviors are some of the most commonly reported behavioral phenotypes in mutant mice. As discussed earlier, many of the tests used to measure these behaviors are easy to construct, conceptually attractive, and require relatively minimal training (of mouse or experimenter). However, precisely because many of these tests measure spontaneous behaviors, they may be relatively sensitive to variations in test procedure and environment. Indeed, it is known that performance on mouse tests for emotion-related behaviors can be modified by a host of procedural variables *(158,159)*. Unfortunately, it is very difficult to ascertain which variables are critical and the degree to which they will reliably affect anxiety- and depression-like behaviors. Studies by Crabbe, Wahlsten, and colleagues have convincingly shown that even ostensibly subtle variations in laboratory environment and test procedure can alter mouse anxiety-related behaviors *(160,161)*. There is also evidence that repeated testing in various tests for anxiety-like behavior markedly alters baseline behaviors and anxiolytic-sensitivity in normal mice *(34,162,163)*. Although formal parametric studies of repeated test exposure have not been conducted for mouse "behavioral despair" tasks, repeated exposure to the mouse FST significantly alters behavior on subsequent exposures in the test, in some *(164)*, but not all, laboratories *(165)*.

A related issue is the potential effect of testing mutant mice on a battery of behavioral assays. Given the cost and effort required to generate sufficient numbers of mutant mice and properly matched controls for behavioral studies, added to the need for replication of findings across different tests, it is often prudent to test the same cohort of mice on multiple tests for emotionality. Paylor and colleagues have recently provided useful information on how specific tests are more or less sensitive to previous test history *(166)*. For these reasons, it is clearly important to carefully consider test order and the potential for previous experience to influence, possibly in a genotype-dependent manner, behavioral phenotypes on tests for anxiety- and depression-like behaviors. Furthermore, given the potential for interlaboratory variations to affect the outcome of mutant studies, a more general recommendation is to realize that "standardized" methods cannot simply be decoded from the methods section of a published paper in a high-profile journal to effective use in one's own laboratory without proper in-house (i.e., pharmacological) validation *(167)*.

4.2. All Mice Are Not Created Equal: The Importance of Genetic Background, Breeding, and Early Life Environment

Numerous genetically distinct strains of mouse are used in studies of anxiety and depression. As noted in Subheading 3.2.1., there are major differences in anxiety-like and depression-related behaviors across the commonly used strains. Some of the more widely reported findings to emerge from these studies is that some strains, such as C57BL/6, exhibit low levels of anxiety- and depression-related behaviors relative to strains such as BALB/cByJ. Other strains, notably 129 substrains, display locomotor abnormalities that complicate their usefulness in mouse mutant studies involving exploration-based tasks *(168–172)*. The upshot of these differences is that the strain, and, therefore, the genetic background, onto which a mutation is backcrossed can affect the detection of an anxiety-like or depression-related phenotype in a mutant mouse. Genetic background can exert effects on a mutant phenotype in a number of ways: (1) by providing an inappropriate baseline level of behavior in wild-type controls, (2) via the unwanted contribution of flanking genes extant from the embryonic stem cell donor strain, and (3) via complex functional interactions between the mutation and background genes *(173–176)*.

Therefore, careful consideration should be given to the choice of background strain for studies of anxiety and depression and, ideally, the mutation should be backcrossed onto a congenic background (Fig. 2) *(177–179)*.

Once a mutation is placed on a suitable genetic background(s), great care should then be taken to minimize unwanted environmental variables during early life. It is increasingly recognized that emotional behaviors in rodents and humans are strongly influenced by perinatal experience *(180)*. If one is to confidently attribute any behavioral differences between adult mutants and controls to an effect of the gene mutation *per se*, it is very important that mutants and controls be reared in the same environment. One should also consider the potential influence of maternal behavior on anxiety-like or depression-related behavior, which, itself, is likely to be strongly influenced by the genotype of the dam *(181–183)*. For these reasons, it is strongly recommended that mutants and nonmutant, wild-type controls are all derived from the same heterozygous by heterozygous mutant breedings and reared together in their littermate groups until weaning (Fig. 2).

5. CONCLUDING REMARKS

It has been estimated that approx 40% of the risk for depression and anxiety disorders is heritable *(184)*. What is clear is that these disorders arise from a combination of environmental and genetic factors, which likely interact in a highly complex manner to influence risk *(185,186)*. Moreover, like other psychiatric diseases, these disorders are genetically complex; i.e., they are influenced by multiple gene variants, each of which can be simplistically thought of as conferring relative susceptibility or resilience (although the direction of this influence may itself be determined by environment) *(187–189)*. Not surprisingly, fear and anxiety-like behaviors in mice also seem to be polygenic and epistatic in nature *(45,190,191)*. Thus, studies in mutant mice are not going to discover the "gene for anxiety" or the "gene for depression." Nonetheless, this approach may provide valid model systems to explore the role of a given gene in molecular pathways that influence depression- and anxiety-related behaviors. That said, although mutant mice are often discussed as "genetic models," the approach has largely been used as a means to study the consequences of altering a specific molecule *ceteris paribus*, much in the same way as a highly selective pharmacological agent would. Indeed, mutant mice are uniquely valuable when tools for dissecting the functional role of a molecule cannot be approached using more traditional approaches, e.g., where pharmacological compounds are unavailable or have poor selectivity for a particular receptor subtype.

On the downside, however, constitutive mutation of a gene raises justifiable concerns over the potential for lifelong, tissue-ubiquitous mutants to invoke developmental alterations that mask the normal function of a molecule or otherwise modify neural systems subserving emotion to muddy the causative link between molecule and behavior. In hindsight, the extent of these adaptations is not surprising, given the significant plasticity of the brain, especially during development. Because of this potential for recruiting compensatory changes to occlude or otherwise alter the normal function of a molecule, it is now widely accepted that phenotypic abnormalities in mutant mice must be carefully interpreted with this caveat in mind. Fortunately, there are a number of emerging approaches that can help mitigate the problem of compensation in mutants. One powerful approach is to restrict mutation of a gene to specific brain regions or cell types. Induction of such "conditional" mutation can also be withheld until after development to further limit the occurrence of adaptive changes. Promising alternatives to mutants are also emerging, including antisense disruption and, more recently, viral-mediated RNA interference of a molecule of interest *(192)*.

Even with these important advances, mutant mice are no research panacea. This valuable tool is most effect when used in conjunction with other techniques to generate converging lines of evidence regarding the role of a molecule in complex neural systems. When viewed as such, we believe that mutant mice will continue to foster understanding of the neural basis of anxiety and depression.

ACKNOWLEDGMENTS

A. Holmes is supported by the Intramural Research Program of the National Institute on Alcohol Abuse and Alcoholism. J. F. Cryan is supported by National Institutes of Mental Health/National Institute on Drug Abuse, grant U01 MH69062.

REFERENCES

1. Andrade L, Caraveo-Anduaga JJ, Berglund P, et al. The epidemiology of major depressive episodes: results from the International Consortium of Psychiatric Epidemiology (ICPE) Surveys. Int J Methods Psychiatr Res 2003;12:3–21.
2. Kessler RC, Berglund P, Demler O, et al. The epidemiology of major depressive disorder: results from the National Comorbidity Survey Replication (NCS-R). JAMA 2003;289:3095–3105.
3. Association AP. Diagnostic and statistical manual of mental disorders: DSM-IV-TR. Washington, DC: American Psychiatric Association, 2000.
4. Levi F, La Vecchia C, Saraceno B. Global suicide rates. Eur J Public Health 2003;13:97–98.
5. Williams LS, Ghose SS, Swindle RW. Depression and other mental health diagnoses increase mortality risk after ischemic stroke. Am J Psychiatry 2004;161:1090–1095.
6. Greenberg PE, Kessler RC, Birnbaum HG, et al. The economic burden of depression in the United States: how did it change between 1990 and 2000? J Clin Psychiatry 2003;64:1465–1475.

7. Holmes A, Heilig M, Rupniak NM, Steckler T, Griebel G. Neuropeptide systems as novel therapeutic targets for depression and anxiety disorders. Trends Pharmacol Sci 2003;24:580–588.
8. Insel TR, Charney DS. Research on major depression: strategies and priorities. JAMA 2003;289:3167–3168.
9. Hamilton M. A rating scale for depression. J Neurol Neurosurg Psychiatry 1970;23:51–56.
10. Geyer MA, Markou A. The Role of Preclinical Models in the Development of Psychotropic Drugs. In: Kupfer DJ, ed. Psychopharmacology: The Fifth Generation of Progress. New York: Raven, 2000.
11. McKinney WT. Overview of the past contributions of animal models and their changing place in psychiatry. Semin Clin Neuropsychiatry 2001;6:68–78.
12. Rodgers RJ. Animal models of 'anxiety': where next? Behav Pharmacol 1997;8:477–96; discussion 497–504.
13. Willner P. Animal models of depression: an overview. Pharmacol Ther 1990;45:425–455.
14. McKinney WT Jr, Bunney WE Jr. Animal model of depression. I. Review of evidence: implications for research. Arch Gen Psychiatry 1969;21:240–248.
15. Bucan M. Abel T. The mouse: genetics meets behaviour. Nat Rev Genet 2002;3:114–123.
16. Cryan JF, Mombereau C. In search of a depressed mouse: utility of models for studying depression-related behavior in genetically modified mice. Mol Psychiatry 2004;9:326–357.
17. Cryan JF, Markou A, Lucki I. Assessing antidepressant activity in rodents: recent developments and future needs. Trends Pharmacol Sci 2002;23:238–245.
18. Holmes A. Targeted gene mutation approaches to the study of anxiety-like behavior in mice. Neurosci Biobehav Rev 2001;25:261–273.
19. Lesch KP, Zeng Y, Reif A, Gutknecht L. Anxiety-related traits in mice with modified genes of the serotonergic pathway. Eur J Pharmacol 2003;480:185–204.
20. Finn DA, Rutledge-Gorman MT, Crabbe JC. Genetic animal models of anxiety. Neurogenetics 2003;4:109–135.
21. Nesse RM. Is depression an adaptation? Arch Gen Psychiatry 2000;57:14–20.
22. File SE, Seth P. A review of 25 years of the social interaction test. Eur J Pharmacol 2003; 463:35–53.
23. Risbrough VB, Brodkin JD, Geyer MA. GABA-A and 5-HT1A receptor agonists block expression of fear-potentiated startle in mice. Neuropsychopharmacology 2003;28:654–663.
24. Crawley JN. Neuropharmacologic specificity of a simple animal model for the behavioral actions of benzodiazepines. Pharmacol Biochem Behav 1981;15:695–699.
25. Zethof TJ, Van der Heyden JA, Tolboom JT, Olivier B. Stress-induced hyperthermia in mice: a methodological study. Physiol Behav 1994;55:109–115.
26. Rodgers RJ, Cao BJ, Dalvi A, Holmes A. Animal models of anxiety: an ethological perspective. Braz J Med Biol Res 1997;30:289–304.
27. Belzung C, Griebel G. Measuring normal and pathological anxiety-like behaviour in mice: a review. Behav Brain Res 2001;125:141–149.
28. Falzone TL, Gelman DM, Young JI, Grandy DK, Low MJ, Rubinstein M. Absence of dopamine D4 receptors results in enhanced reactivity to unconditioned, but not conditioned, fear. Eur J Neurosci 2002;15:158–164.
29. Dulawa SC, Grandy DK, Low MJ, Paulus MP, Geyer MA. Dopamine D4 receptor-knock-out mice exhibit reduced exploration of novel stimuli. J Neurosci 1999;19:9550–9556.
30. Blanchard DC, Griebel G, Blanchard RJ. The Mouse Defense Test Battery: pharmacological and behavioral assays for anxiety and panic. Eur J Pharmacol 2003;463:97–116.
31. van Gaalen MM, Stenzel-Poore MP, Holsboer F, Steckler T. Effects of transgenic overproduction of CRH on anxiety-like behaviour. Eur J Neurosci 2002;15:2007–2015.
32. Santarelli L, Gobbi G, Debs PC, et al. Genetic and pharmacological disruption of neurokinin 1 receptor function decreases anxiety-related behaviors and increases serotonergic function. Proc Natl Acad Sci USA 2001;98:1912–1917.
33. Degroot A, Nomikos GG. Genetic deletion and pharmacological blockade of CB1 receptors modulates anxiety in the shock-probe burying test. Eur J Neurosci 2004;20:1059–1064.

34. Hascoet M, Bourin M, Couetoux du Tertre A. Influence of prior experience on mice behavior using the four-plate test. Pharmacol Biochem Behav 1997;58:1131–1138.
35. Chaki S, Hirota S, Funakoshi T, et al. Anxiolytic-like and antidepressant-like activities of MCL0129 (1-[(S)-2-(4-fluorophenyl)-2-(4-isopropylpiperadin-1-yl)ethyl]-4-[4-(2-met hoxynaphthalen-1-yl)butyl]piperazine), a novel and potent nonpeptide antagonist of the melanocortin-4 receptor. J Pharmacol Exp Ther 2003;304:818–826.
36. Dirks A, Fish EW, Kikusui T, et al. Effects of corticotropin-releasing hormone on distress vocalizations and locomotion in maternally separated mouse pups. Pharmacol Biochem Behav 2002;72:993–999.
37. Crawley JN. Whats wrong with my mouse? Behavioral Phenotyping of Transgenic and Knockout Mice. New York: Wiley-Liss, 2000.
38. Holmes A, Yang RJ, Murphy DL, Crawley JN. Evaluation of antidepressant-related behavioral responses in mice lacking the serotonin transporter. Neuropsychopharmacology 2002;27:914–923.
39. Salas R, Pieri F, Fung B, Dani JA, De Biasi M. Altered anxiety-related responses in mutant mice lacking the beta4 subunit of the nicotinic receptor. J Neurosci 2003;23:6255–6263.
40. Holmes A, Kinney JW, Wrenn CC, et al. Galanin GAL-R1 receptor null mutant mice display increased anxiety-like behavior specific to the elevated plus-maze. Neuropsychopharmacology 2003;28:1031–1044.
41. van Gaalen MM, Steckler T. Behavioural analysis of four mouse strains in an anxiety test battery. Behav Brain Res 2000;115:95–106.
42. Bale TL, Contarino A, Smith GW, et al. Mice deficient for corticotropin-releasing hormone receptor-2 display anxiety-like behaviour and are hypersensitive to stress. Nat Genet 2000; 24:410–414.
43. Coste SC, Kesterson RA, Heldwein KA, et al. Abnormal adaptations to stress and impaired cardiovascular function in mice lacking corticotropin-releasing hormone receptor-2. Nat Genet 2000;24:403–409.
44. Kishimoto T, Radulovic J, Radulovic M, et al. Deletion of crhr2 reveals an anxiolytic role for corticotropin-releasing hormone receptor-2. Nat Genet 2000;24:415–419.
45. Turri MG, Datta SR, DeFries J, Henderson ND, Flint, J. QTL analysis identifies multiple behavioral dimensions in ethological tests of anxiety in laboratory mice. Curr Biol 2001;11:725–734.
46. Lang PJ, Davis M, Ohman A. Fear and anxiety: animal models and human cognitive psychophysiology. J Affect Disord 2000;61:137–159.
47. Kim JJ, Fanselow MS. Modality-specific retrograde amnesia of fear. Science 1992;256:675–677.
48. LeDoux JE. Emotion circuits in the brain. Annu Rev Neurosci 2000;23:155–184.
49. Harding EJ, Paul ES, Mendl M. Animal behaviour: cognitive bias and affective state. Nature 2004;427:312.
50. Eysenck MW. Anxiety and Cognition: A Unified Theory. Hove, England: Psychology Press, 1997.
51. Crestani F, Lorez M, Baer K, et al. Decreased GABAA-receptor clustering results in enhanced anxiety and a bias for threat cues. Nat Neurosci 1999;2:833–839.
52. Cain CK, Blouin AM, Barad M. L-type voltage-gated calcium channels are required for extinction, but not for acquisition or expression, of conditional fear in mice. J Neurosci 2002;22: 9113–9121.
53. El-Ghundi M, O'Dowd BF, George SR. Prolonged fear responses in mice lacking dopamine D1 receptor. Brain Res 2001;892:86–93.
54. Anisman H, Zacharko RM. Multiple neurochemical and behavioral consequences of stressors: implications for depression. Pharmacol Ther 1990;46:119–136.
55. Kessler RC. The effects of stressful life events on depression. Annu Rev Psychol 1997;48:191–214.
56. Sullivan PF, Neale MC, Kendler KS. Genetic epidemiology of major depression: review and meta-analysis. Am J Psychiatry 2000;157:1552–1562.
57. Wong ML, Licinio J. Research and treatment approaches to depression. Nat Rev Neurosci 2001;2:343–351.
58. Nestler EJ, Barrot M, DiLeone RJ, Eisch AJ, Gold SJ, Monteggia LM. Neurobiology of depression. Neuron 2002;34:13–25.

59. Geyer MA, Markou A. Animal Models of Psychiatric Disorders. In: Bloom FE, Kupfer DJ, eds. Psychopharmacology: The Fourth Generation of Progress. New York, Raven Press, 1995, pp. 787–798.
60. Geyer MA, Markou A. The Role of Preclinical Models in the Development of Psychotropic Drugs. In: Bloom FE, Kupfer DJ, eds. Psychopharmacology: The Fifth Generation of Progress. New York, Raven, 2000.
61. Gottesman II, Gould TD. The endophenotype concept in psychiatry: etymology and strategic intentions. Am J Psychiatry 2003;160:636–645.
62. Hyman SE, Fenton WS. Medicine. What are the right targets for psychopharmacology? Science 2003;299:350–351.
63. Inoue K, Lupski JR. Genetics and genomics of behavioral and psychiatric disorders. Curr Opin Genet Dev 2003;13:303–309.
64. Hassler G, Drevets W, Manji H, Charney D. Discovering endophenotypes for major depression. Neuropsychopharmacology, 2004, in press.
65. Porsolt RD. Animal models of depression: utility for transgenic research. Rev Neurosci 2000;11:53–58.
66. Seong E, Seasholtz AF, Burmeister M. Mouse models for psychiatric disorders. Trends Genet 2002;18:643–650.
67. Steru L, Chermat R, Thierry B, Simon P. The tail suspension test: a new method for screening antidepressants in mice. Psychopharmacology (Berl). 1985;85:367–370.
68. Cryan JF, Mombereau C, Vassout A. The tail suspension test as a model for assessing antidepressant activity: Review of pharmacological and genetic studies in mice. Neuroscience and Biobehavioral Reviews, 2004, in press.
69. Bai F, Li X, Clay M, Lindstrom T, Skolnick P. Intra- and interstrain differences in models of "behavioral despair". Pharmacol Biochem Behav 2001;70:187–192.
70. Renard CE, Dailly E, David DJ, Hascoet M, Bourin M. Monoamine metabolism changes following the mouse forced swimming test but not the tail suspension test. Fundam Clin Pharmacol 2003;17:449–455.
71. Liu X, Peprah D, Gershenfeld HK. Tail-suspension induced hyperthermia: a new measure of stress reactivity. J Psychiatr Res 2003;37:249–259.
72. Lucki I, Dalvi A, Mayorga AJ. Sensitivity to the effects of pharmacologically selective antidepressants in different strains of mice. Psychopharmacology (Berl) 2001;155:315–322.
73. Porsolt R, Lenegre A. Behavioral Models of Ddepression. In: Elliott J, Heal D, Marsden C, eds. Experimental Approaches to Anxiety and Depression. London, Wiley, 1992, pp. 73–85.
74. Mombereau C, Kaupmann K, Froestl W, Sansig G, Van Der Putten H, Cryan JF. Genetic and Pharmacological evidence of a role for GABA(B) receptors in the modulation of anxiety- and antidepressant-like behavior. Neuropsychopharmacology 2004;29:1050–1062.
75. Thierry B, Steru L, Chermat R, Simon P. Searching-waiting strategy: a candidate for an evolutionary model of depression? Behav Neural Biol 1984;41:180–189.
76. Dixon AK. Ethological strategies for defence in animals and humans: their role in some psychiatric disorders. Br J Med Psychol 1998;71(Pt 4):417–445.
77. Gilbert P, Allan S. The role of defeat and entrapment (arrested flight) in depression: an exploration of an evolutionary view. Psychol Med 1998;28:585–598.
78. Lucki I. A prescription to resist proscriptions for murine models of depression. Psychopharmacology (Berl) 2001;153:395–398.
79. Weingartner H, Silberman E. Models of cognitive impairment: cognitive changes in depression. Psychopharmacol Bull 1982;18:27–42.
80. Nishimura H, Tsuda A, Oguchi M, Ida Y, Tanaka M. Is immobility of rats in the forced swim test "behavioral despair"? Physiol Behav 1988;42:93–95.
81. West CH, Weiss JM. Effects of antidepressant drugs on rats bred for low activity in the swim test. Pharmacol Biochem Behav 1998;61:67–79.
82. Solberg LC, Horton TH, Turek FW. Circadian rhythms and depression: effects of exercise in an animal model. Am J Physiol 1999;276:R152–R161.

83. Alonso SJ, Damas C, Navarro E. Behavioral despair in mice after prenatal stress. J Physiol Biochem 2000;56:77–82.
84. Tannenbaum B, Tannenbaum GS, Sudom K, Anisman H. Neurochemical and behavioral alterations elicited by a chronic intermittent stressor regimen: implications for allostatic load. Brain Res 2002;953:82–92.
85. Alcaro A, Cabib S, Ventura R, Puglisi-Allegra S. Genotype- and experience-dependent susceptibility to depressive-like responses in the forced-swimming test. Psychopharmacology (Berl) 2002;164:138–143.
86. Kokkinidis L, Zacharko RM, Anisman H. Amphetamine withdrawal: a behavioral evaluation. Life Sci 1986;38:1617–1623.
87. Cryan JF, Hoyer D, Markou A. Withdrawal from chronic amphetamine induces Depressive-Like behavioral effects in rodents. Biol Psychiatry 2003;54:49–58.
88. Porsolt RD, Anton G, Blavet N, Jalfre M. Behavioural despair in rats: a new model sensitive to antidepressant treatments. Eur J Pharmacol 1978;47:379–391.
89. Borsini F, Meli A. Is the forced swimming test a suitable model for revealing antidepressant activity? Psychopharmacology (Berl) 1988;94:147–160.
90. Miyakawa T, Yamada M, Duttaroy A, Wess J. Hyperactivity and intact hippocampus-dependent learning in mice lacking the M1 muscarinic acetylcholine receptor. J Neurosci 2001;21:5239–5250.
91. Overmier JB, Seligman ME. Effects of inescapable shock upon subsequent escape and avoidance responding. J Comp Physiol Psychol 1967;63:28–33.
92. Seligman ME, Maier SF. Failure to escape traumatic shock. J Exp Psychol 1967;74:1–9.
93. Seligman ME, Beagley G. Learned helplessness in the rat. J Comp Physiol Psychol 1975; 88:534–541.
94. Anisman H, Irwin J, Sklar LS. Deficits of escape performance following catecholamine depletion: implications for behavioral deficits induced by uncontrollable stress. Psychopharmacology (Berl) 1979;64:163–170.
95. Anisman H, DeCatanzaro D, Remington G. Escape performance following exposure to inescapable shock: Deficits in motor response maintenance. J Exp Psychol Anim Behav Process 1978;4:197–218.
96. Leshner AI, Remler H, Biegon A, Samuel D. Desmethylimipramine (DMI) counteracts learned helplessness in rats. Psychopharmacology (Berl) 1979;66:207–208.
97. Sherman AD, Petty F. Additivity of neurochemical changes in learned helplessness and imipramine. Behav Neural Biol 1982;35:344–353.
98. Martin P, Soubrie P, Puech AJ. Reversal of helpless behavior by serotonin uptake blockers in rats. Psychopharmacology (Berl) 1990;101:403–407.
99. Weiss JM, Kilts CD. Animal Models of Depression and Schizophrenia. In: Nemeroff CB, Schatzberg AF, eds. Textbook of Psychopharmacology. 2nd ed. American Psychiatric Press, 1998, pp. 88–123.
100. Caldarone BJ, George TP, Zachariou V, Picciotto MR. Gender differences in learned helplessness behavior are influenced by genetic background. Pharmacol Biochem Behav 2000;66: 811–817.
101. Drugan RC, Skolnick P, Paul SM, Crawley JN. A pretest procedure reliably predicts performance in two animal models of inescapable stress. Pharmacol Biochem Behav 1989;33:649–654.
102. Vollmayr B, Henn FA. Learned helplessness in the rat: improvements in validity and reliability. Brain Res Brain Res Protoc 2001;8:1–7.
103. Shanks N, Anisman H. Escape deficits induced by uncontrollable foot-shock in recombinant inbred strains of mice. Pharmacol Biochem Behav 1993;46:511–517.
104. Shanks N, Anisman H. Strain-specific effects of antidepressants on escape deficits induced by inescapable shock. Psychopharmacology (Berl) 1989;99:122–128.
105. Shanks N, Anisman H. Stressor-provoked behavioral changes in six strains of mice. Behav Neurosci 1988;102:894–905.
106. Mogil JS, Wilson SG, Bon K, et al. Heritability of nociception I: responses of 11 inbred mouse strains on 12 measures of nociception. Pain 1999;80:67–82.

107. MacQueen GM, Ramakrishnan K, Croll SD, et al. Performance of heterozygous brain-derived neurotrophic factor knockout mice on behavioral analogues of anxiety, nociception, and depression. Behav Neurosci 2001;115:1145–1153.
108. Willner P, Muscat R, Papp M. An animal model of anhedonia. Clin Neuropharmacol 1992;15(Suppl 1, Pt A):550A–551A.
109. Willner P, Muscat R, Papp M. Chronic mild stress-induced anhedonia: a realistic animal model of depression. Neurosci Biobehav Rev 1992;16:525–534.
110. Willner P. Validity, reliability and utility of the chronic mild stress model of depression: a 10-year review and evaluation. Psychopharmacology (Berl) 1997;134:319–329.
111. Moreau JL, Jenck F, Martin JR, Mortas P, Haefely WE. Antidepressant treatment prevents chronic unpredictable mild stress-induced anhedonia as assessed by ventral tegmental self-stimulation behavior in rats. Eur Neuropsychopharmacol 1992;2:43–49.
112. Matthews K, Forbes N, Reid IC. Sucrose consumption as an hedonic measure following chronic unpredictable mild stress. Physiol Behav 1995;57:241–248.
113. Forbes NF, Stewart CA, Matthews K, Reid IC. Chronic mild stress and sucrose consumption: validity as a model of depression. Physiol Behav 1996;60:1481–1484.
114. Reid I, Forbes N, Stewart C, Matthews K. Chronic mild stress and depressive disorder: a useful new model? Psychopharmacology (Berl) 1997;134:365–367; discussion 371–377.
115. Hatcher JP, Bell DJ, Reed TJ, Hagan JJ. Chronic mild stress-induced reductions in saccharin intake depend upon feeding status. J Psychopharmacol 1997;11:331–338.
116. Harris RB, Zhou J, Youngblood BD, Smagin GN, Ryan DH. Failure to change exploration or saccharin preference in rats exposed to chronic mild stress. Physiol Behav 1997;63:91–100.
117. Nielsen CK, Arnt J, Sanchez C. Intracranial self-stimulation and sucrose intake differ as hedonic measures following chronic mild stress: interstrain and interindividual differences. Behav Brain Res 2000;107:21–33.
118. Lin D, Bruijnzeel AW, Schmidt P, Markou A. Exposure to chronic mild stress alters thresholds for lateral hypothalamic stimulation reward and subsequent responsiveness to amphetamine. Neuroscience 2002;114:925–933.
119. Barr AM, Fiorino DF, Phillips AG. Effects of withdrawal from an escalating dose schedule of d-amphetamine on sexual behavior in the male rat. Pharmacol Biochem Behav 1999;64:597–604.
120. Barr AM, Zis AP, Phillips AG. Repeated electroconvulsive shock attenuates the depressive-like effects of d-amphetamine withdrawal on brain reward function in rats. Psychopharmacology (Berl) 2002;159:196–202.
121. Bielajew C, Konkle AT, Merali Z. The effects of chronic mild stress on male Sprague-Dawley and Long Evans rats: I. Biochemical and physiological analyses. Behav Brain Res 2002;136:583–592.
122. Harkin A, Houlihan DD, Kelly JP. Reduction in preference for saccharin by repeated unpredictable stress in mice and its prevention by imipramine. J Psychopharmacol 2002;16:115–123.
123. Ducottet C, Griebel G, Belzung C. Effects of the selective nonpeptide corticotropin-releasing factor receptor 1 antagonist antalarmin in the chronic mild stress model of depression in mice. Prog Neuropsychopharmacol Biol Psychiatry 2003;27:625–631.
124. Griebel G, Simiand J, Serradeil-Le Gal C, et al. Anxiolytic- and antidepressant-like effects of the non-peptide vasopressin V1b receptor antagonist, SSR149415, suggest an innovative approach for the treatment of stress-related disorders. Proc Natl Acad Sci USA 2002;99:6370–6375.
125. Griebel G, Simiand J, Steinberg R, et al. 4-(2-Chloro-4-methoxy-5-methylphenyl)-N-[(1S)-2-cyclopropyl-1-(3-fluoro-4- methylphenyl)ethyl]5-methyl-N-(2-propynyl)-1, 3-thiazol-2-amine hydrochloride (SSR125543A), a potent and selective corticotrophin-releasing factor(1) receptor antagonist. II. Characterization in rodent models of stress-related disorders. J Pharmacol Exp Ther 2002;301:333–345.
126. Jesberger JA, Richardson JS. Brain output dysregulation induced by olfactory bulbectomy: an approximation in the rat of major depressive disorder in humans? Int J Neurosci 1988;38:241–265.
127. Kelly JP, Wrynn AS, Leonard BE. The olfactory bulbectomized rat as a model of depression: an update. Pharmacol Ther 1997;74:299–316.

128. Lumia AR, Teicher MH, Salchli F, Ayers E, Possidente B. Olfactory bulbectomy as a model for agitated hyposerotonergic depression. Brain Res 1992;587:181–185.
129. van Riezen H, Leonard BE. Effects of psychotropic drugs on the behavior and neurochemistry of olfactory bulbectomized rats. Pharmacol Ther 1990;47:21–34.
130. Neckers LM, Zarrow MX, Myers MM, Denenberg VH. Influence of olfactory bulbectomy and the serotonergic system upon intermale aggression and maternal behavior in the mouse. Pharmacol Biochem Behav 1975;3:545–550.
131. Otmakhova NA, Gurevich EV, Katkov YA, Nesterova IV, Bobkova NV. Dissociation of multiple behavioral effects between olfactory bulbectomized C57Bl/6J and DBA/2J mice. Physiol Behav 1992;52:441–448.
132. Gurevich EV, Aleksandrova IA, Otmakhova NA, Katkov YA, Nesterova IV, Bobkova NV. Effects of bulbectomy and subsequent antidepressant treatment on brain 5-HT2 and 5-HT1A receptors in mice. Pharmacol Biochem Behav 1993;45:65–70.
133. Possidente B, Lumia AR, McGinnis MY, Rapp M, McEldowney S. Effects of fluoxetine and olfactory bulbectomy on mouse circadian activity rhythms. Brain Res 1996;713:108–113.
134. Nesterova IV, Gurevich EV, Nesterov VI, Otmakhova NA, Bobkova NV. Bulbectomy-induced loss of raphe neurons is counteracted by antidepressant treatment. Prog Neuropsychopharmacol Biol Psychiatry 1997;21:127–140.
135. Komori T, Yamamoto M, Matsumoto T, Zhang K, Okazaki Y. Effects of imipramine on T cell subsets in olfactory bulbectomized mice. Neuropsychobiology 2002;46:194–196.
136. Bilkei-Gorzo A, Racz I, Michel K, Zimmer A. Diminished anxiety- and depression-related behaviors in mice with selective deletion of the Tac1 gene. J Neurosci 2002;22:10,046–10,052.
137. Hozumi S, Nakagawasai O, Tan-No K, et al. Characteristics of changes in cholinergic function and impairment of learning and memory-related behavior induced by olfactory bulbectomy. Behav Brain Res 2003;138:9–15.
138. Cryan JF, McGrath C, Leonard BE, Norman TR. Combining pindolol and paroxetine in an animal model of chronic antidepressant action—can early onset of action be detected? Eur J Pharmacol 1998;352:23–28.
139. Cryan JF, McGrath C, Leonard BE, Norman TR. Onset of the effects of the 5-HT1A antagonist, WAY-100635, alone, and in combination with paroxetine, on olfactory bulbectomy and 8-OH-DPAT-induced changes in the rat. Pharmacol Biochem Behav 1999;63:333–338.
140. Stock HS, Hand GA, Ford K, Wilson MA. Changes in defensive behaviors following olfactory bulbectomy in male and female rats. Brain Res 2001;903:242–246.
141. Primeaux SD, Holmes PV. Role of aversively motivated behavior in the olfactory bulbectomy syndrome. Physiol Behav 1999;67:41–47.
142. Mar A, Spreekmeester E, Rochford J. Antidepressants preferentially enhance habituation to novelty in the olfactory bulbectomized rat. Psychopharmacology (Berl) 2000;150:52–60.
143. McNish KA, Davis M. Olfactory bulbectomy enhances sensitization of the acoustic startle reflex produced by acute or repeated stress. Behav Neurosci 1997;111:80–91.
144. van Rijzingen IM, Gispen WH, Spruijt BM. Olfactory bulbectomy temporarily impairs Morris maze performance: an ACTH(4-9) analog accelerates return of function. Physiol Behav 1995;58:147–52.
145. Marcilhac A, Anglade G, Hery F, Siaud P. Olfactory bulbectomy increases vasopressin, but not corticotropin-releasing hormone, content in the external layer of the median eminence of male rats. Neurosci Lett 1999;262:89–92.
146. Marcilhac A, Faudon M, Anglade G, Hery F, Siaud P. An investigation of serotonergic involvement in the regulation of ACTH and corticosterone in the olfactory bulbectomized rat. Pharmacol Biochem Behav 1999;63:599–605.
147. Stock HS, Ford K, Wilson MA. Gender and gonadal hormone effects in the olfactory bulbectomy animal model of depression. Pharmacol Biochem Behav 2000;67:183–191.
148. Holmes PV, Masini CV, Primeaux SD, et al. Intravenous self-administration of amphetamine is increased in a rat model of depression. Synapse 2002;46:4–10.

149. Katkov YA, Otmakhova NA, Gurevich EV, Nesterova IV, Bobkova NV. Antidepressants suppress bulbectomy-induced augmentation of voluntary alcohol consumption in C57Bl/6j but not in DBA/2j mice. Physiol Behav 1994;56:501–509.
150. Sieck MH, Baumbach HD. Differential effects of peripheral and central anosmia producing techniques on spontaneous behavior patterns. Physiol Behav 1974;13:407–425.
151. van Riezen H, Schnieden H, Wren AF. Olfactory bulb ablation in the rat: behavioural changes and their reversal by antidepressant drugs. Br J Pharmacol 1977;60:521–528.
152. Greckschg G, Zhou D, Franke C, et al. Influence of olfactory bulbectomy and subsequent imipramine treatment on 5-hydroxytryptaminergic presynapses in the rat frontal cortex: behavioural correlates. Br J Pharmacol 1997;122:1725–1731.
153. Norrholm SD, Ouimet CC. Altered dendritic spine density in animal models of depression and in response to antidepressant treatment. Synapse 2001;42:151–163.
154. Barr AM, Phillips AG. Withdrawal following repeated exposure to d-amphetamine decreases responding for a sucrose solution as measured by a progressive ratio schedule of reinforcement. Psychopharmacology (Berl) 1999;141:99–106.
155. Lynch MA, Leonard BE. Changes in brain gamma-aminobutyric acid concentrations following acute and chronic amphetamine administration and during post amphetamine depression. Biochem Pharmacol 1978;27:1853–1855.
156. Gilliss B, Malanga CJ, Pieper JO, Carlezon WA Jr. Cocaine and SKF-82958 potentiate brain stimulation reward in Swiss-Webster mice. Psychopharmacology (Berl) 2002;163:238–248.
157. Ikeda K, Moss SJ, Fowler SC, Niki H. Comparison of two intracranial self-stimulation (ICSS) paradigms in C57BL/6 mice: head-dipping and place-learning. Behav Brain Res 2001;126:49–56.
158. Holmes A, Rodgers RJ. Influence of spatial and temporal manipulations on the anxiolytic efficacy of chlordiazepoxide in mice previously exposed to the elevated plus-maze. Neurosci Biobehav Rev 1999;23:971–980.
159. Hogg S. A review of the validity and variability of the elevated plus-maze as an animal model of anxiety. Pharmacol Biochem Behav 1996;54:21–30.
160. Crabbe JC, Wahlsten D, Dudek BC. Genetics of mouse behavior: interactions with laboratory environment. Science 1999;284:1670–1672.
161. Wahlsten D, Metten P, Phillips TJ, et al. Different data from different labs: lessons from studies of gene-environment interaction. J Neurobiol 2003;54:283–311.
162. Holmes A, Iles JP, Mayell SJ, Rodgers RJ. Prior test experience compromises the anxiolytic efficacy of chlordiazepoxide in the mouse light–dark exploration test. Behav Brain Res 2001;122:159–167.
163. Holmes A, Rodgers RJ. Responses of Swiss-Webster mice to repeated plus-maze experience: further evidence for a qualitative shift in emotional state? Pharmacol Biochem Behav 1998;60:473–488.
164. Schramm NL, McDonald MP, Limbird LE. The alpha(2a)-adrenergic receptor plays a protective role in mouse behavioral models of depression and anxiety. J Neurosci 2001;21:4875–4882.
165. Conti AC, Cryan JF, Dalvi A, Lucki I, Blendy JA. cAMP response element-binding protein is essential for the upregulation of brain-derived neurotrophic factor transcription, but not the behavioral or endocrine responses to antidepressant drugs. J Neurosci 2002;22:3262–3268.
166. McIlwain KL, Merriweather MY, Yuva-Paylor LA, Paylor R. The use of behavioral test batteries: effects of training history. Physiol Behav 2001;73:705–717.
167. Wahlsten D, Rustay NR, Metten P, Crabbe JC. In search of a better mouse test. Trends Neurosci 2003;26:132–136.
168. Holmes A, Wrenn CC, Harris AP, Thayer KE, Crawley JN. Behavioral profiles of inbred strains on novel olfactory, spatial and emotional tests for reference memory in mice. Genes Brain Behav 2002;1:55–69.
169. Rodgers RJ, Davies B, Shore R. Absence of anxiolytic response to chlordiazepoxide in two common background strains exposed to the elevated plus-maze: importance and implications of behavioural baseline. Genes Brain Behav 2002;1:242–251.

170. Voikar V, Koks S, Vasar E, Rauvala H. Strain and gender differences in the behavior of mouse lines commonly used in transgenic studies. Physiol Behav 2001;72:271–281.
171. Cook MN, Bolivar VJ, McFadyen MP, Flaherty L. Behavioral differences among 129 substrains: implications for knockout and transgenic mice. Behav Neurosci 2002;116:600–611.
172. Contet C, Rawlins JN, Deacon RM. A comparison of 129S2/SvHsd and C57BL/6JOlaHsd mice on a test battery assessing sensorimotor, affective and cognitive behaviours: implications for the study of genetically modified mice. Behav Brain Res 2001;124:33–46.
173. Holmes A, Lit Q, Murphy DL, Gold E, Crawley JN. Abnormal anxiety-related behavior in serotonin transporter null mutant mice: the influence of genetic background. Genes Brain Behav 2003;2:365–380.
174. Nadeau JH. Modifier genes and protective alleles in humans and mice. Curr Opin Genet Dev 2003;13:290–295.
175. Hood HM, Belknap JK, Crabbe JC, Buck KJ. Genomewide search for epistasis in a complex trait: pentobarbital withdrawal convulsions in mice. Behav Genet 2001;31:93–100.
176. Gerlai R. Gene-targeting studies of mammalian behavior: is it the mutation or the background genotype? Trends Neurosci 1996;19:177–181.
177. Crawley JN, Belknap JK, Collins A, et al. Behavioral phenotypes of inbred mouse strains: implications and recommendations for molecular studies. Psychopharmacology (Berl) 1997; 132:107–124.
178. Wolfer DP, Crusio WE, Lipp HP. Knockout mice: simple solutions to the problems of genetic background and flanking genes. Trends Neurosci 2002;25:336–340.
179. Wong GT. Speed congenics: applications for transgenic and knock-out mouse strains. Neuropeptides 2002;36:230–236.
180. Meaney MJ. Maternal care, gene expression, and the transmission of individual differences in stress reactivity across generations. Annu Rev Neurosci 2001;24:1161–1192.
181. Anisman H, Zaharia MD, Meaney MJ, Merali Z. Do early-life events permanently alter behavioral and hormonal responses to stressors? Int J Dev Neurosci 1998;16:149–164.
182. Francis DD, Szegda K, Campbell G, Martin WD, Insel TR. Epigenetic sources of behavioral differences in mice. Nat Neurosci 2003;6:445–446.
183. Bale TL, Picetti R, Contarino A, Koob GF, Vale WW, Lee KF. Mice deficient for both corticotropin-releasing factor receptor 1 (CRFR1) and CRFR2 have an impaired stress response and display sexually dichotomous anxiety-like behavior. J Neurosci 2002;22:193–199.
184. Kendler KS. Twin studies of psychiatric illness: an update. Arch Gen Psychiatry 2001;58: 1005–1014.
185. Murphy DL, Uhl GR, Holmes A, et al. Experimental gene interaction studies with SERT mutant mice as models for human polygenic and epistatic traits and disorders. Genes Brain Behav 2003;2:350–364.
186. Rutter M, Silberg J. Gene-environment interplay in relation to emotional and behavioral disturbance. Annu Rev Psychol 2002;53:463–490.
187. McGuffin P, Riley B, Plomin R. Genomics and behavior. Toward behavioral genomics. Science 2001;291:1232–1249.
188. Caspi A, McClay J, Moffitt TE, et al. Role of genotype in the cycle of violence in maltreated children. Science 2002;297:851–854.
189. Caspi A, Sugden K, Moffitt TE, et al. Influence of life stress on depression: moderation by a polymorphism in the 5-HTT gene. Science 2003;301:386–389.
190. Singer JB, Hill AE, Burrage LC, et al. Genetic dissection of complex traits with chromosome substitution strains of mice. Science 2004;304:445–448.
191. Talbot CJ, Radcliffe RA, Fullerton J, Hitzemann R, Wehner JM, Flint J. Fine scale mapping of a genetic locus for conditioned fear. Mamm Genome 2003;14:223–230.
192. Hommel JD, Sears RM, Georgescu D, Simmons DL, DiLeone RJ. Local gene knockdown in the brain using viral-mediated RNA interference. Nat Med 2003;9:1539–1544.

193. El-Ghundi M, O'Dowd BF, Erclik M, George SR. Attenuation of sucrose reinforcement in dopamine D1 receptor deficient mice. Eur J Neurosci 2003;17:851–862.
194. Dixon AK, Huber C, Lowe DA. Clozapine promotes approach-oriented behavior in male mice. J Clin Psychiatry 1994;55(Suppl B):4–7.
195. Nonogaki K, Strack AM, Dallman MF, Tecott LH. Leptin-independent hyperphagia and type 2 diabetes in mice with a mutated serotonin 5-HT2C receptor gene. Nat Med 1998;4:1152–1156.
196. Karolyi IJ, Burrows HL, Ramesh TM, et al. Altered anxiety and weight gain in corticotropin-releasing hormone-binding protein-deficient mice. Proc Natl Acad Sci USA 1999;96:11,595–11,600.
197. Boutrel B, Franc B, Hen R, Hamon M, Adrien J. Key role of 5-HT1B receptors in the regulation of paradoxical sleep as evidenced in 5-HT1B knock-out mice. J Neurosci 1999;19:3204–3212.
198. Boutrel B, Monaca C, Hen R, Hamon M, Adrien J. Involvement of 5-HT1A receptors in homeostatic and stress-induced adaptive regulations of paradoxical sleep: studies in 5-HT1A knock-out mice. J Neurosci 2002;22:4686–4692.
199. El Yacoubi M, Bouali S, Popa D, et al. Behavioral, neurochemical, and electrophysiological characterization of a genetic mouse model of depression. Proc Natl Acad Sci USA 2003;100: 6227–6232.
200. Kafkafi N, Pagis M, Lipkind D, et al. Darting behavior: a quantitative movement pattern designed for discrimination and replicability in mouse locomotor behavior. Behav Brain Res 2003; 142:193–205.
201. Mizoguchi K, Yuzurihara M, Ishige A, Sasaki H, Tabira T. Chronic stress impairs rotarod performance in rats: implications for depressive state. Pharmacol Biochem Behav 2002;71:79–84.
202. Nonogaki K, Abdallah L, Goulding EH, Bonasera SJ, Tecott LH. Hyperactivity and reduced energy cost of physical activity in serotonin 5-HT(2C) receptor mutant mice. Diabetes 2003;52:315–320.
203. Grippo AJ, Beltz TG, Johnson AK. Behavioral and cardiovascular changes in the chronic mild stress model of depression. Physiol Behav 2003;78:703–710.
204. Ballard TM, Pauly-Evers M, Higgins GA, et al. Severe impairment of NMDA receptor function in mice carrying targeted point mutations in the glycine binding site results in drug-resistant nonhabituating hyperactivity. J Neurosci 2002;22:6713–6723.
205. Cheeta S, Ruigt G, van Proosdij J, Willner P. Changes in sleep architecture following chronic mild stress. Biol Psychiatry 1997;41:419–427.
206. Estape N, Steckler T. Cholinergic blockade impairs performance in operant DNMTP in two inbred strains of mice. Pharmacol Biochem Behav 2002;72:319–334.
207. Contarino A, Dellu F, Koob GF, et al. Reduced anxiety-like and cognitive performance in mice lacking the corticotropin-releasing factor receptor 1. Brain Res 1999;835:1–9.
208. van Gaalen MM, Stenzel-Poore M, Holsboer F, Steckler T. Reduced attention in mice overproducing corticotropin-releasing hormone. Behav Brain Res 2003;142:69–79.

13
Mutant Mouse Models of Bipolar Disorder
Are There Any?

Anneloes Dirks, Lucianne Groenink, and Berend Olivier

Summary

Bipolar disorder (also known as manic–depressive illness) is distinctive among psychiatric illnesses in that it is characterized by spontaneously alternating episodes of depression and mania. Over the years, extensive research into the pathophysiology of bipolar disorder has resulted in a growing understanding of the cellular, biochemical, and molecular changes associated with bipolar disorder and its treatment. However, given its unique nature, developing an animal model for bipolar disorder in which all aspects of the illness are emulated is challenging. Indeed, fully validated animal models of bipolar disorder are not available and a variety of models are used to represent a single manic or depressive episode, with some models possibly representing the progressive nature of the disorder. Nonetheless, targeted mutations of specific neurotransmitter systems, including receptors and transporters, as well as genetic manipulations of cellular signaling pathways, produce a variety of changes in affective-like behavior, with most changes consistent with manic-like behavior. As such, these mutant mouse models (with their own limitations) could contribute to the research of the underlying brain mechanisms of mania and/or bipolar disorder. In this chapter, we present an overview of neurochemical, neuroendocrine, and behavioral changes in bipolar disorder, and of the available mutant mouse models in which some aspects of the disorder are emulated. The mutant mouse models include targeted overexpression or knockout/knock-down of genes coding for corticotropin-releasing factor, glucocorticoid receptor, serotonin transporters, and dopamine transporters, and of genes involved in intracellular signaling pathways.

Key Words: Bipolar disorder; validity; animal model; mutant mice; dopamine transporter; serotonin transporter; CRF; BNDF; ERK.

1. INTRODUCTION

Although I was frequently depressed, when I felt good I felt "better than good."
—Suzanne J. Fiala, MD *(1)*

Bipolar disorder (also know as manic–depressive illness) is distinctive among psychiatric illnesses in that it is characterized by spontaneously alternating episodes of depression and mania. Few disorders throughout history have been described with such consistency as bipolar disorder. Symptoms that characterize the disorder can be found in medical literature throughout the centuries, from the ancient Greeks to the present day (reviewed in ref. 2). Perhaps the first person who described mania and depression (melancholia) as two different phenomenological states of one and the same disease was the Greek physician, Aretaeus of Cappadocia, in the 1st century AD. Emil Kraepelin, however, was the first to segregate the psychotic disorders and to draw boundaries around their clinical diagnoses, distinguishing between manic–depressive psychosis (bipolar disorder) and dementia praecox (schizophrenia) (*see* ref. 2).

Over the years, extensive research into the pathophysiology of bipolar disorder have resulted in a growing understanding of the cellular, biochemical, and molecular changes associated with bipolar disorder and its treatment (*see* refs. 3–5). Mutant mice with genetically altered expression of a specific protein (either a receptor, transporter, enzyme, or signal transduction molecule) *(6)* have only been recently added to the list of available tools in this line of research. These mutant mouse models could provide further insight into the underlying pathophysiological and/or etiological mechanisms of the disorder.

In this chapter, we describe the pathophysiology of bipolar disorder and the available mutant mouse models relevant to bipolar disorder.

2. BIPOLAR DISORDER

2.1. General Description of Disorder

In the Diagnostic and Statistical Manual of Mental Disorders, 4th edition (DSM-IV) *(7)*, several bipolar disorders are distinguished (e.g., Bipolar I and Bipolar II). These disorders comprise both manic episodes and the presence or history of major depressive episodes, i.e., at least 2 wk of depressed mood or loss of interest accompanied by at least four additional symptoms of depression, including sleep disturbances; changes in appetite or weight; psychomotor changes, such as agitation and retardation; decreased energy; difficulty thinking, concentrating, or making decisions; sense of worthlessness or guilt; and thoughts of death, suicidal ideation, or suicide attempts. A manic episode is defined by a distinct period during which mood is abnormally and persistently elevated (euphoria), expansive (unceasing and indiscriminate enthusiasm for interpersonal, sexual, or occupational interactions), or irritable; together with inflated self-esteem or grandiosity, decreased need for sleep, pressure of speech, flight of ideas, distractibility, increased involvement in goal-directed activities or psychomotor agitation, and excessive involvement in pleasurable activities with a high potential for painful consequences (according to the DSM-IV). The disturbance must be sufficiently severe to cause marked impairment in social or occupational functioning, to require hospitalization, or be characterized by psychotic features.

Bipolar I disorder, the most classic form of bipolar disorders, is characterized by a sequence of manic or mixed episodes (symptoms of both a manic episode and major depressive episode occurring nearly every day for at least a 1-wk period) with major depressive episodes. The course of Bipolar II disorder is similar but more unstable; major depressive episodes alternate with hypomanic episodes (i.e., manic episodes that are not severe enough to cause a marked impairment in social or occupational functioning or to require hospitalization, with the absence of psychotic features) *(8)*. The current concept of bipolar disorder is thought of as a continuum of phenotypes that range from very mild forms of depression with brief hypomania to more severe forms in which criteria for severity may be the speed of the cycles or the presence of psychotic features *(8)*.

The lifetime prevalence of bipolar disorder in community samples ranges from 0.4 to 1.6% *(7)*. Recent epidemiological studies, however, suggest that bipolarity may affect at least 5% of the general population *(9)*. Furthermore, bipolar disorder is a condition that is frequently associated with one or more comorbid psychiatric disorders, including anxiety disorders (e.g., refs. *10–13*). The onset of the illness usually occurs before the age of 25 yr *(8)*. Manic episodes typically begin suddenly, frequently after psychosocial stressors, with a rapid escalation of symptoms over a few days.

Behaviorally, the core symptoms of bipolar disorders, i.e., euphoric, expansive, or irritable mood, can all be regarded as symptoms that are associated with disinhibition *(14)*. Indeed, neuropsychologically, patients with bipolar disorder have been found to exhibit pronounced cognitive impairments (e.g., refs. *15* and *16*), and, during manic episodes, to be indistinguishable from schizophrenia patients on a variety of information processing tasks (e.g., refs. *17* and *18*). Prepulse inhibition (PPI) is often used to assess information processing in human and rodent studies, and it is used as an operational measure for a process called sensorimotor gating, by which excess or trivial stimuli are screened or "gated out" of awareness, so that an individual can focus attention on the most important aspects of the environment *(19)*. Thus, PPI refers to a general reduction in the processing of and distraction by irrelevant or repetitive stimuli. The only study published on sensorimotor gating deficits in bipolar disorder patients so far revealed significantly reduced PPI in bipolar patients compared with control subjects, depending on the time interval between onset of prepulse and onset of startle stimulus, and found no differences between bipolar disorder and schizophrenia patients *(14)*.

2.2. Treatment

Lithium and valproate are the most commonly used mood stabilizers in the treatment of bipolar disorders (*see* refs. *20–23*). Lithium is a monovalent cation, and valproate is a branched-chain fatty acid *(24)*. The full clinical beneficial effects of lithium and valproate require weeks of treatment and the maintenance of blood drug levels within a narrow therapeutic window *(25)*. The anticonvulsants, carbamazepine and lamotrigine, may also be useful as mood stabilizers in the more prevalent "soft" spectrum of bipolar disorder, including Bipolar II disorder, whereas lithium may be indicated in a smaller group of patients with a "classic" bipolar picture (Bipolar I disorder) (e.g., refs. *13, 20,* and *21*).

2.3. Pathophysiology

A number of abnormalities in the dopaminergic, serotonergic, noradrenergic, and other neurotransmitter and neuropeptide systems have been demonstrated in bipolar disorder (*see* refs. *3, 26,* and *27*). Dopamine has been proposed to be especially involved in the pathophysiology of mania, and changes in dopaminergic neurotransmission have been reported as consistent neurobiological abnormalities in bipolar disorder *(28)*. Aside from alterations in central neurotransmitter systems, dysregulation of the hypothalamus–pituitary–adrenal axis (HPA axis), the major pathway by which stress exerts its effects on the brain and other organ systems, is commonly observed in bipolar disorder (reviewed in refs. *3* and *4*). Among the various peptide and neurotransmitter systems that have been implicated in the regulation of stress responses, the corticotropin-releasing factor (CRF; also referred to as corticotropin-releasing hormone) *(29)* system takes a prominent position *(30–34)*. Although there is no evidence to support changes in concentrations of cerebrospinal fluid CRF in mania (e.g., refs. *35* and *36*), it is likely that central CRF systems are hyperactive in mania because such patients exhibit marked HPA-axis hyperactivity *(37)*. In line with this assumption, are the findings in rats that marked changes are observed in CRF concentrations and *CRF* messenger RNA expression in several brain areas (including hypothalamus) after subchronic or chronic treatment with valproate or lithium, with the effects of valproate being more widespread *(37,38)*. Only valproate, but not lithium, altered plasma corticosterone concentrations and only after subchronic, but not acute or chronic, administration *(37,38)*.

Despite the devastating impact of the disorder and the relatively well-characterized clinical picture, with accompanying neurochemical, neuroendocrine, and behavioral changes, little is known for certain about the etiology or pathophysiology of bipolar disorder *(26)*. One theory proposes that altered expression of critical proteins arising directly or indirectly from a series of susceptibility genes (*see* refs. *39–43*) predisposes to a dysregulation of signaling in regions of the brain, resulting in the periodic loss of homeostasis and the clinical appearance of mania and/or depression *(3,44)*. Indeed, a variety of family epidemiological data has long argued for a strong genetic component (*see* ref. *43*). Twin studies of bipolar disorder are generally consistent in observing greater concordance among monozygotic twins than dizygotic twins (for recent review, *see* ref. *45*). Furthermore, recurrence risk of bipolar disorder for first-degree relatives of bipolar probands is 8.7%, whereas the risk for unipolar depression is 14.1% (*see* ref. *45*). An alternative theory states that a major component of the pathophysiology of bipolar disorder may stem from discordant biological rhythms that ultimately may drive the periodic recurrent nature of the disorder (*see* ref. *26*).

3. MODELING BIPOLAR DISORDER

Because of the intriguing alternation of manic and depressive episodes, it is a challenge to develop a valid animal model for bipolar disorder. As outlined earlier in Section 2., depressive symptoms include anhedonia, lack of motivation, loss of appetite, insomnia, motor retardation or agitation, fatigue, cognitive impairment, and suicidal thoughts. Manic behaviors include motor hyperactivity, excessive energy and speech, grandiosity, aggressiveness, loss of appetite, and insomnia.

The validity of animal models of psychiatric disorders can be assessed based on how these models are in agreement with three major criteria: face, construct, and predictive validity (*see* refs. *6, 26, 28,* and *46*). Face validity tests the similarities in behavioral features between the model and the modeled disorder. For example, face validity of most animal models of mania usually focus primarily on insomnia and hyperactivity as symptoms that resemble mania. Predictive validity assesses the specificity of clinically effective drugs (in the case of bipolar disorder, lithium, valproate, and carbamazepine) to a therapeutic-like effect in the behavioral model. Construct validity tests possible common mechanisms that can explain both the model and the modeled disorder.

As described by Einat and colleagues *(47)*, an ideal model for bipolar disorder should include spontaneous behavior that moves back and forth between increased and decreased expression of the model behavior and that may be similar to human mania or depression (face validity). For predictive validity, the model behavior should also be normalized by chronic but not acute treatment with the mood stabilizers, lithium and valproate, and the anticonvulsants, carbamazepine and lamotrigine (e.g., refs. *20–23*), but should induce manic-like behavior in response to treatment with antidepressants (e.g., refs. *48* and *49*). Additionally, a well-validated model should be based on one of the mechanistic theories of the disease (construct validity).

Modeling of bipolar disorder may be a much more difficult task compared with other somatic diseases, because the diagnosis of bipolar disorder is based on reports of symptoms rather than specific measures *(47)*. Specific features of bipolar disorder, including euphoria, racing thoughts, depressed mood, guilt, and/or suicidal thoughts can be fully appreciated only in humans, wherein the cortical mantle has evolved to a greater extent than in other species *(40)*. However, other endophenotypes of bipolar disorder (e.g., mania) have their equivalents in, for instance, the mouse, thus, potentially allowing for phenotypes that can be modeled in the laboratory *(40)*. The traditional strategy to induce psychopathology in animals is by using pharmacological, behavioral, or genetic manipulations. Animal models for bipolar disorder based on pharmacological interventions are recently reviewed by Machado-Vieira and Einat and colleagues *(28,47)*. For example, an acute treatment with appropriate doses of psychostimulants, including amphetamine and cocaine, can produce a range of mania-like behaviors, including hyperactivity, heightened sensory awareness, alertness, and changes in sleep patterns (*see* refs. *26, 40,* and *47*). However, traditional animal models for bipolar disorder have concentrated on modeling a single manic episode. Some models have focused on the hyperactivity feature of the disorder; others are based on measures of motivation and reward or the effects of sleep deprivation (for review, *see* ref. *50*). Although available models can replicate specific aspects of mania or depression, no model incorporates both components in alternating fashion, and, therefore, the models fail to mimic the hallmark of the illness *(28)*. Given the unique nature of bipolar disorder, it will be very difficult, not to say impossible, to develop an animal model for bipolar disorder in which all aspects of the illness will be present. It will be more feasible to model particular aspects of the disorder in mutant mouse models.

One of the most important advances in understanding psychiatric disorders has been the development of mice with genetically altered expression of a specific protein, be it a receptor, transporter, enzyme, or signal transduction molecule *(6)*. Indeed, mutant mouse strains demonstrate a variety of different phenotypes, including biochemical and behavioral changes, and the screening of available strains may result in new and valid animal models for several psychiatric disorders *(26)*. Mutant mice with phenotypes that resemble a psychiatric disorder may also offer new clues about the mechanisms of the disease.

4. MUTANT MOUSE MODELS FOR BIPOLAR DISORDER

Currently, fully validated animal models of bipolar disorder are not available and a variety of models are used to represent a single manic or depressive episode, with some models possibly representing the progressive nature of the disorder *(26)*. Nonetheless, targeted mutations of specific neurotransmitter systems, including receptors and transporters, as well as genetic manipulations of cellular signaling pathways produce a variety of changes in affective-like behavior, with most changes consistent with manic-like behavior. In general, the choices of the genes targeted in these mutant mouse models are based on the already existing knowledge of the pathophysiology of bipolar disorder or the effects of mood stabilizers on intracellular events and processes. The mutant mouse models include targeted overexpression or knockout (KO)/knock-down (KD) of genes coding for CRF, glucocorticoid receptor (GR), serotonin transporters (5-HTT), and dopamine transporters (DAT), and of genes involved in intracellular signaling pathways, including extracellular signal-regulated kinase (ERK)-1 and brain-derived neurotrophic factor (BDNF). Table 1 summarizes the parallels between phenotypic abnormalities in mice with targeted mutations and human bipolar disorder.

4.1. Corticotropin-Releasing Factor

As stated earlier, it is likely that central CRF systems are hyperactive in mania because patients exhibit marked HPA-axis hyperactivity *(37)*. Furthermore, it has been hypothesized that the mood-stabilizing effects of valproate may be mediated in part by alterations in CRF neuronal activity *(38)*. As such, transgenic mice with a hyperactive CRF system could potentially be representative for bipolar disorder. Indeed, transgenic mice overexpressing CRF (CRF-OE) exhibit features similar to depressive illness, including increased heart rate and body temperature, decreased heart rate variability, and altered HPA-axis regulation reflected in increased basal plasma corticosterone concentrations, and nonsuppression of corticosterone secretion in response to dexamethasone *(51–53)*. Interestingly, CRF-OE mice show sensorimotor gating deficits (i.e., impaired PPI) commonly associated with

Table 1
Parallels Between Phenotypic Abnormalities in Mice With Targeted Mutations and Human Bipolar Disorder

	Mania	Depression	Anxiety	Impulsivity/ aggression	Prepulse inhibition	Miscellaneous	References
Bipolar disorder	Yes	Yes	Comorbid	High	↓	Alternating between moods Sleep disturbances Hyperactive HPA axis Cognitive impairments	3,14 26,27

	Locomotor activity								
Mutant mouse model	Home cage	Open field	After psychostimulant	Depression-related behavior	Anxiety-related behavior	Impulsivity/ aggression	Prepulse inhibition	Miscellaneous	References
CRF-OE	=	↓					↓	Hyperactive HPA axis	51–54,56
GRov	=	=	↑	↑	↑			Increased emotional "liability" Normal HPA axis	57
DAT KO	=	↑				=	↓	Perserverative motor patterns	62–66
DAT KD	=	↑					=	Perserverative motor patterns	67,68
5-HTT KO	↓	↓		=/↑[a]	=/↑[a]	↓	=	Normal learning/memory	73,74,76–78
ERK-1 KO		↑		↓	=/↑	↑	=	Normal learning/memory	87,89,90
BDNF+/−	=/↑		↑	=	=/↑	↑	=	Normal learning/memory	101,102, 104–108
Emx-BDNF KO		↓			=		=	Impaired learning/memory	109
TrkB KO		↑		=	=	↑			112,113

[a] Depending on background strain. ↑, Increased vs wild-type mice; ↓, decreased vs wild-type mice; =, no difference. CRF-OE, overexpressing corticotropin-releasing factor; GRov, glucocorticoid receptor-overexpressing mice; DAT, dopamine transporter; KO, knockout; KD, knock-down; 5-HTT, serotonin transporter; ERK, extracellular signal-regulated kinase; BDNF, brain-derived neurotrophic factor.

schizophrenia *(54)* that can be reversed by the typical antipsychotic, haloperidol, and the atypical antipsychotics, clozapine and risperidone *(55)*. However, these mutant mice display reduced levels of locomotor activity in the open field when compared with wild-type littermates *(52,56)*. Moreover, several behaviors, physiological measures, and pharmacological interventions relevant to mania, including psychostimulant-induced hyperactivity, impulsive behavior, sleep, and the effects of mood stabilizers and carbamazepine, have not yet been examined in these mice. Therefore, although these CRF-OE mice show a clear depression-like phenotype with possible psychotic features, future research should also be aimed at establishing a mania-like phenotype in the CRF-OE mice.

4.2. Glucocorticoid Receptor

"Emotional instability" or "liability" refers to a greater likelihood of switching from one emotional state to another and/or having a wider range of emotional intensity, in contrast to consistency in mood *(57)*. Bipolar disorder is a good example of uncontrolled emotional liability whereby, as noted earlier, patients exhibit extreme negative and positive emotions and can switch readily between mania and depression. The molecular mechanisms that control the range and stability of emotions are still unknown.

Recently, transgenic mice were generated that overexpress the GR specifically in the forebrain, including the hippocampus, prefrontal cortex, striatum, amygdala, and hypothalamus *(57)*. GR is one of the two distinct nuclear receptor types for corticosteroids and, together with mineralocorticoid receptors, mediates, in a coordinate manner, the steroid control of the HPA-axis activity and behavior *(58)*. These GR-overexpressing mice display significant increases in anxiety-like behaviors in elevated plus maze and light/dark box as well as increased depressant-like behaviors in the forced-swim test relative to wild-type mice. These changes in behavior seem to be extremely sensitive to treatment with antidepressants. Furthermore, although general locomotor activity in an open field did not differ, the GR-overexpressing mice demonstrated hyperlocomotion after cocaine administration. There were no neuroendocrine differences between genotypes under basal conditions or after mild stress.

Thus, mice overexpressing GR in the forebrain have a consistently wider than normal range of reactivity in both positive and negative emotionality tests, which is associated, in specific brain regions, with increased expression of genes relevant to emotionality (i.e., CRF, 5-HTT, norepinephrine transporters, DAT, and serotonin [5-HT] 1A receptor). The authors conclude that GR overexpression in forebrain causes higher "emotional lability" secondary to a unique pattern of molecular regulation and suggest that natural variations in GR gene expression can contribute to the fine-tuning of emotional stability or lability and may play a role in bipolar disorder *(57)*.

4.3. Dopamine Transporter

Dysregulation of the dopamine system is thought to underlie several psychiatric disorders, including schizophrenia and bipolar disorder. Several studies have focused on DAT as a possible candidate gene in association studies in patients with bipolar disorder *(59,60)*. The plasma membrane DAT is responsible for the rapid uptake of dopamine into presynaptic terminals after its release *(61)*.

One method to study the behavioral consequences of a chronically dysregulated dopamine system is by studying DAT-null mutant mice. Mice lacking DAT are characterized by high extracellular dopamine levels and spontaneous hyperlocomotion *(62–66)*. Their locomotor activity is characterized by repetitive, perseverative straight movements in the periphery of the test enclosure *(65)*. DAT KO mice were easily aroused by novelty and always responded with hyperlocomotion, which interfered with habituation to the testing environment, exploratory behavior in an open field, and the coping response to forced-swimming stress *(64)*. Aggressive behaviors were not modified in DAT KO mice compared with wild-type controls *(64)*. However, DAT KO mice exhibited a consistent deficit in sensorimotor gating, as measured by the PPI of the startle response, without any significant alteration in startle reactivity *per se (65)*.

Recently, a second kind of DAT mutant mouse has been created. Unlike the full DAT KO, these KD mice have an approx 90% loss of DAT *(67)*. These mice also have a chronic hyperdopaminergic tone (e.g., a 70% increase in extracellular striatal DA), but they do not show gross physical changes. Similar to the DAT KO mice, the DAT KD mice have normal home cage activity, but are hyperactive in a novel open field and display perseverative motor patterns *(67,68)*. Furthermore, in the novel-object exploration test, the hyperactivity of the DAT KD occurred specifically in response to novel stimuli, possibly caused by a decreased response habituation *(67)*. If indeed a dysregulated dopamine system underlies some of the key symptoms of mania, and if the DAT KD mice model a dysregulated DA system, drugs that successfully treat manic symptoms should attenuate the hyperactivity displayed by the DAT KD mutant mice. Indeed, valproate significantly attenuated the hyperactivity in the DAT KD mice but had no effect on motor behavior in the wild-type controls *(68)*. Valproate also diminished the degree of perseverative straight locomotor patterns in DAT KD mice *(68)*. Unlike the full DAT KO, DAT KD mice do not show deficits in PPI compared with wild-type controls *(68)*.

When the dopamine system is functioning normally, as in wild-type mice, behavioral organization is intact, and normal motor activity and gating can occur. If, however, the DA system is perturbed, in this case through the hyperdopaminergic state produced by DAT mutations, there seems to be a disruption in the processes that are involved in both the modulation of incoming stimuli and/or the appropriate expression of motor responses *(65)*. If indeed the hyperdopaminergic state of the DAT KO, more so for the DAT KD but mice, is consistent with a manic state in humans with bipolar disorder, than the DAT KD mice seem to provide a model of some aspects of manic behavior. With the limited models of bipolar disorder available, the DAT KD mice might provide a promising vehicle to screen for new psychiatric therapies to treat mania and its related symptoms *(68)*.

4.4. Serotonin Transporter

Available data on the role of the 5-HTT suggests that a dysfunction in brain serotonergic system activity contributes to the vulnerability to affective disorders (e.g., refs. *69–71*). The 5-HTT is the major site of 5-HT reuptake into the presynaptic neuron, and it has been shown that an insertion/deletion polymorphism in the promoter region (*5-HTTLPR*) and a variable number of tandem repeats polymorphism in the second intron may affect gene-transcription activity (*see* ref. *72*). The available evidence of a link between genetic variation of the 5-HTT and affective disorders prompted the generation of 5-HTT KO mice *(73,74)*.

5-HTT KO mice (5-HTT$^{-/-}$) showed, as could be expected, an absence of 5-HT reuptake, leading to a decreased rate of synaptic 5-HT clearance and profound increase in basal levels of forebrain extracellular 5-HT, accompanied by modest compensatory alterations in dopamine or norepinephrine function *(75)*, as well as marked changes within the serotonergic system itself (reviewed in ref. *74*). In the open field, exploratory behavior, measured as horizontal and vertical activity and the percentage of time spent in the center (often used as an index of anxiety), was generally lower in 5-HTT$^{-/-}$ mice than wild-type controls *(76)*. Furthermore, 5-HTT$^{-/-}$ mice showed significantly lower levels of home cage activity as compared with 5-HTT$^{+/-}$ and 5-HTT$^{+/+}$ littermate controls *(77)*. In concordance with the open-field behavior, male and female 5-HTT$^{-/-}$ mice showed robust phenotypic abnormalities as compared with 5-HTT$^{+/+}$ littermates, suggestive of increased anxiety-like behavior in the elevated plus maze, light/dark-exploration test, and emergence test *(76)*. It should be noted that the behavioral phenotype strongly depends on the background strain of these mutant mice (*see* ref. *74*). The effects of the genetic background on which the 5-HTT null mutation was placed were even more pronounced in tests for antidepressant activity. 5-HTT KO mice on a C57BL/6J background showed normal baseline performance on both the tail-suspension and forced-swim tests, whereas, in contrast, 5-HTT$^{-/-}$ mice on a 129S6 background showed an antidepressant-like decrease in immobility in the tail-suspension test, but an increase in immobility in the forced-swim test *(78)*. Furthermore, male 5-HTT$^{-/-}$ mice showed reduced aggression in the resident–intruder test compared with wild-type littermate controls, reflected by a longer latency to attack an intruder, and attacking the intruder less frequently *(77)*. The amount of time 5-HTT mutant mice spent in nonaggressive, social investigative behavior was similar to that of wild-type controls, implying a specific inhibition of aggressive responses, rather than a more general social deficit, in 5-HTT KO mice *(77)*.

Taken together, although human association studies suggest a link between bipolar disorder and the 5-HTT, 5-HTT$^{-/-}$ mice do not show an evident bipolar disorder-like phenotype with, among others, hyperactivity and heightened aggression/impulsivity. It should be noted that gene linkage and association studies yielded conflicting results regarding the involvement of the 5-HTT in bipolar disorder (for reviews, *see* refs. *72, 79,* and *80).* Thus, although valuable in research investigating the role of the 5-HT receptor in, for instance, treatment response, these 5-HTT KO mice seem not to be a good animal model for bipolar disorder.

4.5. ERK–MAP Kinase-Signaling Pathway

Extensive research into the pathophysiology of bipolar disorder, using increasingly accurate methods and tools, has resulted in increasing knowledge of the cellular, biochemical, and molecular changes associated with bipolar disorder and its treatment (*see* refs. *3–5).* A number of theories have been developed based on this knowledge, suggesting that bipolar disorder or the effects of mood stabilizers may be related to a number of intracellular events and processes (*see* refs. *3–5,* and *81).* One such novel hypothesis regarding the pathophysiology of bipolar disorder suggests a possible involvement of the ERK–mitogen-activated protein (MAP) kinase-signaling pathway (e.g., ref. *82).*

Cellular signaling pathways interact at various levels, thereby forming complex signaling networks that allow neurons to receive, process, and respond to information, and to modulate the signal generated by multiple different neurotransmitter and neuropeptide systems (*see* ref. *26).* Intracellular signaling cascades are critically involved in regulating complex psychological, motoric, and cognitive processes, as well as diverse neurovegetative functions, such as appetite and wakefulness—all of which represent dimensions of bipolar disorder *(26).* Recent research showed that the ERK-signaling cascade—a pathway that plays a major role in regulating synaptic plasticity and long-term neuronal adaptations (for review, *see* ref. *83)*—is a biochemical target for effective treatment for bipolar disorder. Genetic manipulations of the BDNF–ERK kinase pathway produce a variety of changes in affective-like behavior, with most changes consistent with manic-like behavior *(26).*

4.5.1. Extracellular Signal-Regulated Kinase-1

The putative role of the ERK-signaling cascade in bipolar disorder has been extensively reviewed by Einat and colleagues *(26).* In the context of bipolar disorder, the emerging role for ERK kinases in a wide variety of forms of synaptic plasticity and memory formation is interesting (e.g., refs. *84–86).* Prolonged activation of postsynaptic receptors results in translocation of ERK kinases to the cell nucleus, resulting in phosphorylation and activation of transcription factors, such as cyclic adenosine monophosphate response element-binding protein *(87).* These nuclear events initiate cell-specific gene expression programs necessary for synaptic remodeling and long-term changes in synaptic efficacy. Lithium and valproate, at therapeutically relevant concentrations, robustly activate the ERK–MAP kinase cascade in rats *(82),* as well as in a human neuroblastoma cell line *(88).*

The most abundant ERK kinases in the brain are the products of the *ERK-1* and *ERK-2* genes. Mice with a targeted disruption of the *ERK-1* gene were generated on two different background strains, i.e. C57BL/6 *(87)* and 129/SvImJ *(89).* The ERK-1 KO mice seem to be mildly hyperactive in the open field compared with their wild-type littermates *(87,89,90)*. More specifically, these mutant mice showed increased horizontal activity in the open field *(87,89)*, but displayed no changes in vertical activity in one study *(89)*, and significantly increased vertical activity in the other study *(87)*. Furthermore, whereas wild-type mice showed clear habituation from d 1 to 3 in the open field, both in horizontal and in vertical activity, ERK-1 mutants retained a significantly higher activity on d 3, both in horizontal activity and in vertical activity *(87)*.

Preliminary data suggest that when the ERK-1 mice are given access to a running wheel in the home cage, the mutant mice show increased activity that persists for at least 2 wk *(90)*. Furthermore, ERK-1 KO mice and their wild-type controls performed comparably on the accelerating rotarod, indicating that the mutant mice were not impaired in terms of motor coordination or motor learning *(89)*. These mutant mice were also tested for their reactivity to novel environmental stimuli using acoustic startle, and for information processing by means of PPI. There were no differences in the amount of acoustic startle in response to a 120-dB stimulus, and PPI did not differ between genotypes *(89)*.

Regarding cognitive functioning, no differences were observed in the acquisition or retention (24 h and 2 wk after training) of either contextual or cue fear conditioning between the ERK-1 KO mice and their wild-type littermate controls *(87,89)*. ERK-1 mutants of both background strains were tested in a passive avoidance paradigm. In this test, animals learn to suppress their natural tendency to seek out dark areas over well-lit areas after their entry into a dark compartment is paired with a mild foot shock. ERK-1 mutants on 129/Sv background were not significantly different from littermate controls 24 h after training in step-through latency, the measure of passive avoidance *(89)*. As such, the authors conclude that hippocampus- and amygdala-dependent emotional learning does not depend critically on the activity of ERK-1 *(89)*. However, in a similar passive avoidance procedure, long-term retention, tested 24 h after training, was significantly better in ERK-1-deficient mice on C57BL/6 background compared with their wild-type littermates *(87)*. There was no significant difference between ERK-1 KO and wild-type mice in step-through latency at 30 min. Therefore, depending on the background strain, in vivo loss of ERK-1 may or may not result in an enhancement of memory consolidation, whereas learning and short-term memory seem to be normal.

ERK-1 mutant mice exhibited longer periods of swimming and less immobility in the forced-swim test, suggesting less depressive-like behavior in these mice *(90)*. Furthermore, these mice do not exhibit any change in anxiety, as measured by the elevated plus maze. However, time spent in the center of the open field was less than that of wild-type mice, which may suggest a higher level of anxiety, or less risk taking behavior *(90)*.

Taken together, these results imply that the most prominent behavioral effect of the lack of ERK-1 activity is a heightened level of activity that is exhibited in a novel environment as well as in the home cage.

4.5.2. Brain-Derived Neurotrophic Factor

The findings that, in rodents, stress decreases levels of BDNF in brain regions associated with depression has resulted in increased focus on BDNF as being a putative important factor in the etiology and pathophysiology of major depression and bipolar disorder *(91–94)*. BDNF, an upstream activator of the ERK–MAP kinase pathway, is a member of the neurotrophin family. This family of peptides is important for the development, maintenance, and function of the nervous system (for reviews, see refs. *94–97*). Chronic administration of lithium or valproate increased the expression of BDNF in the rat brain, suggesting that these mood stabilizers may produce a neurotrophic effect mediated by the upregulation of BDNF in the brain *(98)*. The lithium- and valproate-induced increases are most prominent in frontal cortex, in contrast to the effects exerted on BDNF expression by antidepressants that produce their greatest effects in the hippocampus *(26)*. BDNF influences phenotype, structural plasticity, and perhaps survival of central 5-HT neurons (e.g., ref. *99*) as well as certain parts of the dopaminergic systems within the brain, in particular D_3 receptor-expressing neurons (see ref. *100*). Indeed, mice with a targeted mutation of the *BDNF* gene show alterations in postsynaptic 5-HT receptor expression *(101)*, despite normal levels of 5-HT and its metabolites in neocortex, hippocampus, or hypothalamus *(101,102)*. Furthermore, these mutant mice also show increased dopamine concentrations in striatum *(103,104)*, although, in a more recent study, normal tissue levels of dopamine and its metabolites were observed in several brain areas, including striatum *(102)*.

Spontaneous (i.e., basal) locomotion in BDNF-heterozygous mice was not different from wild-type controls *(101,102,104)*, although this finding is not unequivocal *(105)*. With respect to bipolar disorder, it is interesting that locomotor activity was significantly increased after acute administration of amphetamine *(104)*.

Locomotion-related measures are not the only behaviors with face validity to model mania, and some other behaviors that were reported for BDNF-mutant mice may also demonstrate manic-like behavior. For instance, heterozygous $BDNF^{+/-}$ mice develop enhanced inter-male aggressiveness in a resident–intruder paradigm in early adulthood *(101)*. The $BDNF^{+/-}$ mice seemed agitated and intolerant of the intruder and engaged in offensive behaviors (e.g., chasing, tail rattling) at the onset of testing, whereas wild-type mice explored their environment and engaged in social interest behaviors directed at the intruder before they displayed aggressive behaviors. In addition, the severity and duration of biting attacks were greater in $BDNF^{+/-}$ mutant mice *(101)*. The authors suggested that $BDNF^{+/-}$ mice are more aggressive and possibly more irritable and impulsive than wild-type mice. Furthermore, they showed that partial disruptions in BDNF expression in the mutant mice lead to alterations in 5-HT neuronal functioning and in behavior associated with this neurotransmitter system, such as the increased aggressiveness *(101)*. In support of these findings, the heightened aggressiveness in $BDNF^{+/-}$ mice can be ameliorated by chronic treatment with the selective 5-HT reuptake inhibitor, fluoxetine, a strategy that augments 5-HT neurotransmission in the brain *(101)*.

In behavioral tests investigating anxiety (elevated plus/zero maze, light–dark box), fear-associated learning, anhedonia (sucrose preference), behavioral despair (Porsolt forced-swim test), and spatial learning (Morris water maze), BDNF$^{+/-}$ mice were indistinguishable from wild-type littermates *(102,106,107)*. However, the fact that BDNF mutants spent more time at the center (and not the periphery) of the open field may be interpreted as an increased risk-taking behavior, another common characteristic of mania *(26,101)*. In contrast, conditional BDNF mutants, besides being hyperaggressive, have increased levels of anxiety in a novel cage and light/dark-exploration test *(108)*.

Mutant mice with reduced BDNF expression in the forebrain (Emx-BDNF KO) that develop in the absence of BDNF in the dorsal cortex, hippocampus, and parts of the ventral striatum and amygdala, showed impairments of specific forms of learning *(109)*. These mutants failed to learn the Morris water-maze task, a hippocampal-dependent visuospatial learning task. Freezing during all phases of cued-contextual fear conditioning, a behavioral task designed to study hippocampal-dependent associative learning, was enhanced. These mice learned a brightness discrimination task well, but were impaired in a more difficult pattern-discrimination task. Forebrain-restricted BDNF KO mice did not exhibit altered sensory processing and gating, as measured by the acoustic startle response or PPI of the startle response. Although they were less active in an open-field arena, they did not show alterations in anxiety, as measured in the elevated plus maze, black/white chamber, or mirrored-chamber tasks. Combined, these data indicate that absence of forebrain BDNF does not disrupt acoustic sensory processing or alter baseline anxiety, but selectively affects learning *(109)*. Therefore, although these mice show clear learning deficits, the mania-like phenotype, as observed in heterozygous BDNF-mutant mice, is not present in Emx-BDNF KO mice, indicating that other brain areas, such as the striatum, might be more important in mania-related behavior than forebrain structures.

The biological activity of BDNF is mediated by the TrkB receptor, a receptor tyrosine kinase *(110,111)*. BDNF function is also studied in vivo using mice with a targeted disruption of the *TrkB* gene. Mutant mice have been generated in which the KO of the *TrkB* gene is restricted to the forebrain neocortex and hippocampus and occurs only during postnatal development *(112)*. These TrkB$^{CaMKII-CRE}$ mice showed a stereotyped hyperlocomotion with reduced explorative activity (significantly less time in the center field), and impulsive reactions to novel stimuli *(113)*. The TrkB-mutant mice did not exhibit depression-like behaviors, such as increased "despair" in the forced-swim test, increased anxiety in the elevated zero maze, or neophobia in the novel-object test *(113)*. The behavioral findings observed in TrkB$^{CaMKII-CRE}$ mice are in accordance with observations in BDNF-heterozygous mice. Similar to TrkB$^{CaMKII-CRE}$ mice, BDNF-heterozygous mice show hyperlocomotion and a trend to "neomania" in a novel-object test, but, at the same time, an evident lack of depression-like behavior in the forced-swim test as well as in the learned-helplessness paradigm *(113)*.

Taken together, the BDNF-ERK pathway mutants are promising animal models for bipolar disorder or mania. Affective-like behavioral changes in mouse strains with genetic alterations of the ERK-signaling pathway include changes in anxiety and neophobia, changes in aggressiveness and food consumption, changes in locomotor patterns, and an altered locomotor response to acute and chronic psychostimulant administration. These behaviors can all be considered to have commonalities with bipolar disorder. However, the effect of mood stabilizers needs to be assessed in these mutant mouse models, to also establish predictive validity.

4.6. Glycogen Synthase Kinase-3β

The enzyme glycogen synthase kinase (GSK)-3 is a direct target of lithium (*see* ref. *114*). GSK-3 has an essential role in a number of signaling pathways and regulates the function of a diverse number of proteins, notably transcription factors and cytoskeletal elements. The most important functions of the enzyme with respect to bipolar disorder may be critical effects on cellular resilience and neuronal plasticity (reviewed in ref. *114*). GSK-3β transgenic mice, generated on a FVB/N genetic background by expressing GSK-3β(S9A) under the control of the Thy-1 promoter cassette, showed no differences in total sleep–wake state durations when compared with nontransgenic wild-type littermates, but did show clear aberrant sleep fragmentation *(115)*. To our knowledge, sleep patterns are, however, the only behaviors investigated in these transgenic mice thus far. Future studies should investigate several behavioral and physiological measures relevant to bipolar disorder in these mice.

5. CONCLUSIONS

The increasing knowledge regarding the pathophysiology of bipolar disorder brought about a number of new mechanistic theories of the disorder, mostly relating either the disease or the therapeutic effects of its treatment to alterations in, particularly, second messenger and intracellular transmission pathways *(47)*. Of the available mutant mouse models for bipolar disorder, genetic alterations in the BDNF–ERK pathway are the most promising. It should be noted, however, that although BDNF is an upstream activator of the ERK–MAP kinase-signaling pathway, ERK-1 KO mice show a clearly different phenotype than heterozygous BDNF mutants. For instance, hyperactivity is increased in ERK-1 KOs, whereas it is unaffected in BDNF heterozygotes. Furthermore, risk-taking behaviors seem to be increased in BDNF mutants and decreased in ERK-1 KO mice.

One should always take into account that because of a lifelong overproduction or absence of a gene product, compensatory adaptations may occur in the brains of these mice. Given the possibilities for developmental compensation and the influence of epigenetic and environmental factors on subsequent behavior in adult mice, it is clear that the lack of a specific protein during the course of early development may result in alterations in molecular pathways and neural circuits relevant to behavior *(6)*.

Given its unique nature, developing an animal model for bipolar disorder in which all aspects of the illness are emulated, including both depressive and manic episodes, is challenging, if not impossible. As clearly demonstrated in this chapter, the available mutant mouse models for bipolar depression only can replicate specific aspects of the disorder, but not the entire spectrum of the disorder. Furthermore, as with many other psychiatric disorders, bipolar disorder is a polygenic disorder and not a monogenic disorder, thus, downsizing the contribution of the mutant mouse models in which the expression of one gene is changed, as animal models for bipolar disorder even further. However, it is clear that whereas bipolar disorder may not be as suitable for modeling in the laboratory as other disorders, many useful paradigms exist that enable researchers to investigate various aspects of bipolar disorder in genetically modified animals. If one adopts the approach proposed by Lenox and colleagues, to investigate endophenotypes relevant to bipolar disorder in mutant mouse models *(40)*, then the available mutant mouse models prove to be very useful and will assist in the development of a greater understanding regarding the pathophysiology of this devastating illness.

REFERENCES

1. Fiala SJ. Normal is a place I visit. JAMA 2004;291:2924–2926.
2. Angst J, Marneros A. Bipolarity from ancient to modern times: conception, birth and rebirth. J Affect Disord 2001;67:3–19.
3. Manji HK, Lenox RH. The nature of bipolar disorder. J Clin Psychiatry 2000;61:42–57.
4. Gould TD, Quiroz JA, Singh J, Zarate CA, Manji HK. Emerging experimental therapeutics for bipolar disorder: insights from the molecular and cellular actions of current mood stabilizers. Mol Psychiatry 2004;9:734–755.
5. Quiroz JA, Singh J, Gould TD, Denicoff KD, Zarate CA, Manji HK. Emerging experimental therapeutics for bipolar disorder: clues from the molecular pathophysiology. Mol Psychiatry 2004;9:756–776.
6. Cryan JF, Mombereau C. In search of a depressed mouse: utility of models for studying depression-related behavior in genetically modified mice. Mol Psychiatry 2004;9:326–357.
7. American Psychiatric Association. Diagnostic and statistical manual of mental disorders (DSM-IV). 4th ed. Text Revision. Washington DC: American Psychiatric Association, 2000.
8. Thomas P. The many forms of bipolar disorder: a modern look at an old illness. J Affect Disord 2004;79:3–8.
9. Akiskal HS, Bourgeois ML, Angst J, Post R, Moller HJ, Hirschfeld R. Re-evaluating the prevalence of and diagnostic composition within the broad clinical spectrum of bipolar disorders. J Affect Disord 2000;59:S5–S30.
10. Chen YW, Dilsaver SC. Comorbidity of panic disorder in bipolar illness: evidence from the Epidemiologic Catchment Area Survey. Am J Psychiatry 1995;152:280–282.
11. MacKinnon DF, McMahon FJ, Simpson SG, McInnis MG, DePaulo JR. Panic disorder with familial bipolar disorder. Biol Psychiatry 1997;42:90–95.
12. McElroy SL, Altshuler LL, Suppes T, et al. Axis I psychiatric comorbidity and its relationship to historical illness variables in 288 patients with bipolar disorder. Am J Psychiatry 2001;158: 420–426.
13. Alda M. The phenotypic spectra of bipolar disorder. Eur Neuropsychopharmacol 2004;14: S94–S99
14. Perry W, Minassian A, Feifel D, Braff DL. Sensorimotor gating deficits in bipolar disorder patients with acute psychotic mania. Biol Psychiatry 2001;50:418–424.

15. Bearden CE, Hoffman KM, Cannon TD. The neuropsychology and neuroanatomy of bipolar affective disorder: a critical review. Bipolar Disord 2001;3:106–150.
16. Clark L, Iversen SD, Goodwin GM. A neuropsychological investigation of prefrontal cortex involvement in acute mania. Am J Psychiatry 2001;158:1605–1611.
17. Addington J, Addington D. Facial affect recognition and information processing in schizophrenia and bipolar disorder. Schizophr Res 1998;32:171–181.
18. Tam WC, Sewell KW, Deng H. Information processing in schizophrenia and bipolar disorder: a discriminant analysis. J Nerv Ment Dis 1998;186:597–603.
19. Braff DL, Geyer MA. Sensorimotor gating and schizophrenia. Human and animal model studies. Arch Gen Psychiatry 1990;47:181–188.
20. Kahn D, Chaplan R. The "good enough" mood stabilizer: a review of the clinical evidence. CNS Spectr 2002;7:227–237.
21. Holtzheimer PE, Neumaier JF. Treatment of acute mania. CNS Spectr 2003;8:917–920; 924–928.
22. Keck PE Jr, Nelson EB, McElroy SL. Advances in the pharmacologic treatment of bipolar depression. Biol Psychiatry 2003;53:671–679.
23. Calabrese JR, Kasper S, Johnson G, et al. International Consensus Group on bipolar i depression treatment guidelines. J Clin Psychiatry 2004;65:569–579.
24. Owens MJ, Nemeroff CB. Pharmacology of valproate. Psychopharmacol Bull 2003;37(Suppl 2):17–24.
25. Gould TD, Chen G, Manji HK. Mood stabilizer psychopharmacology. Clin Neurosci Res 2002;2:193–212.
26. Einat H, Manji HK, Gould TD, Du J, Chen G. Possible involvement of the ERK signaling cascade in bipolar disorder: behavioral leads from the study of mutant mice. Drug News Perspect 2003;16:453–463.
27. Post RM, Speer AM, Hough CJ, Xing G. Neurobiology of bipolar illness: implications for future study and therapeutics. Ann Clin Psychiatry 2004;15:85–94.
28. Machado-Vieira R, Kapczinski F, Soares JC. Perspectives for the development of animal models of bipolar disorder. Prog Neuropsychopharmacol Biol Psychiatry 2004;28:209–224.
29. Hauger RL, Grigoriadis DE, Dallman MF, Plotsky PM, Vale WW, Dautzenberg FM. International Union of Pharmacology. XXXVI. Current status of the nomenclature for receptors for corticotropin-releasing factor and their ligands. Pharmacol Rev 2003;55:21–26.
30. Dunn AJ, Berridge CW. Physiological and behavioral responses to corticotropin-releasing factor administration: is CRF a mediator of anxiety or stress responses? Brain Res Brain Res Rev 1990;15:71–100.
31. Owens MJ, Nemeroff CB. Physiology and pharmacology of corticotropin-releasing factor. Pharmacol Rev 1991;43:425–473.
32. Koob GF, Heinrichs SC, Merlo Pich E, et al. The role of corticotropin-releasing factor in behavioural responses to stress. Ciba Found Symp 1993;172:277–295.
33. Koob GF, Heinrichs SC. A role for corticotropin releasing factor and urocortin in behavioral responses to stressors. Brain Res 1999;848:141–152.
34. Holsboer F. The rationale for corticotropin-releasing hormone receptor (CRH-R) antagonists to treat depression and anxiety. J Psychiatr Res 1999;33:181–214.
35. Berrettini WH, Nurnberger JI, Zerbe RL, Gold PW, Chrousos GP, Tamao T. CSF neuropeptides in euthymic bipolar patients and controls. Br J Pharmacol 1987;150:208–212.
36. Banki CM, Karmacsi L, Bissette G, Nemeroff CB. Cerebrospinal fluid neuropeptides in mood disorder and dementia. J Affect Disord 1992;25:39–45.
37. Gilmor ML, Skelton KH, Nemeroff CB, Owens MJ. The effects of chronic treatment with the mood stabilizers valproic acid and lithium on corticotropin-releasing factor neuronal systems. J Pharmacol Exp Ther 2003;305:434–439.
38. Stout SC, Owens MJ, Lindsey KP, Knight DL, Nemeroff CB. Effects of sodium valproate on corticotropin-releasing factor systems in rat brain. Neuropsychopharmacology 2001;24:624–631.

39. Niculescu AB III, Segal DS, Kuczenski R, Barrett T, Hauger RL, Kelsoe JR. Identifying a series of candidate genes for mania and psychosis: a convergent functional genomics approach. Physiol Genomics 2000;4:83–91.
40. Lenox RH, Gould TD, Manji HK. Endophenotypes in bipolar disorder. Am J Med Genet 2002;114:391–406.
41. Prathikanti S, MacMahon FJ. Genome scans for susceptibility genes in bipolar affective disorder. Ann Med 2001;33:257–262.
42. Glahn DC, Bearden CE, Niendam TA, Escamilla MA. The feasibility of neuropsychological endophenotypes in the search for genes associated with bipolar affective disorder. Bipolar Disord 2004;6:171–182.
43. Tsuang MT, Taylor L, Faraone SV. An overview of the genetics of psychotic mood disorders. J Psychiatr Res 2004;38:3–15.
44. Mathews CA, Reus VI. Genetic linkage in bipolar disorder. CNS Spectr 2003;8:891–904.
45. Smoller JW, Finn CT. Family, twin, and adoption studies of bipolar disorder. Am J Med Genet 2003;123C:48–58.
46. Geyer MA, Markou A. Animal Models of Psychiatric Disorders. In: Bloom FE, Kupfer DJ, eds. Psychopharmacology: The Fourth Generation of Progress. New York: Raven, 1995, pp. 787–798.
47. Einat H, Belmaker RH, Manji HK. New approaches to modeling bipolar disorder. Psychopharmacol Bull 2003;37:47–63.
48. El Mallakh RS, Karippot A. Use of antidepressants to treat depression in bipolar disorder. Psychiatr Serv 2002;53:580–584.
49. Goldberg JF, Truman CJ. Antidepressant-induced mania: an overview of current controversies. Bipolar Disord 2003;5:407–420.
50. Nestler EJ, Gould E, Manji H, et al. Preclinical models: status of basic research in depression. Biol Psychiatry 2002;52:503–528.
51. Dirks A, Groenink L, Bouwknecht JA, et al. Overexpression of corticotropin-releasing hormone in transgenic mice and chronic stress-like autonomic and physiological alterations. Eur J Neurosci 2002;16:1751–1760.
52. Groenink L, Pattij T, de Jongh R, et al. 5-HT$_{1A}$ receptor knockout mice and mice overexpressing corticotropin-releasing hormone in models of anxiety. Eur J Pharmacol 2003;463:185–197.
53. Groenink L, Dirks A, Verdouw PM, et al. HPA-axis dysregulation in mice overexpressing corticotropin-releasing hormone. Biol Psychiatry 2002;51:875–881.
54. Dirks A, Groenink L, lutje Schipholt M, et al. Reduced startle reactivity and plasticity in transgenic mice overexpressing corticotropin-releasing hormone. Biol Psychiatry 2002;51:583–590.
55. Dirks A, Groenink L, Westphal KGC, et al. Reversal of startle gating deficits in transgenic mice overexpressing corticotropin-releasing factor by antipsychotic drugs. Neuropsychopharmacology 2003;28:1790–1798.
56. Dirks A, Groenink L, Verdouw P, et al. Behavioral analysis of transgenic mice overexpressing corticotropin-releasing hormone in paradigms emulating aspects of stress, anxiety and depression. Int J Comp Psychol 2001;14:123–135.
57. Wei Q, Lu XY, Liu L, et al. Glucocorticoid receptor overexpression in forebrain: a mouse model of increased emotional lability. Proc Natl Acad Sci USA 2004;101:11,851–11,856.
58. De Kloet ER. Stress in the brain. Eur J Pharmacol 2000;405:187–198.
59. Kelsoe JR. Recent progress in the search for genes for bipolar disorder. Curr Psychiatry Rep 1999;1:135–140.
60. Greenwood TA, Alexander M, Keck PE, et al. Evidence for linkage disequilibrium between the dopamine transporter and bipolar disorder. Am J Med Genet 2001;105:145–151.
61. Giros B, Caron MG. Molecular characterization of the dopamine transporter. Trends Pharmacol Sci 1993;14:43–49.
62. Giros B, Jaber M, Jones SR, Wightman RM, Caron MG. Hyperlocomotion and indifference to cocaine and amphetamine in mice lacking the dopamine transporter. Nature 1995;379:606–612.

63. Gainetdinov RR, Jones SR, Caron MG. Functional hyperdopaminergia in dopamine transporter knock-out mice. Biol Psychiatry 1999;46:303–311.
64. Spielewoy C, Roubert C, Hamon M, Nosten-Bertrand M, Betancur C, Giros B. Behavioural disturbances associated with hyperdopaminergia in dopamine-transporter knockout mice. Behav Pharmacol 2000;11:279–290.
65. Ralph RJ, Paulus MP, Fumagalli F, Caron MG, Geyer MA. Prepulse inhibition deficits and perseverative motor patterns in dopamine transporter knock-out mice: differential effects of D1 and D2 receptor antagonists. J Neurosci 2001;21:305–313.
66. Morice E, Denis C, Giros B, Nosten-Bertrand M. Phenotypic expression of the targeted null-mutation in the dopamine transporter gene varies as a function of the genetic background. Eur J Neurosci 2004;20:120–126.
67. Zhuang X, Oosting RS, Jones SR, et al. Hyperactivity and impaired response habituation in hyperdopaminergic mice. Proc Natl Acad Sci U S A 2001;98:1982–1987.
68. Ralph-Williams RJ, Paulus MP, Zhuang X, Hen R, Geyer MA. Valproate attenuates hyperactive and perseverative behaviors in mutant mice with a dysregulated dopamine system. Biol Psychiatry 2003;53:352–359.
69. Dean B. The neurobiology of bipolar disorder: findings using human postmortem central nervous system tissue. Aus N Z J Psychiatry 2004;38:135–140.
70. Lesch KP, Bengel D, Heils A, et al. Association of anxiety-related traits with a polymorphism in the serotonin transporter gene regulatory region. Science 1996;274:1527–1530.
71. Stockmeier CA. Involvement of serotonin in depression: evidence from postmortem and imaging studies of serotonin receptors and the serotonin transporter. J Psychiatr Res 2003;37:357–373.
72. Murphy DL, Lerner A, Rudnick G, Lesch KP. Serotonin transporter: gene, genetic disorders, and pharmacogenetics. Mol Interv 2004;4:109–123.
73. Bengel D, Murphy DL, Andrews AM, et al. Altered brain serotonin homeostasis and locomotor insensitivity to 3,4-methylenedioxymethamphetamine ("Ecstasy") in serotonin transporter-deficient mice. Mol Pharmacol 1998;53:649–655.
74. Holmes A, Murphy DL, Crawley JN. Abnormal behavioral phenotypes of serotonin transporter knockout mice: parallels with human anxiety and depression. Biol Psychiatry 2003;54:953–959.
75. Montanez S, Owens WA, Gould GG, Murphy DL, Daws LC. Exaggerated effect of fluvoxamine in heterozygote serotonin transporter knockout mice. J Neurochem 2003;86:210–219.
76. Holmes A, Yang RJ, Lesch KP, Crawley JN, Murphy DL. Mice lacking the serotonin transporter exhibit 5-HT(1A) receptor-mediated abnormalities in tests for anxiety-like behavior. Neuropsychopharmacology 2003;28:2077–2088.
77. Holmes A, Murphy DL, Crawley JN. Reduced aggression in mice lacking the serotonin transporter. Psychopharmacology 2002;161:160–167.
78. Holmes A, Yang RJ, Murphy DL, Crawley JN. Evaluation of antidepressant-related behavioral responses in mice lacking the serotonin transporter. Neuropsychopharmacology 2002;27:914–923.
79. Anguelova M, Benkelfat C, Turecki G. A systematic review of association studies investigating genes coding for serotonin receptors and the serotonin transporter: I. Affective disorders. Mol Psychiatry 2003;8:591
80. Lotrich FE, Pollock BG. Meta-analysis of serotonin transporter polymorphisms and affective disorders. Psychiatr Genet 2004;14:121–129.
81. Ackenheil M. Neurotransmitters and signal transduction processes in bipolar affective disorders: a synopsis. J Affect Disord 2001;62:101–111.
82. Einat H, Yuan P, Gould TD, et al. The role of the extracellular signal-regulated kinase signaling pathway in mood modulation. J Neurosci 2003;23:7311–7316.
83. Pearson G, Robinson F, Beers Gibson T, et al. Mitogen-activated protein (MAP) kinase pathways: regulation and physiological functions. Endocr Rev 2001;22:153–183.
84. Adams JP, Sweatt JD. Molecular psychology: roles for the ERK MAP kinase cascade in memory. Annu Rev Pharmacol Toxicol 2002;42:135–163.
85. Mazzucchelli C, Brambilla R. Ras-related and MAPK signalling in neuronal plasticity and memory formation. Cell Mol Life Sci 2000;57:604–611.

86. Sweatt JD. Mitogen-activated protein kinases in synaptic plasticity and memory. Curr Opin Neurobiol 2004;14:311–317.
87. Mazzucchelli C, Vantaggiato C, Ciamei A, et al. Knockout of ERK1 MAP Kinase enhances synaptic plasticity in the striatum and facilitates striatal-mediated learning and memory. Neuron 2002;34:807–820.
88. Yuan PX, Huang LD, Jiang YM, Gutkind JS, Manji HK, Chen G. The mood stabilizer valproic acid activates mitogen-activated protein kinases and promotes neurite growth. J Biol Chem 2001;276:31,674–31,683.
89. Selcher JC, Nekrasova T, Paylor R, Landreth GE, Sweatt JD. Mice lacking the ERK1 isoform of MAP Kinase are unimpaired in emotional learning. Learn Mem 2001;8:11–19.
90. Engel SE, Nekrasova T, Einat H, et al. Role of the extrcellular regulated kinase (ERK)-1 in the regulation of the behaviors associated with mood. Program No. 755. 7. 2003 Abstract Viewer/Itinerary Planner. Washington, DC: Society for Neuroscience, 2003. Online.
91. Neves-Pereira M, Mundo E, Muglia P, King N, Macciardi F, Kennedy JL. The brain-derived neurotrophic factor gene confers susceptibility to bipolar disorder: evidence from a family-based association study. Am J Hum Genet 2002;71:651–655.
92. Duman RS. Role of neurotrophic factors in the etiology and treatment of mood disorders. Neuromolecular Med 2004;5:11–25.
93. Hashimoto K, Shimizu E, Iyo M. Critical role of brain-derived neurotrophic factor in mood disorders. Brain Res Brain Res Rev 2004;45:104–114.
94. Lang UE, Jockers-Scherubl MC, Hellweg R. State of the art of the neurotrophin hypothesis in psychiatric disorders: implications and limitations. J Neural Transm 2004;111:387–411.
95. Bibel M, Barde YA. Neurotrophins: key regulators of cell fate and cell shape in the vertebrate nervous system. Genes Dev 2000;14:2919–2937.
96. Huang EJ, Reichardt LF. Neurotrophins: roles in neuronal development and function. Annu Rev Neurosci 2001;24:677–736.
97. Horch HW. Local effects of BDNF on dendritic growth. Rev Neurosci 2004;15:117–129.
98. Fukumoto T, Morinobu S, Okamoto Y, Kagaya A, Yamawaki S. Chronic lithium treatment increases the expression of brain-derived neurotrophic factor in the rat brain. Psychopharmacology 2001;158:100–106.
99. Mamounas LA, Blue ME, Siuciak JA, Altar CA. Brain-derived neurotrophic factor promotes the survival and sprouting of serotonergic axons in rat brain. J Neurosci 1995;15:7929–7939.
100. Guillin O, Griffon N, Diaz J, et al. Brain-derived neurotrophic factor and the plasticity of the mesolimbic dopamine pathway. Int Rev Neurobiol 2004;59:425–444.
101. Lyons WE, Mamounas LA, Ricaurte GA, et al. Brain-derived neurotrophic factor-deficient mice develop aggressiveness and hyperphagia in conjunction with brain serotonergic abnormalities. Proc Natl Acad Sci USA 1999;96:15,239–15,244.
102. Chourbaji S, Hellweg R, Brandis D, et al. Mice with reduced brain-derived neurotrophic factor expression show decreased choline acetyltransferase activity, but regular brain monoamine levels and unaltered emotional behavior. Brain Res Mol Brain Res 2004;121:28–36.
103. Dluzen DE, Story GM, Xu K, Kucera J, Walro JM. Alterations in nigrostriatal dopaminergic function within BDNF mutant mice. Exp Neurol 1999;160:500–507.
104. Dluzen DE, Gao X, Story GM, Anderson LI, Kucera J, Walro JM. Evaluation of nigrostriatal dopaminergic function in adult +/+ and +/– BDNF mutant mice. Exp Neurol 2001;170:121–128.
105. Kernie SG, Liebl DJ, Parada LF. BDNF regulates eating behavior and locomotor activity in mice. EMBO J 2000;19:1290–1300.
106. Montkowski A, Holsboer F. Intact spatial learning and memory in transgenic mice with reduced BDNF. NeuroReport 1997;8:779–782.
107. MacQueen GM, Ramakrishnan K, Croll SD, et al. Performance of heterozygous brain-derived neurotrophic factor knockout mice on behavioral analogues of anxiety, nociception, and depression. Behav Neurosci 2001;115:1145–1153.
108. Rios M, Fan G, Fekete C, et al. Conditional deletion of brain-derived neurotrophic factor in the postnatal brain leads to obesity and hyperactivity. Mol Endocrinol 2001;15:1748–1757.

109. Gorski JA, Balogh SA, Wehner JM, Jones KR. Learning deficits in forebrain-restricted brain-derived neurotrophic factor mutant mice. Neuroscience 2003;121:341–354.
110. Barbacid M. Structural and functional properties of the TRK family of neurotrophin receptors. Ann NY Acad Sci 1995;766:442–458.
111. Purcell AL, Carew TJ. Tyrosine kinases, synaptic plasticity and memory: insights from vertebrates and invertebrates. Trends Neurosci 2003;26:625–630.
112. Minichiello L, Korte M, Wolfer DP, et al. Essential role for TrkB receptors in hippocampus-mediated learning. Neuron 1999;24:401–414.
113. Zorner B, Wolfer DP, Brandis D, et al. Forebrain-specific TrkB-receptor knockout mice: behaviorally more hyperactive than "depressive." Biol Psychiatry 2003;54:972–982.
114. Gould TD, Zarate CA, Manji HK. Glycogen synthase kinase-3: a target for novel bipolar disorder treatments. J Clin Psychiatry 2004;65:10–21.
115. Vaarties KG, Ahnaou A, Huysmans H, Moechars D, Taymans JM, Drinkenburg WH. Mice overexpressing glycogen synthase kinase 3 beta (GSK-3b) show normal total sleep-wake duration with altered body activity and body temperature. Program No. 930. 19. 2003 Abstract Viewer/Itinerary Planner. Washington, DC: Society for Neuroscience, 2003. Online.

INDEX

A

Acetylcholine, 194–196, 201–203, 216, 248
Acoustic startle response, *see* Startle response
Adenine phosophoribosyltransferase (APRT), 30, 31
Additive summation, 65
 causal logic, 65
Aggressive behavior, *see* Behavior
Agoraphobia, 240
AGTR2 gene, *see* Gene
α-7 nicotinic acetylcholine receptor, 201–203
Alprazolam, *see* Anxiolytics
Alzheimer's disease, 19, 70, 75, 88, 221–223, 231
APPV717F, 230
Amphetamine, *see* Psychostimulant
Amygdala,
 anxiety and fear, 18, 156, 272, 276, 278
 cognitive function, 245, 251
 long-term potentiation, 110
 startle response, 196,
Animal models, *see also* Mouse models, Rat strains, 3, 4, 7, 16, 18, 19, 25, 46, 73, 76, 80, 87, 89, 94, 96, 103, 105, 116, 127, 128, 151–154, 156, 167, 168, 177, 178, 181, 182, 190, 193–195, 221, 222, 224, 233, 237, 238, 265, 269, 270, 279, 280
Antidepressant, 27, 228, 247–251, 269, 272, 274, 277
Antipsychotic, 178, 184, 193–199, 203, 207, 212, 214, 216, 272
 atypical, 194, 196, 198, 203, 207, 214, 216, 272
 chlorpromazine, 196
 clozapine, 178, 194, 197, 203, 212–214, 217, 272
 conventional, 193, 194, 197, 203, 216
 haloperidol, 178, 197, 202, 203, 208, 212–214, 217, 229, 272
 imipramine, 197, 227–229
 olanzapine, 216
 quetiapine, 197, 216
Antisense oligonucleotide, 222
Anxiety, 18, 19, 27, 31, 33, 35, 36, 75, 79, 105–108, 112, 114, 115, 155, 156, 161, 163, 165, 166, 185, 221–224, 226, 227, 230, 232, 233, 237–245, 252–254, 267, 271, 272, 274, 276, 278, 279, *see also* Anxiogenic
Anxiogenic, 163, 165, 222, 241
Anxiolytics,
 alprazolam, 228, 229
 benzodiazepine, 79, 80, 227, 228, 255
 chlordiazepoxide, 227, 228
 clobazam, 227, 229
 clonazepam, 228, 229
 diazepam, 33, 197, 227–229
 flumazenil, 228, 229
 fluoxetine, 227–229, 250, 277
 m-chlorophenyl piperazine (mCPP), 228, 229
 moclobemide, 228, 229
 phenelzine, 228
Apomorphine, 194, 196–199, 208, 209, *see also* Dopamine agonist
Apoptosis, 108, 133, 181, 214, 215

287

APPV717F, *see* Alzheimer's Disease
Area X, *see FOXP2*
ARHGEF6 gene, *see* Gene
Arylsulfatase A (*ASA*), *see* Gene
Associationism, 12, 19, 20
Ataxia, 28, 87–92, 95, 110, 112, 211, 212
Ataxin-1, *see* Gene
Attention, 4, 8, 71, 75, 76, 156, 161, 163, 165, 181, 182, 184–186, 245, 246, *see also* Attention deficit hyperactivity disorder
Attention deficit hyperactivity disorder (ADHD), 71, 76, *see also* Hyperactivity
Antipsychotic drugs, medication, 178, 185, 193–198, 203, 207, 212, 214, 216, 272
Audiogenic seizures, 105
Autism, 18, 19, 36, 37, 69, 70, 75, 151–158, 160, 161, 163, 165, 167–169
Autistic, 103, *see also* Autism

B

BALB/c mouse, *see* Mouse, strains
Basolateral amygdala, *see* Amygdala
Behavior, *see also* Endophenotypes
  aggressive, 34, 109, 159, 269, 273, 274, 277–279
  animal, 3–10, 12–15, 18–20, 25–27, 30–39, 54–66, 69–73, 75, 76, 79, 80, 90, 93, 95, 101–107, 109–119, 136, 139, 140, 142, 144–145, 151–153, 155–158, 161–163, 165, 166, 168
  defensive, 223–227, 233, 243, 250
  exploratory, 5, 105, 109, 112, 114, 117, 123, 162–163, 165, 166, 231, 233, 273, 274
  human, 3–5, 9, 10, 14, 20, 45, 49, 54–66, 168
  learned helplessness, 237, 248, 249, 278
  measures of,
    chase-flight test, 226, 227
    contextual fear conditioning, 108, 112, 117, 195, 231–233, 276, 278
    contextual defense, 226
    delayed spatial alternation task, 216
    discrimination learning, 13
    fear conditioning, 18, 29, 105, 109, 116, 232, 233, 245
    food preference, 75
    forced swim test, 19, 217, 237, 247, 272–274, 276, 278
    four-plate test, 237, 243
    hotplate, 107
    immobility, 224, 231, 248, 251, 274, 276
    light/dark exploration, 274, 278
    locomotor activity, 114, 165, 198, 208, 211, 213, 232, 243, 271–273, 277
    mirrored chamber test, 241, 278
    Morris water maze, 16, 29, 72, 106, 109, 111, 116, 154, 163, 164, 251, 278
    open field test, 116, 165, 241
    operant conditioning, 3, 11–13, 17, 161
    passive avoidance, 29, 72, 105, 112, 114, 251, 276
    plus-maze, 165
    radial maze, 106, 109, 114, 166, 205, 231
    rotarod, 90–92, 96, 105, 107–109, 112, 117, 243, 276
    shock-probe burying, 237, 243
    staircase test, 241
    tail suspension test (TST), 237, 247, 248, 251, 274
    T-maze, 210, 229–229
Behavioral despair, 237, 247, 248, 252, 278
Behaviorism, 3, 8, 10, 14, 19, 20, 56
Benzodiazepine, *see* Anxiolytics
β-hexosaminidase, 29, 30
Bipolar disorder, 16, 19, 204, 265–275, 277, 279, 280, *see also* Psychosis
  lithium treatment, 267–269, 275, 277, 279

Index 289

Brain, 4, 15, 88, 157, 167, 222
   behavior, 20, 70, 75, 102, 104, 137
   emotion, 18, 250, 251, 277
   gene expression, 104, 107, 108,
      111, 112, 114, 128, 139, 142,
      144, 145, 154, 187, 204, 206,
      213, 215, 216, 249, 268, 270,
      272, 274, 275, 277, 278
   intelligence, 16
   lesion, 17, 128, 155, 198, 207, 216
   function, 15, 16, 107, 108, 118,
      128, 130, 132, 137, 138, 157,
      160, 178, 180, 185, 196, 205,
      222, 265, 277
   pathology, 70, 110–113, 167, 181,
      184, 186, 187, 195, 201, 204,
      206–208, 214, 215, 238
   structure, 3, 27, 28, 32, 102, 104, 105,
      113, 114, 118, 127, 132, 137–
      142, 154, 156, 178, 180–182,
      185, 187, 196, 201, 211, 213,
      222, 254, 268, 272, 277, 278
     caudate nucleus, 137, 139,
        141, 145
     cerebellum, 28, 92, 108, 137,
        140, 141, 204, 205, 215
     corpus callosum, 33, 113
     entorhinal cortex, 207
     frontal cortex, 140, 161, 167,
        194, 196, 198, 201, 204,
        207–210, 213–216, 245,
        272, 277
     hippocampus, 17, 38, 75, 104,
        109–112, 114, 115, 118,
        119, 155, 156, 194, 196,
        198, 201–203, 205–210,
        213–216, 245, 251, 272,
        276–278
     inferior olives, 140, 142
     nucleus accumbens, 161, 207,
        209, 215
     prefrontal cortex, 140, 161,
        167, 196, 204, 207–210,
        213, 245, 272
     temporolimbic cortices, 206–208
     thalamus, 139–142, 194, 196,
        201, 214

Brain-derived neurotropic factor,
   see Brain, gene expression
Broca's area, 132, 137
Buspirone, 27, 197, 227–228, *see also*
   Serotonin

C

C57BL mouse, *see* Mouse, strains
C57BL/6J mouse, *see* Mouse,
   strains, C57BL
CA3 pyramidal cell, *see* Hippocampus
Cabib, S, 32
Caudate nucleus, *see* Brain, structure
CD-1, *see* Mouse strains
Cell adhesion molecule L1, *see* Gene
Cerebellum, *see* Brain
Chaperones, *see* Molecular pathways
Chase-flight test, *see* Behavior,
   measures
Chlordiazepoxide, *see* Anxiolytics
Chlorpromazine, *see* Antipsychotics
Chronic mild stress, 237, 249
Clayton, NS, 73
Clobazam, *see* Anxiolytics
Clonazepam, *see* Anxiolytics
Clozapine, *see* Antipsychotics
Cocaine, 228, 229, 269, 272
Coffin-Lowry syndrome, 101, 108, 109
Cognition, 7, 80, 101, 117, 118, 140,
   182, 188, 210, 233, 245
Cognitive function, 29, 72, 103, 114,
   118, 151, 168, 276
Cognitive psychology, 3 14–16, 20
Coherence, 51, 52
Compensation, 28, 31, 39, 119, 187,
   254, 279
COMT gene, *see* Gene
Conflict, 237, 240–243
Congenic, 33–34, 199, 253,
   see also Mouse
Consciousness, 3, 4, 6, 8–10, 14, 15,
   19, 20
Contextual fear conditioning, *see*
   Behavior, measures of
Contextual defense, *see* Behavior,
   measures of

Conventional antipsychotic, *see* Antipsychotic
Corpus callosum, *see* Brain, structure
Corticotropin-releasing factor (CRF), 239, 265, 268, 270
Corticotropin-releasing hormone receptor (CRHR), 35
Crabbe, JC, 19, 73, 252
CRASH syndrome, 101, 113
Cue dependent, *see* Fear conditioning

D

Darwin, C, 3–6, 9, 12–16, 19
DBA/2J mouse, *see* Mouse strains
Defensive behavior, *see* Behavior
Delayed spatial alternation task, *see* Behavior, measures of
Dementia, 16, 70, 101, 102, 266, *see also* Schizophrenia
Dendritic spines, 90, 104, 154, 203–205
Depression, 19, 31, 37, 69–71, 75, 104, 181, 231, 237–239, 245–254, 265–269, 271, 272, 277, 278, 280
  symptoms, 268
  manic-depression, 16
Developmental disorders, 127, 137, 155, 213, *see also* Autism
Diazepam, *see* Anxiolytics
Direct observation, 46–48, 50, 51
Discrimination
  learning, *see* Behavior, measures of
  drug, 211
Dopamine,
  receptor (D2), 34, 187, 193, 196, 198, 214
  transporter (DAT), 265, 270, 271, 273, *see also* Neurotransmitters, Dopamine
Dopaminergic, *see also* Neurotransmitters, Dopamine
  hypersensitivity, 187
  transmission, 172, 193–195, 197, 198, 207–210, 216, 218, 268, 273, 277
Down syndrome, 16, 18, 102
Drosophila, 92–94, 96, 109, 155, 168
Drug discrimination, *see* Discrimination, drug

D-Serine, *see* Glycine, agonist
DSM-IV, 70, 71, 157, 158, 238, 245–247, 250, 266
Dysgenesis, 33
Dyslexia, 26, 128, 136, 137

E

Elevated plus-maze, *see* Behavior, measures of
Emotionality, 232, 252, 272
Endophenotype, 71, 199, 201, *see also* Intermediate traits
Entorhinal Cortex, *see* Brain, structure
Epilepsy, 104, 115, 154, 155, 167, 169, 180, 187
Epistasis, 36
Estrogen receptor, 34
Ethology, 12, 221
Evolution, 3, 7, 9, 12, 13, 19, 20, 178, 245
  cognitive function, 151, 168
  vocal learning, 127, 129–131, 140, 142–144
Exploratory behavior, *see* Behavior
Extracellular signal-regulated kinase (ERK), 155, 270, 271, 275–277, 279

F

False-positives, 252
Familial AD, *see* Alzheimer's disease
Fear, 8, 9, 18, 79, 109, 156, 195, 221–224, 230–233, 239, 240, 245, 253
Fear conditioning, *see*, Behavior, measures of, fear conditioning
Fimbria-fornix, 202
Fitch, WT, 131
Flumazenil, *see* Anxiolytics
Fluoxetine, *see* Anxiolytics
FMR1, *see* Gene
Food preference paradigm, *see* Behavior, measures of
Forced swim test, *see* Behavior, measures of
Four-plate test, *see* Behavior, measures of

Index

FOXP2, see Gene
   area X, 230
FOXC1, see Gene
   forkhead, 133, 135
Fragile X, *see FMR1*
Fragile X mental retardation gene 1
   (*FXR1*), *see* Gene
Fragile X mental retardation gene 2
   (*FXR2*), *see* Gene
Fragile X syndrome, 36, 102–104,
   107, 152, 153
FRAXE syndrome, 107
Frontal cortex, *see* Brain, structure

G

GABA receptor, 72, 79, 80, 200, 201,
   203–205, 217
Galantamine, 217
γ-aminobutyric acid (GABA), 72,
   79, 80, 200, 201, 205, 217, 248
GDI1, see Gene
Gene,
   AGTR2, 101
      mutations,
         in X-linked MR, 112
         in mice, 112
   ARHGEF6, 101
      mutations,
         in X-linked MR 114, 115,
         117, 120
   arylsulfatase A (*ASA*), 28
   ataxin-1, 87–97
   cell adhesion molecule L1, 101, 113
   COMT, 187
   FMR1, 101–107, 152–154, 168
   FOXP2, 127–129, 132–145
   FOXC1, 133, 135
   FXR1, 102–106, 154
   FXR2, 106, 107, 154
   GDI1, 109, 110, 117, 119
   methyl-CpG-binding protein 2
      (*MECP2*), 101, 110–112,
      152–155, 168
   NF1, 101, 116, 117, 152, 155, 156, 168
   PAH, 101, 115
   PRODH, 186–187
   reelin (*RELN*), 168, 203–206
   ribosomal S6 kinase 2 (*RSK2*),
      101, 108, 109

Gene targeting, 32, 155, 177, 178,
   180, 184
Generalized anxiety disorder,
   see Anxiety
Genetic background, 32, 33, 70,
   104–106, 111, 113, 117, 178,
   186, 237, 252, 253, 274, 279
Genetic disorders, 3, 18, 29, 152, *see
   also* Specific genetic disorders
Genetic lesion, 27, 32, 36, 183
Genetic mouse model, 69, 70, 73,
   101, 177, 179, 180, 182, 194
Genetic susceptibility, 29, 36, 37, 39
Genetics, 3, 15, 16, 18, 20, 22, 31,
   35–38, 69, 70, 72, 73, 79, 80, 87,
   94, 103, 107–110, 112–116, 130,
   168, 181, 182
   linkage, 36, 152, 153, 183, 201, 274
Gerlai, R, 17, 32, 73
Glucocorticoid receptor (GR), 265,
   270–272
Glutamate, *see* Neurotransmitters
Glycine, 211, 216, 217
   agonist,
      D-serine, 210
Glycogen synthase kinase (GSK)-
   3β, 279
Griebel, G, 225

H

Habits, 3, 5, 9
Habituation, 54, 114, 162, 164, 165,
   195, 213, 225, 251, 273, 276
Haloperidol, *see* Antipsychotics
Hayes, SC, 56
Hinde, RA, 11
Hippocampus, *see* Brain, structure
   CA3 pyramidal cell, 201
Holmes, A, 33
Holmes, SJ, 9
Hotplate, *see* Behavior, measures of
Human genetics, 103, 107–110, 112–
   116, 181, 182, *see also* Genetics
Huntington's disease, 72, 88, 94,
   185, 198
Hydrocephalus, 101, 113
Hyperactivity, 92, 105, 112, 153, 187,
   198, 250, 251, 263, 268–270,
   272–274, 279, *see also* Attention
   deficit hyperactivity disorder

Hypoactivity, 112, 116, 155, 248
Hyponeophagia, 237, 243
Hypoplasia, 101, 113
Hypothalamic–periaqueductal grey, 228
Hypothalamic–pituitary–adrenal axis (HPA), 209, 268, 270–272
Hypoxanthine phosophoribosyl-transferase (HPRT), 30, 31, 39

I

Ibotenic Acid, 208–210
Imipramine, *see* Antipsychotic
Immobility, *see* Behavior, measures of
Imprinting, 12, 92
Impulsivity, 71, 75, 102, 242, 270, 271, 274
Inbred strains, 32, 34, 71, 72, 79, 198, 199, 222, *see also* Mouse, strains
Inclusions, nuclear, 90, 92, 93, 96
Inferior olives, *see* Brain, structure of
Intelligence, 3, 6–9, 13–15, 17, 102, 128, 185
Intermediate traits, 69–71, 79, 80, 157, *see also* Endophenotype
Intracranial self-stimulation, 246, 250

J

James, W, 6, 7
Jennings, HS, 8

K

Kantor, JR, 49, 56
KE family, 135–138
Knock-in, *see* Mouse, knock-in
Knockout, *see* Mouse, knockout
Köhler, W, 10–12, 14, 17

L

Language, 45, 69, 127–133, 135–138, 142–145, 151, 158, 160, 167, 195
Latent inhibition, 185
Learned helplessness, *see* Behavior

Learning, 3, 4, 8, 9, 11–20, 33, 34, 72, 73, 75, 89, 101, 106, 109, 111, 112, 114, 117, 119, 127, 128, 136, 140, 142–144, 153–158, 162–164, 167, 178, 180, 182, 185, 186, 194, 196, 203, 221, 230–233, 245, 251, 271, 276, 278
  deficit, 116
  disabilities, 3 19, 20, 116, 155
  reward, 76, 79, 106, 161, 162, 210, 246, 250, 251, 269
Lesch-Nyhan syndrome, 28, 31
Leucine zipper, 133, 135
Light/dark exploration, *see* Behavior, measures of
Linkage, *see* Genetics
Lithium, *see* Bipolar disorder
Locomotor activity, *see* Behavior, measures of
Loeb, J, 8, 9
Long-term depression, 104
Long-term potentiation (LTP), 17, 18, 38, 104, 108, 110, 112, 114, 156
Lorenz, K, 12
Lowe syndrome (OCRL), 28, 29
Lubbock, J, 6, 7

M

Maier, NRF, 10, 11
Mania, 265, 266, 268–274, 277–279, *see also* Bipolar disorder
Manipulation, 46–48, 50, 72, 103, 131, 177, 224, 237, 239, 249
m-chlorophenyl piperazine (mCPP), *see* Anxiolytics
Mental disorders, 15, 16, 70, 102, 157, 266
Mental retardation, 3, 15, 49, 92, 101, 113, 153, 154, 163, 167, 168, 187
Metachromatic leukodystrophy, 28, 29
Methyl-CpG-binding protein 2 (MECP2), *see* Gene
Milacemide, 211, 217
Mind, 5, 6, 8–10, 14, 15
  theory of, 151, 158
Mirrored chamber test, *see* Behavior, measures of
Mitogen-activated protein (MAP), 275, 277, 279

Index

Moclobemide, *see* Anxiolytics
Molecular,
    genetics, 18, 20, 69, 70, 79, 80,
        see also Genetics
    pathways,
        chaperones, 90–93
Mood, 27, 193, 266, 267, 269, 270,
        272, 275, 277, 279
    disorders, 19, 102, 237–239
    Morgan, CL, 6, 7
    Morris water maze, *see* Behavior,
        measures of
Motor
    dysfunction, 96, 112
    impairments, 102
Mouse,
    model,
        knock-in, 72, 92, 93, 104, 117, 239
        knockout, 3, 18, 19, 22, 25–28, 30–
            32, 34–37, 39, 49, 72, 76, 88,
            104–108, 110, 113, 114, 119,
            145, 154, 180, 239, 242, 243,
            245, 248, 265, 270, 271
        transgenic, 18, 19, 26, 27, 30, 32,
            38, 49, 70, 75, 88–96, 104,
            106, 112, 154, 222, 230–
            232, 239, 270, 272, 279
    NR1 deficient, 194, 213–216
    strains,
        BALB/c, 35, 217, 249, 253
        C57BL/, 32–35, 104, 105, 108,
            111, 113, 199, 206, 249,
            253, 274–276
        CD-1, 32, 249
        DBA/2, DBA/2J, 32, 35, 194,
            199, 202, 203
Mouse defense test battery, 221,
    223, 229, 237, 243
Mutant mice, 18, 30, 45, 104, 109–
    112, 115, 116, 119, 120, 145,
    153–156, 184, 186, 194, 213,
    214, 216, 237, 239, 241, 243–
    245, 247, 250–254, 265, 266,
    270, 272–274, 276–278
    PD-APP, 221, 230–233
Mutation, 16, 18, 19, 26–29, 32–39,
    75, 87–90, 95–97, 107–119, 128,
    135, 138, 139, 152–154, 156,
    168, 178, 181, 182, 186, 187,
    230, 242, 253, 254, 274, 277

N

Naloxone, 197
Neophobia, 158, 161–163, 278, 279
Neuritic plaques, 230
Neurodegeneration, 87–89, 93, 94,
    96, 214, 215
Neurodegenerative disease, 49, 87–
    88, 128, 154, 215
Neurodevelopmental disorders,
    136, 213
Neurofibromatosis type 1, 116,
    153, 155
Neuronal nitric oxide synthase
    (nNOS), 34, 76
Neuroscience, 3, 14, 15, 18, 20, 130,
    144, 239
Neurotransmitter, 38, 110, 180,
    193–195, 199, 201, 207, 214,
    217, 265, 268, 270, 275, 277
    dopamine, 34, 172, 181, 187, 193–
        197, 207, 214, 242, 268, 270,
        273, 274, 277
        agonists,
            apomorphine, 194, 196–
                199, 208, 209
        receptors, 34, 178, 181, 182, 187
    glutamate, 91, 96, 104, 187, 195,
        196, 205, 208, 211, 214
    serotonin (5HT), 27, 31, 196, 210,
        239, 247
        agonist, 227, 228
        transporter (5HTT), 265, 270–
            272, 274
NF1, *see* Gene
Nitric oxide synthase, 34, 76, 217
N-methyl-D-aspartic acid (NMDA),
    194, 197, 211–217
    receptor, 18, 38, 116
        hypofunction (NRH), 194, 211
Novelty,
    seeking, 242
    suppressed feeding, 243
NR1-deficient mouse, *see* Mouse,
    NR1-deficient
Nucleus accumbens, Brain, structure
Null allele, 32, 33, 35, 111, 180

O

Observation,
  direct, 47
  indirect, 48–50
Obsessive–compulsive disorder, 185, 196, 198, 221
Olanzapine, see Antipsychotic
Olfactory bulbectomy, 237, 250–251
Open field test, see Behavior, measures of
Operant behavior, 54, 55, 58
Operant conditioning, see Behavior, measures of

P

PAH, see Gene
Passive avoidance, see Behavior, measures of
Pavlov, IP, 9–12, 14, 16, 18
PCP, 197, 211, 212, 214–217
PD-APP, see Mutant mice
Peptide, 94, 200, 230, 268
Phencyclidine, 194, 197
Phenelzine, see Anxiolytics
Phenotype, 19, 25–37, 39, 79, 87, 92, 94, 101, 103, 104, 106, 107, 109–112, 117, 119, 136, 140, 154, 155, 180, 183, 187, 202, 205, 206, 241–243, 245, 248, 253, 272, 274, 277–279
Phenylalanine hydroxylase, 102, 115
Phenylketonuria, 19, 101, 102, 115
Phobia, 166, 240
Phosphorylation, 95–97, 143, 203, 275
Prozac, see Antidepressant
Pleiotropy, 180
Popping behavior, 217
Predictive validity, 73, 241, 269, 279
Prefrontal Cortex, see Brain, structures
Prepulse inhibition (PPI), 34, 35, 105, 107, 108, 113, 154, 165, 166, 185, 186, 194–200, 205, 209, 216, 267
Problem solving, 11, 14
PRODH, see Gene
Proline, 186, 187
Proteasome, 90–94
Psychiatric disorder, 36, 64, 71, 177, 178, 181, 270
Psychopathology, 3, 16, 18, 45, 46, 64, 66, 80, 193–197, 207, 210, 212, 218, 269
Psychosis, 28, 37, 183, 193–195, 200, 202, 204, 211, 231, 266
Psychostimulant,
  amphetamine, 32, 187, 193, 194, 208
  addiction and withdrawal, 251, 269
  locomotor behavior, 32, 209, 210, 277
Purkinje cells, 28, 87, 89–96, 140, 142

Q

Quantitative trait loci (QTL), 28, 79, 83, 199, 200, 245
Quetiapine, see Antipsychotic

R

Radial maze, see Behavior, measures of
Rat strains,
  Brown Norway, 200
  Sprague Dawley, 198, 202, 208–210
  Wistar, 200, 212
Reelin (*RELN*), see Gene
Reflex, 4, 9, 105
  startle, 154, 156, 195, 198
Reflexive behavior, 54, 55, 59, 64
Reliability, 247, 249, 250
Remote observation, 46–48
Reticular nucleus, see Thalamus
Rett syndrome (RTT), 18, 101, 110–112, 152, 154, 155, 158
Reward, see Learning, reward
Ribosomal S6 kinase 2 (*RSK2*), see Gene
Risk
  allele, 180, 181
  assessment, 224–227, 239, 240, 243
RNA interference, 222, 254
Romanes, G, 6, 10, 13
Rotarod, see Behavior, measures of

S

SCA1, 87–96
Schizophrenia, 16, 18–20, 35, 37, 38, 69–71, 76, 102, 165, 177, 180–187, 193–217
Schizotypal, 36, 197, 198
Schneirla, TC, 10, 11
Scopolamine, 76, 214
Seizures, 18, 28, 92, 102, 105, 110, 112, 154, 155, 167, 185, 217
Sensorimotor gating, 35, 165, 182, 185, 186, 196, 198, 199, 209, 267, 270, 273
Sensory inhibition, 200–203
Serotonin (5HT), 27, 227, 228, 239, 247
  transporter (5HTT), 31, 196, 210, 265, 270–272, 274
Sandhoff disease, 29, 30
Sherrington, CS, 8, 9, 16
Shock-probe burying, *see* Behavior, measures of
Skinner, BF, 10–12, 14, 55, 56, 63
Sleep,
  architecture, 246, 248
  patterns, 19, 233, 266, 269, 271, 272, 279
Social,
  interaction, 75, 111, 151, 157–159, 163, 212, 241
  transfer, 75
Song system, 142
Spalding, D, 5, 6
Speech disorders, 25, 28, 103, 110, 112, 127–133, 135–139, 143–145, 158, 160, 266, 269
Sprague Dawley, *see* Rat strains
Staircase test, *see* Behavior, measures of
Startle reflex, 154, 166, 195, 198
Startle response (also acoustic startle response), 105, 106, 108, 113, 185, 194–197, 199, 209, 216, 273, 278
Stereotypic behavior, 110–111, 113
Strain differences, 32, 198–200, 209, 211, 243, 249
Stress, 75, 108, 161, 167, 208, 209, 211, 216, 217
Stress-induced hyperthermia, 237, 241, 243
Substitute responding, 57, 61
Substitute stimulation, 57–60, 63, 64
Suicide, 238, 245, 246, 266
Synaptogenesis, 168, 203
Syntax, 129, 132, 136, 142

T

Tail suspension test (TST), *see* Behavior, measures of
Tay-Sachs disease, 28–30
Temporolimbic cortices, *see* Brain, structure
Thalamus, *see* Brain, structure
Theory of mind, 151, 158
Thompson, RF, 17
Thorndike, EL, 7, 8, 10, 11, 14
Threat, 223–228, 239–241, 245
T-maze, *see* Behavior, measures of
Topiramate, 217
Transcriptional profiling, 187
Transformation, 46–48
Transgenic, *see* Mouse model
Tuberous sclerosis, 15, 152, 155
22q11,
  deletion syndrome, 183, 184, 186, 187
  schizophrenia susceptibility locus, 182, 186, 187

U

Ubiquitin, 91–95

V

Validity, 19, 47, 51, 52, 63, 72, 73, 80, 230, 233, 241, 250, 269, 277, 279
Valproate, 267–270, 273, 275, 277, *see also* Mood disorders
Vasopressin, 75, 156, 250
Velocardiofacial syndrome, 36
Ventral hippocampus, *see* Hippocampus
Verbal dyspraxia, 136
Vocal learning, 128, 140, 142–144

W

Washburn, MF, 9
Water maze, *see* Morris water maze
Watson, JB, 10–12, 14
Weight,
 body, 50, 108, 111, 112, 246
 brain, 110–112
Wistar, *see* Rat strains
Wistar-Kyoto, *see* Rat strains, Wistar

Working memory, 109, 114, 119, 164, 182–186, 205, 207, 210, 216, 246
Wundt, W, 7, 9, 10

Y

Yerkes, RM, 8
Yohimbine, 228–229

Z

Zinc finger, 133, 135